Phase Separation in Living Cells

Riki Kurokawa
Editor

Phase Separation in Living Cells

Benefits and Risks

Springer

Editor
Riki Kurokawa
School of Medicine
Saitama Medical University
Hidaka, Saitama, Japan

ISBN 978-981-99-4885-7 ISBN 978-981-99-4886-4 (eBook)
https://doi.org/10.1007/978-981-99-4886-4

This Springer imprint is published by the registered company Springer Nature Singapore Pte Ltd.
The registered company address is: 152 Beach Road, #21-01/04 Gateway East, Singapore 189721, Singapore

Preface

There have been emerging demands to appreciate biological actions of phase separation in divergent aspects of science. This concise book presents comprehensive understanding of recent progress of phase separation. A key molecule of phase separation is intrinsically disordered region (IDR)/proteins (IDPs), mostly comprised of RNA-binding proteins (RBPs). Phase separation is a physicochemical process that induces a single solution of solvent and IDPs into two phases in which one phase contains just solvent and another has IDPs in solvent. Multivalent interactions among IDRs induce phase separation. This process generates condensates of IDP solution. One of the major roles of phase separation is to generate the condensates, membraneless organelles including stress granule, Cajal body, and nucleolus. Phase separation is also involved in fundamental biological functions like transcription and translation. Accidentally, IDPs form aggregation to harm cells of the central nervus system. These aggregations potentially cause neurodegenerative diseases like Alzheimer's disease and amyotrophic lateral sclerosis (ALS). For the risk, regulatory systems are developed to repress accidental aggregation formation in living cells.

This book presents the leading edge of the young field of phase separation. We plan to show just active areas of the phase separation studies. The book consists of four parts: Part I Physics and Chemistry has topics of structural biology of RBP FUS/TLS and chaperon for phase separation, computational approach of phase separation, and chemistry of G-quadruplex of DNA and RNA. Part II Molecular Biology has topics from molecular mechanism of IDR sequence phase separation and potential cellular function of these labile fibers, formation of membraneless organelles, and roles of noncoding RNAs in phase separation. Part III Biology has topics from nuclear pore complex, developmental biology regarding force-dependent remodeling, and neurobiology and higher order brain functions. Part IV Medicine has topics from condensates and cancer therapy, pathogenesis of neurodegenerative diseases based on IDPs, responses of microglia to amyloid, and hot topic of role of phase separation in COVID-19. These four parts give a rise to novel insights into phase separation in living cells. These are thoroughly new. Previously, separate publications regarding each area of phase separation have been appeared, but our book presents these topics in a single volume. This should be a big benefit for readers. Very recently, coronavirus takes advantage of phase separation on its infections. Therefore, emerging needs and opportunity for knowing phase separation have been arising. Timely, this book should satisfy these needs.

Hidaka, Saitama, Japan Riki Kurokawa

Contents

Part I
Physics and Chemistry

Chapter 1
FUS Aggregation by Shear Stress on Pipetting and Its Suppression by Non-coding RNA

Masato Katahira

Abstract Fused in sarcoma/translocated in liposarcoma (FUS/TLS) is a multitasking RNA/DNA-binding protein. FUS aggregation is implicated in various neurodegenerative diseases. We revealed the large conformational change of FUS from a compact to an extended structure upon binding of non-coding RNA transcribed from the upstream region of the cyclin D1 gene. In the course of this study, we found by chance that FUS transforms into the amorphous aggregation state as an instant response to the shear stress caused by usual pipetting even at a low FUS concentration, 100 nM. It was also revealed that the non-coding RNA can suppress the transformation of FUS into aggregates. A small fragment (13 nucleotides) of the full length non-coding RNA (602 nucleotides) was shown to be functional for both the conformational change of FUS and suppression of the transformation of FUS into aggregates. The suppressive effect of RNA on FUS aggregation turns out to be sequence-dependent. These results suggested that the non-coding RNA could be a prospective suppressor of FUS aggregation caused by mechanistic stress in cells. Our finding might pave the way for more research on the role of RNAs as aggregation inhibitors, which could facilitate the development of therapies for neurodegenerative diseases.

Keywords FUS · Non-coding RNA · ALS · LLPS · Neurodegenerative diseases · Aggregation

---

M. Katahira (✉)
Institute of Advanced Energy, Kyoto University, Uji, Kyoto, Japan
e-mail: katahira.masato.6u@kyoto-u.ac.jp

R. Kurokawa (ed.), *Phase Separation in Living Cells*,
https://doi.org/10.1007/978-981-99-4886-4_1

1.1 FUS Protein and Long Non-coding RNA That Binds to FUS

1.1.1 FUS Protein Implicated in Neurodegenerative Diseases

Fused in sarcoma (FUS), also known as translocated in liposarcoma (TLS), is an RNA- and DNA-binding protein, which regulates various biological processes (Wang et al. 2008; Takahama et al. 2013; Yu and Reed 2015; Zhang et al. 2018; Lagier-Tourenne et al. 2010). FUS has been considered as a molecular link between apparently different human diseases such as cancer and neurodegenerative diseases (Campos-Melo et al. 2014; Doi et al. 2004, 2008, 2010; Kwiatkowski et al. 2009; Vance et al. 2009). FUS was found as the major component of nuclear polyglutamine (polyQ) aggregates in a Huntington disease (HD) cell model (Doi et al. 2004), where FUS was converted from a soluble form to insoluble aggregates (Doi et al. 2008). Then, FUS was also found to be a member of the PolyQ aggregates in other diseases including spinocerebellar ataxia (SCA) types 1, 2, and 3 and dentatorubral-pallidoluysian atrophy (DRPLA) (Doi et al. 2010). It was also found that FUS mutations were present in amyotrophic lateral sclerosis (ALS) (Kwiatkowski et al. 2009; Vance et al. 2009) and frontotemporal lobar degeneration (FTLD) patients (Neumann et al. 2009). It was found that those mutations accelerate the FUS transition into an insoluble form (Patel et al. 2015). Although these neurodegenerative diseases have different manifestations, FUS aggregation is associated with all of them (Lagier-Tourenne et al. 2010). This suggests a common pathway for their neuropathologies.

1.1.2 Several Different States of FUS

FUS is known to take on different states such as dispersed, liquid droplet, gel, and fibril ones depending on factors such as pH, ionic strength, protein concentration, thermal stress, shear stress, and RNA presence (St George-Hyslop et al. 2018; Maharana et al. 2018; Schwartz et al. 2013; Shen et al. 2020). It was found that FUS-containing droplets yielded by liquid–liquid phase separation (LLPS) play a key role in the assembly of membrane-less organelles such as stress granules (Molliex et al. 2015). It was also found that high concentration of RNA can suppress LLPS of FUS (Maharana et al. 2018). It was revealed that FUS can interchange between a dispersed phase, liquid droplets, and a reversible gel and that liquid droplets can be converted into irreversible gels and fibrils through aging or pathological conditions (St George-Hyslop et al. 2018; Alberti and Hyman 2016).

FUS comprises a low complexity domain (LC), the first arginine-glycine-glycine rich motif (RGG1), an RNA recognition motif (RRM), the second RGG (RGG2), a Zinc finger domain (ZnF), and the third RGG (RGG3) (Fig. 1.1a) (Hamad et al.

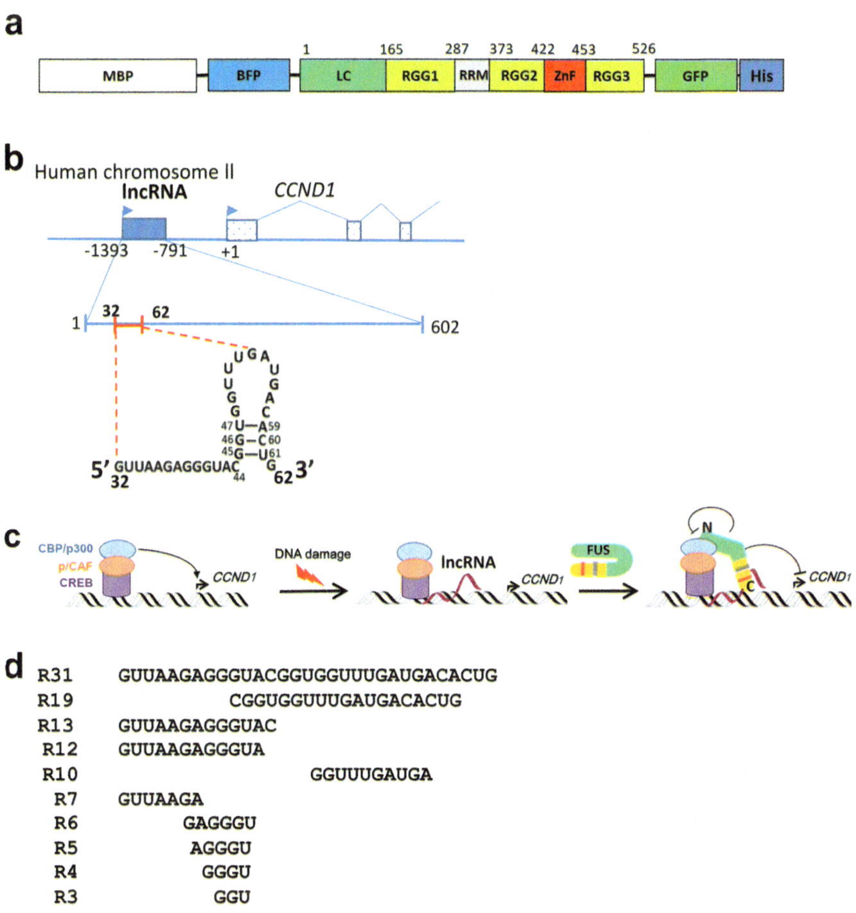

Fig. 1.1 (**a**) Schematic illustration of the protein used in this study, MBP-BFP-FUS-GFP-6xHis. Residue numbers of FUS are indicated. (**b**) The localization, length, sequence, and secondary structure of lncRNA (Yoneda et al. 2016). (**c**) The mechanism of the repression of *CCND1* by FUS on DNA damage (Wang et al. 2008). A compact-to-extended conformational change of FUS is hypothesized on binding lncRNA, which enables FUS to interact with CBP/p300 and repress their histone acetyltransferase activities, resulting in the repression of *CCND1*. (**d**) The names and sequences of fragments of R31 of lncRNA. (Reproduced and modified from Hamad et al. 2020a)

2020a). The LC and RGGs are known to be intrinsically disordered regions (IDRs). RGG, RRM, and ZnF are well known nucleic acid binding domains.

1.1.3 Long Non-coding RNA That Binds to FUS

It was reported that DNA damage triggers the transcription of long non-coding RNA (lncRNA) of 602 nucleotides from the 5′ upstream region of the cyclin D1 gene (*CCND1*) (Wang et al. 2008) (Fig. 1.1b). Subsequently, FUS is recruited to the 5′ upstream region of *CCND1* through binding to lncRNA (Fig. 1.1c). Binding of lncRNA also relieves the N-terminal region of FUS from masking by the C-terminal region of FUS, allowing the N-terminal region of FUS to bind to CREB-binding protein (CBP)/E1A-binding protein P300 (p300) and to repress the histone acetyltransferase HAT activity of CBP/p300 allosterically (Fig. 1.1c). A model was suggested in which lncRNA serves as a ligand for FUS, causing an allosteric effect to release it from an inactive conformation (Wang et al. 2008) (Fig. 1.1c). This allows gene-specific FUS–CBP/p300 interactions resulting in inhibition of the HAT activity of CBP/p300 and the repression of *CCND1* transcription (Fig. 1.1c). Later, we found that a portion of lncRNA (nucleotide numbers 32–62) comprising 31 nucleotides, R31, is sufficient for binding of FUS. It was demonstrated that the 5′ half of R31 takes on a single-stranded structure and that the 3′ half forms a stem-loop structure (Fig. 1.1b) (Yoneda et al. 2016).

1.2 Large Conformational Change of FUS from a Compact to an Extended Structure upon Binding of Non-coding RNA

1.2.1 FRET Analysis

In the model presented in Fig. 1.1c, upon binding of non-coding RNA, large conformational change of FUS from a compact to an extended structure is expected. To investigate the structural changes of FUS upon addition of various nucleic acids, we performed fluorescence resonance energy transfer (FRET) assay. FRET is a methodology that can determine the distance between two chromophores, a donor and acceptor. If the distance between the donor–acceptor pair is less than 10 nm, FRET efficiency is high, whereas FRET efficiency is poor if the distance exceeds more than 10 nm (Sahoo 2011).

For the FRET assay, we introduced blue fluorescent protein (BFP) and green fluorescent protein (GFP) to the N- and C-termini of FUS, respectively (Fig. 1.1a) (Hamad et al. 2020a). Additionally, maltose binding protein (MBP) and a 6xHis-tag were attached for the enhancement of the solubility and purification, respectively (Fig. 1.1a). We expect that FUS takes on a compact form in a free form, in which GFP and BFP are supposed to be close, and shows high FRET efficiency (Fig. 1.2a, left) (Hamad et al. 2020a). In a RNA-bound form, however, we expect that FUS takes on an extended form, in which GFP and BFP are supposed to be far apart and shows low FRET efficiency (Fig. 1.2a, right) (Hamad et al. 2020a).

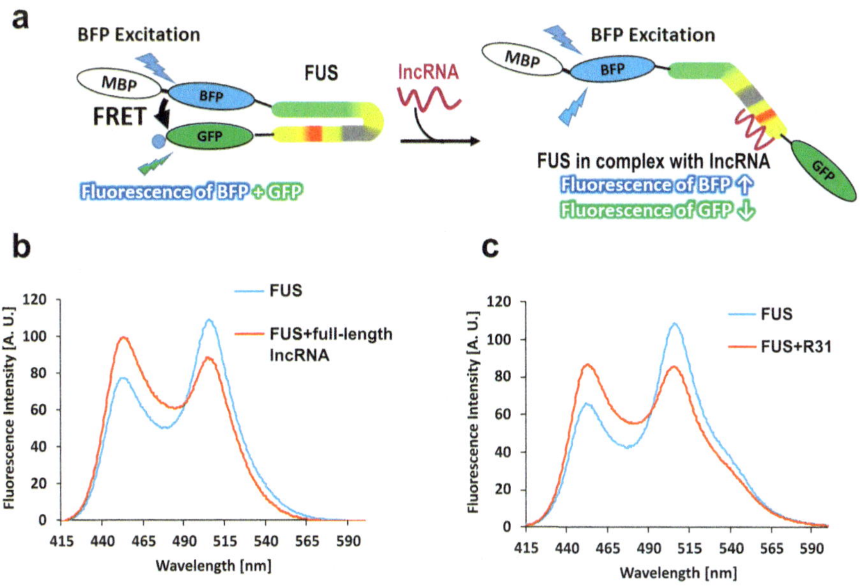

Fig. 1.2 (**a**) The strategy for the detection of the compact-to-extended conformational change of FUS by FRET. (**b**) Fluorescence spectrum of FUS fusion protein (MBP-BFP-FUS-GFP-6xHis) (100 nM), excited at 402 nm, in either the absence (blue) or presence (red) of an equimolar amount of full-length lncRNA of 602 nt. (**c**) Fluorescence spectrum of FUS in either the absence (blue) or presence (red) of an equimolar amount of R31 of lncRNA. (Reproduced and modified from Hamad et al. 2020a)

MBP-BFP-FUS-GFP-6xHis protein was successfully expressed in E. coli cells and purified by Ni-affinity chromatography followed by size-exclusion chromatography (Hamad et al. 2020a). Hereafter, MBP-BFP-FUS-GFP-6xHis protein is simply called FUS.

The conformational change of FUS caused by binding of the full-length lncRNA (602 nucleotides) was examined. Figure 1.2b shows the fluorescence spectrum of FUS excited at 402 nm. Due to FRET, in addition to the peak of BFP at around 453 nm, a peak of GFP at around 506 nm was observed (Fig. 1.2b) (Hamad et al. 2020a). When the full-length lncRNA was added, the fluorescence intensity at 506 nm decreased, while that at 453 nm increased (Fig. 1.2b) (Hamad et al. 2020a). This means the reduction in FRET, which indicates the increase in the BFP-GFP distance. It is suggested that FUS undergoes the compact-to-extended conformational change on binding to lncRNA. This enables FUS to interact with CBP/p300 to repress the expression of the *CCND1* gene (Fig. 1.1c) (Hamad et al. 2020a).

Then, the conformational change caused by R31 (Fig. 1.1d) was examined. The spectral change of FUS caused by R31 turned out to be very similar to that by the full-length lncRNA (Fig. 1.2c) (Hamad et al. 2020a). This indicates that a shorter RNA, R31, is sufficient to cause the same extent of conformational change as the full-length lncRNA. It was also revealed that even a much shorter RNA, R13 (Fig 1.1d), can cause the similar conformational change of FUS to a certain extent, while the conformational change caused by R10 was rather smaller (Hamad et al. 2020a). Furthermore, the conformational change caused by U_{13} comprising 13 uracil residues (Fig 1.1d) was much smaller. Thus, it turned out that the conformation change of FUS by RNA is sequence-dependent (Hamad et al. 2020a).

1.2.2 AFM Analysis

We also directly visualized a dynamical conformational change of a single molecule of FUS by high-speed atomic force microscopy (HS-AFM) (Hamad et al. 2020b). As described above, all domains of FUS are IDRs except for the RRM and ZnF ones (Hoell et al. 2011). In order to help understanding the AFM images, MBP-BFP-FUS-GFP-6xHis protein was used. MBP, BFP, and GFP are known to form a globular structure and thus are expected to be clearly visualized as spheres on HS-AFM. Therefore, when FUS takes on a compact globular structure, observation of maximally four spheres is expected. When FUS takes on an extended flexible structure and is visualized just faintly due to its IDR characteristics, the two spheres in proximity on HS-AFM must be MBP and BFP located at the N-terminus of FUS, and the remaining visible sphere must be GFP located at the C-terminus of FUS. The attachment of the three proteins facilitated understanding of the conformational behavior of FUS as demonstrated below.

Although the full length lncRNA comprises 602 nucleotides, a fragment of 31 nucleotides, R31 (Fig. 1.1d), was shown to be enough for appropriate binding to FUS (Yoneda et al. 2016). Hereafter, R31 is simply called lncRNA.

In the absence of lncRNA, it was revealed that FUS takes on a compact conformation presumably due to the electrostatic interaction between its N- and C-terminal regions (Fig. 1.3) (Hamad et al. 2020b). In the compact conformation, the N-terminal region is supposed to be masked by the C-terminal region (Fig. 1.1c) (Hamad et al. 2020a). Then, the compact-to-extended conformational change was successfully captured (Fig. 1.3) (Hamad et al. 2020b). In Fig. 1.3, lncRNA was added at time 0 s. A FUS molecule (and a fusion protein molecule also) in Fig. 1.3 forms a compact structure for a moment and then, at a certain time, FUS quickly transformed into an extended structure. The distance between BFP and GFP is a measure of the extension of a FUS molecule. When the compact-to-extended conformational change was caused by lncRNA, the BFP-GFP distance dramatically changed from ca. 10 nm to 27 nm (Hamad et al. 2020b). In the extended conformation, the N-terminal region of FUS is supposed to be exposed, which enables FUS to interact with CBP/p300 for interference of their acetyltransferase activities (Fig. 1.1c) (Hamad et al. 2020a). Thus, we succeeded in visualizing the conformational transition of FUS in response to binding of lncRNA.

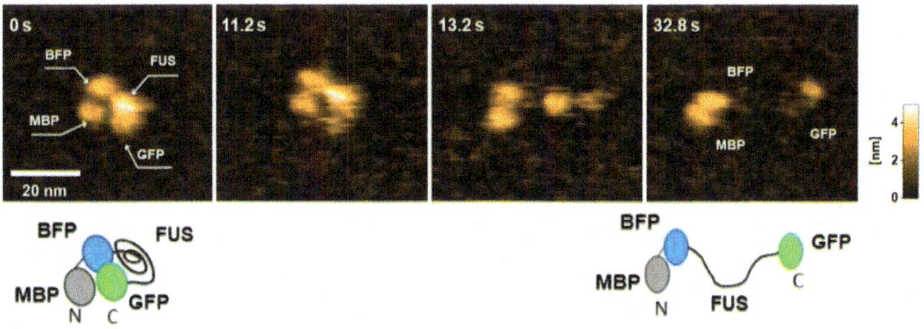

Fig. 1.3 Consecutive HS-AFM images showing the moment of the conformational transition of FUS (and FUS fusion, MBP-BFP-FUS-GFP-6xHis, also) in the presence of lncRNA. A 2 μL droplet of the 5 nM fusion protein was placed on the APTES-mica surface. 14-fold equimolar amount of lncRNA to the fusion protein was added to the solution in the chamber at time 0. 0.2 s/frame. (Reproduced and modified from Hamad et al. 2020b with permission from the Royal Society of Chemistry)

FRET and HS-AFM clearly revealed the large conformational change of FUS from a compact to an extended structure upon binding of non-coding RNA. It was also shown that this effect of RNA is sequence-dependent.

1.3 FUS Aggregation by Shear Stress on Pipetting

1.3.1 Fluorescence Spectroscopy Analysis

In the course of the study of the conformational change of FUS by lncRNA, we found by chance that mechanistic shear stress caused by pipetting can induce FUS aggregation (Hamad et al. 2021). We noticed that upon pipetting of MBP-BFP-FUS-GFP-6xHis, which is simply called FUS in this section, both the BFP and GFP fluorescence intensities decreased. To examine the effect of pipetting more quantitatively, 45 strokes of pipetting were applied to a FUS protein sample and fluorescence spectra were measured every three strokes of pipetting (the interval between measurements was set to 60 s) (Fig. 1.4) (Hamad et al. 2021). The sample volume was 150 μL and the pipetting volume was set to 140 μL. We supposed that the observed reduction in the fluorescence intensity for the entire wavelength range is caused by aggregation of FUS. Aggregates may precipitate or stay at the bottom of the cuvette and do not contribute to the fluorescence spectrum because the light does not pass the bottom of the cuvette. In order to confirm this idea, the concentration of the protein in the supernatant of the sample was examined. The concentration of the supernatant of the sample after 30 strokes of pipetting turned out to be lower than that of the sample without pipetting by ca. 40%. This reduction in concentration is qualitatively consistent with the reduction in the fluorescence intensity, which supports our idea (Hamad et al. 2021).

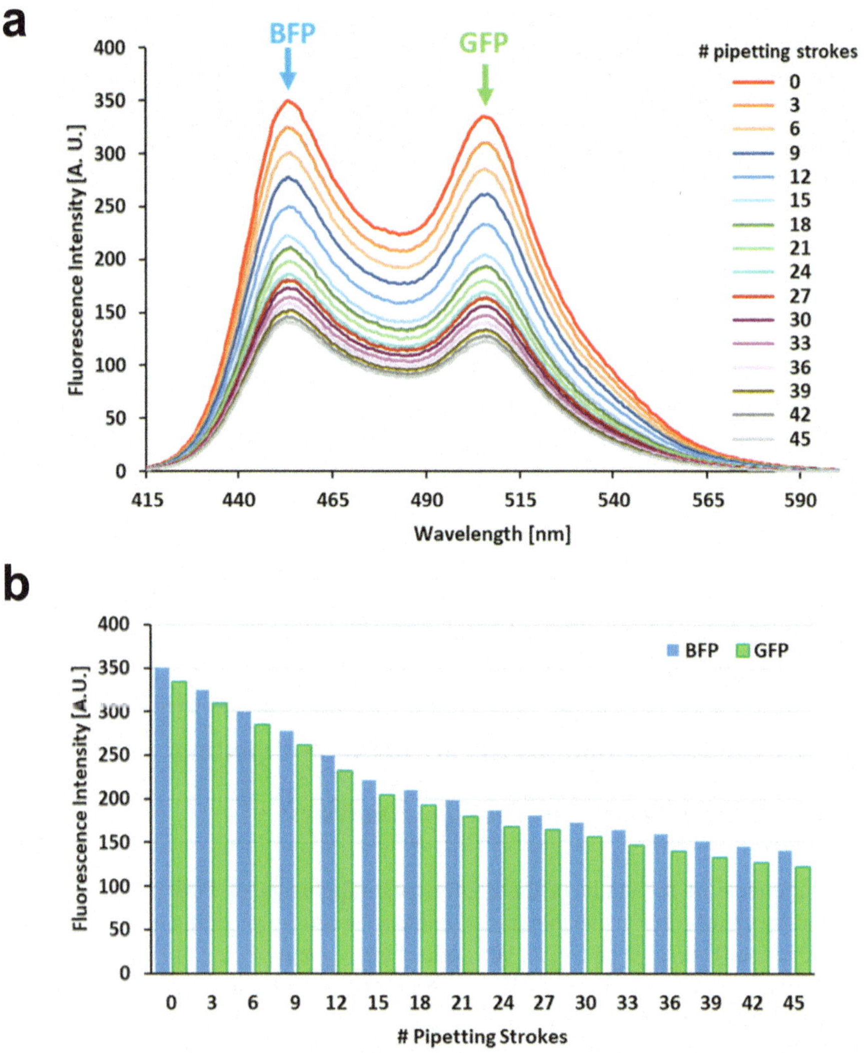

Fig. 1.4 (**a**) Overlaid fluorescence spectra of 100 nM FUS fusion protein (MBP-BFP-FUS-GFP-6xHis). The spectra were measured every 60 s, during which the sample (150 μL) was mixed by three strokes of pipetting (pipetting volume 140 μL). The total number of pipetting strokes is indicated on the right. (**b**) Bar graph showing the BFP (at 453 nm) and GFP (at 506 nm) fluorescence intensities of the fluorescence spectra shown in (**a**). The averages of two independent experiments ± standard deviation (S.D.) are shown. (Reproduced from Hamad et al. 2021)

Pipetting can subject protein molecules to shear stress due to the velocity gradient. It is supposed that shear stress is most prominent for molecules close to the surface of the pipette tip (Bekard et al. 2011). The unequal force distribution on a protein molecule might induce some conformational change that leads to aggregation, as discussed in Sect. 1.3.4.

1.3.2 Fluorescence Microscopy Analysis

In order to further verify our idea, the effect of pipetting on FUS was visualized by fluorescence microscopy. Another fusion protein, streptavidin recognition sequence (Strep)-GFP-FUS, which is simply called FUS in this section, was constructed in order to exclude the effects of MBP and BFP on FUS aggregation. We prepared four protein samples with different numbers of strokes of pipetting (0, 15, 30, and 45 strokes). For the P0 sample (100 nM), no pipetting was applied. The P15, P30, and P45 samples underwent 15, 30, and 45 strokes of pipetting, respectively. All samples were examined by fluorescence microscopy (Fig. 1.5a) (Hamad et al. 2021). The number of particles larger than 0.002 mm^2 was counted. More FUS particles were formed as the number of strokes of pipetting increased (Fig. 1.5a, b) (Hamad et al. 2021).

Particles can be either droplets formed due to LLPS or aggregates. The shape of a droplet is known to be completely round (St George-Hyslop et al. 2018). The shapes of particles turned out to be mostly not round but irregular (data not shown), indicating that most particles are not droplets but amorphous aggregates. We also examined the nature of the particles by using 1,6-hexanediol, which is known to dissolve droplets formed by LLPS (Kroschwald et al. 2017). Ten percent 1,6-hexanediol was added to each sample, and then the particles were counted again. It was confirmed that the number of FUS particles that are resistant to 1,6-hexanediol treatment and thus are supposed to be not droplets but aggregates increased as the number of pipetting strokes increased (Fig. 1.5c) (Hamad et al. 2021). Thus, our idea was supported.

1.3.3 TEM Analysis

Then, we investigated the appearance of the pipetting-induced FUS aggregates by transmission electron microscopy (TEM) to determine whether the formed aggregates are amorphous or take on a particular structure such as amyloid fibrils. A protein sample was examined before pipetting, the P0 sample, and after 30 strokes of pipetting, the P30 sample. For the P0 sample, aggregates were rarely found (Fig. 1.6a) (Hamad et al. 2021). However, for the P30 sample, many aggregates were found and they were mostly amorphous (Fig. 1.6b) (Hamad et al. 2021).

Thus, our idea that the FUS aggregates are formed by pipetting was solidly confirmed with three independent methods: fluorescence spectroscopy, fluorescence microscopy, and TEM.

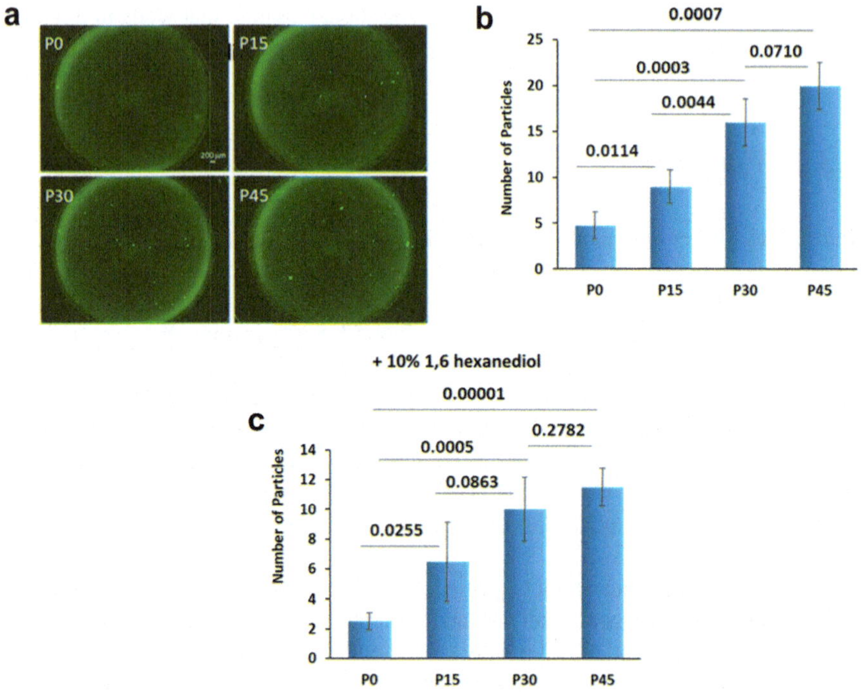

Fig. 1.5 (**a**) Representative fluorescence microscope images of FUS fusion protein (Strep-GFP-FUS). The 100 nM Strep-GFP-FUS solution was subjected to pipetting 0, 15, 30, and 45 strokes (P0, P15, P30, and P45, respectively) before measurement. (**b**) A bar graph showing the number of particles >0.002 mm^2 observed in (**a**). The number of particles was determined with Fiji software. The bar graphs show the averages of three independent experiments ± standard deviation (S.D.). *p* values are indicated. (**c**) A bar graph showing the numbers of particles >0.002 mm^2 observed after addition of 10% 1,6-hexanediol, which is known to disrupt the liquid–liquid phase separation, to the samples shown in (**a**). (Reproduced from Hamad et al. 2021)

1.3.4 The Mechanism of the FUS Aggregation on Pipetting

MBP-BFP-FUS-GFP-6xHis was used for fluorescent spectroscopy and TEM, while strep-GFP-FUS was used for fluorescence microscopy. The aggregation was found on pipetting for both MBP-BFP-FUS-GFP-6xHis and strep-GFP-FUS. Thus, it is not likely that either MBP or BFP is the origin of the aggregation. Additionally, it was reported that GFP does not aggregate on shearing (Duerkop et al. 2018). Therefore, it is suggested strongly that FUS is the origin of the aggregation.

Here, we demonstrated that the usual pipetting can induce aggregation of FUS. The more the number of strokes of pipetting is, the more the number of aggregates is. We assumed that the shear stress caused by pipetting is a driving force for the formation of aggregates. The shear stress on the protein structure and the subsequent induction of aggregation have been studied for decades (Bekard et al. 2011; Hill et al. 2006; Xie et al. 2007; Di Stasio and De Cristofaro 2010; Ramstack et al. 1979).

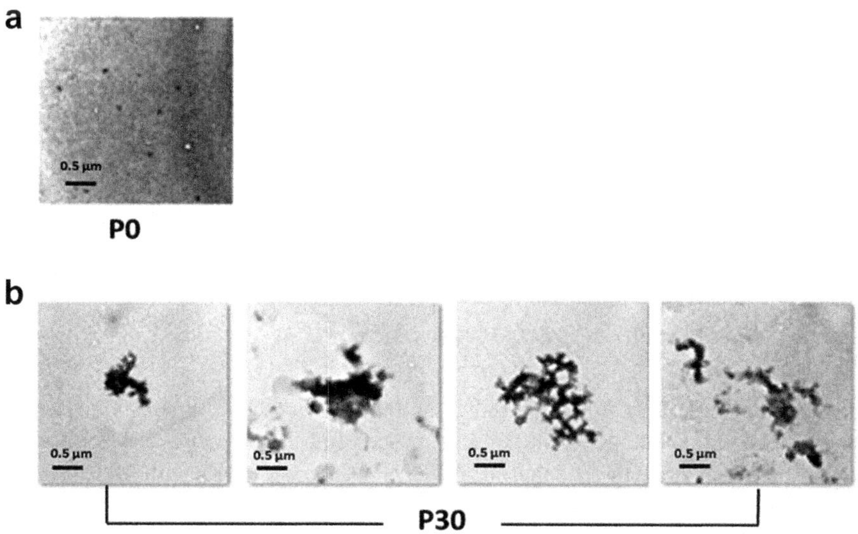

Fig. 1.6 (**a**) A representative TEM image of FUS fusion protein (MBP-BFP-FUS-GFP-6xHis) before pipetting (P0). (**b**) TEM images showing FUS aggregates formed after 30 strokes of pipetting (P30). (Reproduced from Hamad et al. 2021)

Regarding the mechanism of shear stress-induced protein aggregation, there is a general consensus that mechanical perturbation of a protein molecule often results in structural destabilization of the native conformation, leading to the exposure of hydrophobic residues. Solvent-exposed hydrophobic groups become nucleated via hydrophobic interactions and subsequently aggregate (Bekard et al. 2011). Shear stress has been reported to induce aggregation of the whey protein beta-lactoglobulin (Hill et al. 2006; Akkermans et al. 2006). Therefore, it is possible that shear stress caused by pipetting induced the transition to the aggregate state.

1.3.5 *The Relationship with Other Studies Including LLPS One*

We found that the formation of aggregates by shear stress caused by pipetting occurred at a low FUS concentration, even at 100 nM. This is in contrast to the situation that a higher FUS concentration, 1–5 μM, is usually needed for the formation of droplets due to LLPS (Maharana et al. 2018; Shen et al. 2020; Wang et al. 2018). FUS aggregates were instantly formed by shear stress caused by pipetting (Fig. 1.4). This is also in contrast to the situation that a longer incubation time, minutes to an hour, is usually needed for the formation of droplets due to LLPS (Wang et al. 2018; Alberti et al. 2018).

Recently, shear-mediated formation of solid fibers of FUS was reported (Shen et al. 2020). Shear stress was applied to FUS at a rather low concentration of 100 nM in our case, while it was applied to a LLPS form of FUS in their study. This may be related to the formation of different species, amorphous aggregates or fibers.

1.4 Suppression of FUS Aggregation by Long Non-coding RNA in a Sequence-Dependent Manner

1.4.1 Fluorescence Spectroscopy Analysis

It was interesting to find that the reduction in the fluorescence intensity caused by shear stress on pipetting was not observed when full-length lncRNA was added to the FUS fusion protein (Hamad et al. 2021). When the full-length lncRNA was added to FUS, the pattern of the fluorescence spectrum changed due to the compact-to-extended conformational change of FUS induced by lncRNA (Fig. 1.7a), as described in Sect. 1.2.1. When strokes of pipetting were applied, however, the spectrum rarely changed (Fig. 1.7a). The decrease in the fluorescence intensity caused by pipetting for all wavelengths, which was observed in the absence of RNA (Fig. 1.4a), was not seen. This indicates that lncRNA can protect FUS from aggregation caused by the shear stress of pipetting.

As described in Sect. 1.2.1., we examined the extent of the conformational change of FUS caused by lncRNA and its fragments on the basis of the change in FRET. Then, we found that the extent of the conformational change of FUS is greater for full-length lncRNA, R31, R19, and R13 than for R10, R7, R5, and R4 (see Fig. 1.1d for sequences) (Hamad et al. 2020a). When either R31 or R13 was added to the FUS fusion protein, the decrease in the fluorescence intensity caused by pipetting for all wavelengths was not seen (Fig. 1.7b, c, e). This indicates that R31 and R13 can prevent the aggregation of FUS. Then, we examined the suppressive effect of a non-specific counterpart of R13, U_{13} comprising 13 uracil residues. When U_{13} was added to the FUS fusion protein, the fluorescence intensity continued to decrease with increasing number of strokes of pipetting (Fig. 1.7d, f). This indicates that U_{13} cannot protect FUS from aggregation by pipetting. Thus, it was revealed that suppression of FUS aggregation by lncRNA is sequence-specific.

1.4.2 Fluorescence Microscopy Analysis

We further confirmed the difference between R13 and U_{13} as to the suppression of FUS aggregation by fluorescence microscopy. Thirty strokes of pipetting were applied for 100 nM Strep-GFP-FUS fusion protein in the absence of RNA, or in the presence of either R13 or U_{13}. Samples were examined by fluorescence

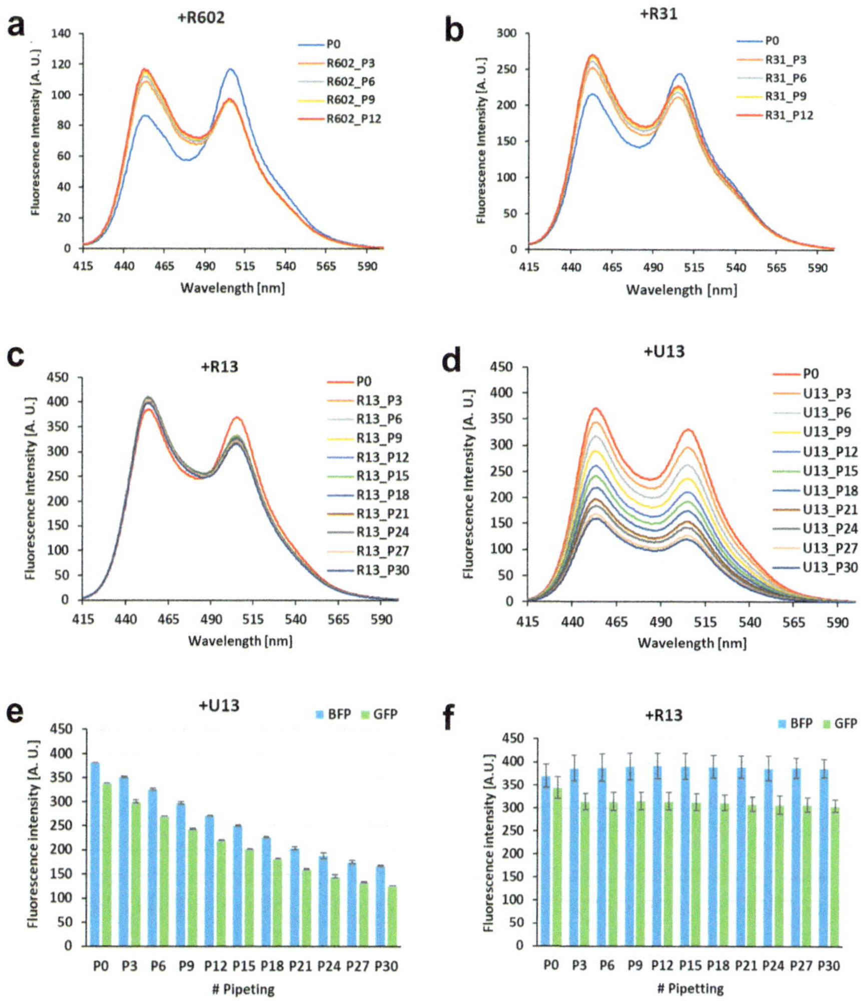

Fig. 1.7 (**a–d**) The fluorescence spectrum of 100 nM FUS fusion protein (MBP-BFP-FUS-GFP-6xHis) without pipetting is shown as P0. An equimolar amount of either the full-length lncRNA (R602) (**a**), R31 (**b**), R13 (**c**), or U_{13} (**d**) was added to the protein solution. Then, fluorescence spectra were measured after every three stokes of pipetting: cumulative numbers of strokes are indicated, e.g., R602_P6 for six cumulative strokes of pipetting after the addition of R602 RNA. (**e**) A bar graph for BFP and GFP fluorescence intensities, at 453 and 506 nm, respectively, of the fluorescence spectra shown in (**c**). (**f**) The similar bar graph of the fluorescence spectra shown in (**d**). (Reproduced from Hamad et al. 2021)

microscopy, and particles larger than 0.002 mm^2 were counted (Fig. 1.8a). Then, droplets due to LLPS were dissolved by adding 10% 1,6-hexanediol, and the particles were counted again (Fig. 1.8b). In this case, only FUS aggregates are counted. These experiments revealed that FUS aggregates were significantly less

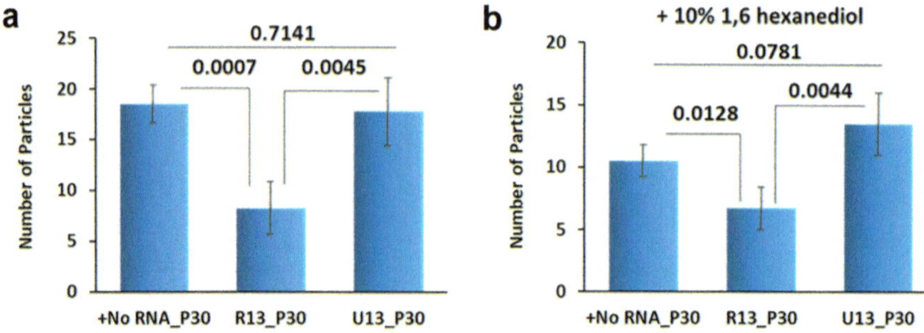

Fig. 1.8 (**a**) A bar graph showing the numbers of particles >0.002 mm^2 in fluorescence microscope images of 100 nM FUS fusion protein (Strep-GFP-FUS) after the addition of an equimolar amount of either no RNA, R13, or U_{13} and subsequent application of 30 strokes of pipetting. The bar graphs show the averages of three independent experiments ±S.D. p values are indicated. (**b**) A bar graph showing the numbers of particles >0.002 mm^2 in fluorescence microscope images after addition of 10% 1,6-hexanediol to the samples of (**a**). (Reproduced and modified from Hamad et al. 2021)

in the presence of R13 than in the presence of U_{13}. Thus, it is confirmed that the suppression of FUS aggregation by R13 is sequence-specific.

1.4.3 The Mechanism of Suppression of FUS Aggregation by Long Non-coding RNA in a Sequence-Dependent Manner

It was revealed that the prion-like domain (residues 1–239), comprising LC and a part of RGG1, and RGG2 are essential for aggregation (Sun et al. 2011). Previously, we revealed that lncRNA and its fragments bind to the C-terminal region of FUS comprising RGG2, ZnF, and RGG3 (Yoneda et al. 2016). Therefore, there is the possibility that RNA bound to the C-terminal region of FUS masks the interface required for the formation of aggregates, which results in the prevention of aggregate formation.

The cation–π interaction between the LC domain and RGG domains was reported to be critical for LLPS of FUS (Wang et al. 2018; Qamar et al. 2018). It is supposed that RNA bound to the C-terminal region of FUS neutralizes the cations and reduces the cation–π interaction. The reduction of the interaction may prevent FUS not only from LLPS but also from aggregation.

Specific RNAs bind to FUS with higher affinity than non-specific ones. This might be a reason why the suppressive effect on aggregation of FUS is RNA sequence-dependent. It might also be the case that non-specific RNA does not necessarily bind to the interface needed for the formation of aggregates, which results in a lower suppressive effect. The correlation between the extent of conformational change of FUS caused by each RNA and resistance to aggregation by each

RNA may also imply that the FUS conformation induced by RNA is unfavorable for the formation of aggregates.

1.5 Conclusions

We found that the shear stress caused by pipetting instantly induces the transition of FUS to amorphous aggregates even at low FUS concentration. The non-coding RNA we previously identified (Wang et al. 2008; Yoneda et al. 2016), lncRNA, can suppress this transition in a sequence-dependent manner. Our finding might serve for the development of therapies for neurodegenerative diseases such as ALS and HD by using RNA as aggregation inhibitors.

Acknowledgments This work was conducted in collaboration with Professors N. Hamad, T. Mashima, T. Nagata, Y. Yamaoki, and K. Kondo of Kyoto University, Professors R. Yoneda and R. Kurokawa of Saitama medical University, Professors H. Watanabe and T. Uchihashi of Nagoya University, Professor M. So of Osaka University, and Professor T. Oyoshi of Shizuoka University.

This work was supported by JSPS KAKENHI, Japan (20H03192, 22H05596, 23H02419 and 23H04069 to M. K., 20 K06524 to T. N., and 19 K16054 and 22 K05314 to Y. Y.), AMED, Japan (JP23fk0410048 to M. K. and 22ak0101097 to T. N.), and the joint usage/research programs of the Institute of Advanced Energy, Kyoto University (ZE2023A-12).

References

Akkermans C et al (2006) Shear pulses nucleate fibril aggregation. Food Biophys 1:144–150

Alberti S, Hyman AA (2016) Are aberrant phase transitions a driver of cellular aging? BioEssays 38:959–968

Alberti S, Gladfelter A, Mittag T (2018) Leading edge primer considerations and challenges in studying liquid-liquid phase separation and biomolecular condensates. Cell 176:419–434

Bekard IB, Asimakis P, Bertolini J, Dunstan DE (2011) Review the effects of shear flow on protein structure and function. Biopolymers 95:733–745

Campos-Melo D, Droppelmann CA, Volkening K, Strong MJ (2014) RNA-binding proteins as molecular links between cancer and neurodegeneration. Biogerontology 15:587–610

Di Stasio E, De Cristofaro R (2010) The effect of shear stress on protein conformation: physical forces operating on biochemical systems: the case of von Willebrand factor. Biophys Chem 153: 1–8

Doi H et al (2004) Identification of ubiquitin-interacting proteins in purified polyglutamine aggregates. FEBS Lett 571:171–176

Doi H et al (2008) RNA-binding protein TLS is a major nuclear aggregate-interacting protein in huntingtin exon 1 with expanded polyglutamine-expressing cells. J Biol Chem 283:6489–6500

Doi H, Koyano S, Suzuki Y, Nukina N, Kuroiwa Y (2010) The RNA-binding protein FUS/TLS is a common aggregate-interacting protein in polyglutamine diseases. Neurosci Res 66:131–133

Duerkop M, Berger E, Dürauer A, Jungbauer A (2018) Impact of cavitation, high shear stress and air/liquid interfaces on protein aggregation. Biotechnol J 13:1800062

Hamad N, Mashima T, Yamaoki Y, Kondo K, Yoneda R, Oyoshi T, Kurokawa R, Katahira M (2020a) RNA sequence and length contribute to RNA-induced conformational change of TLS/FUS. Sci Rep 10:2629
Hamad N, Watanabe H, Uchihashi T, Kurokawa R, Nagata T, Katahira M (2020b) Direct visualization of the conformational change of FUS/TLS upon binding to promoter-associated non--coding RNA. Chem Commun 56:9134–9137
Hamad N, Yoneda R, So M, Kurokawa R, Nagata T, Katahira M (2021) Non-coding RNA suppresses FUS aggregation caused by mechanistic shear stress on pipetting in a sequence-dependent manner. Sci Rep 11:9523
Hill EK, Krebs B, Goodall DG, Howlett GJ, Dunstan DE (2006) Shear flow induces amyloid fibril formation. Biomacromolecules 7:10–13
Hoell JI et al (2011) RNA targets of wild-type and mutant FET family proteins. Nat Struct Mol Biol 18:1428
Kroschwald S, Maharana S, Simon A (2017) Hexanediol: a chemical probe to investigate the material properties of membrane-less compartments. Matters 3:e201702000010
Kwiatkowski TJ et al (2009) Mutations in the FUS/TLS gene on chromosome 16 cause familial amyotrophic lateral sclerosis. Science 323:1205–1208
Lagier-Tourenne C, Polymenidou M, Cleveland DW (2010) TDP-43 and FUS/TLS: emerging roles in RNA processing and neurodegeneration. Hum Mol Genet 19:R46–R64
Maharana S et al (2018) RNA buffers the phase separation behavior of prion-like RNA binding proteins. Science 360:918–921
Molliex A et al (2015) Phase separation by low complexity domains promotes stress granule assembly and drives pathological fibrillization. Cell 163:123–133
Neumann M et al (2009) A new subtype of frontotemporal lobar degeneration with FUS pathology. Brain 132:2922–2931
Patel A et al (2015) A liquid-to-solid phase transition of the ALS protein FUS accelerated by disease mutation. Cell 162:1066–1077
Qamar S et al (2018) FUS phase separation is modulated by a molecular chaperone and methylation of arginine cation-π interactions. Cell 173:720–734.e15
Ramstack JM, Zuckerman L, Mockros LF (1979) Shear-induced activation of platelets. J Biomech 12:113–125
Sahoo H (2011) Förster resonance energy transfer – a spectroscopic nanoruler: principle and applications. J Photochem Photobiol C: Photochem Rev 12:20–30
Schwartz JC, Wang X, Podell ER, Cech TR (2013) RNA seeds higher-order assembly of FUS protein. Cell Rep 5:918–925
Shen Y et al (2020) Biomolecular condensates undergo a generic shear-mediated liquid-to-solid transition. Nat Nanotechnol 15:841–847
St George-Hyslop P et al (2018) The physiological and pathological biophysics of phase separation and gelation of RNA binding proteins in amyotrophic lateral sclerosis and fronto-temporal lobar degeneration. Brain Res 1693:11–23
Sun Z et al (2011) Molecular determinants and genetic modifiers of aggregation and toxicity for the ALS disease protein FUS/TLS. PLoS Biol 9:e1000614
Takahama K et al (2013) Regulation of telomere length by G-quadruplex telomere DNA- and TERRA-binding protein TLS/FUS. Chem Biol 20:341
Vance C et al (2009) Mutations in FUS, an RNA processing protein, cause familial amyotrophic lateral sclerosis type 6. Science 323:1208–1211
Wang X et al (2008) Induced ncRNAs allosterically modify RNA-binding proteins in cis to inhibit transcription. Nature 454:126–130
Wang J et al (2018) A molecular grammar governing the driving forces for phase separation of prion-like RNA binding proteins. Cell 174:688–699.e16
Xie Y, Wang F, Puscheck EE, Rappolee DA (2007) Pipetting causes shear stress and elevation of phosphorylated stress-activated protein kinase/jun kinase in preimplantation embryos. Mol Reprod Dev 74:1287–1294

Yoneda R et al (2016) The binding specificity of Translocated in LipoSarcoma/FUsed in Sarcoma with lncRNA transcribed from the promoter region of cyclin D1. Cell Biosci 6:4

Yu Y, Reed R (2015) FUS functions in coupling transcription to splicing by mediating an interaction between RNAP II and U1 snRNP. Proc Natl Acad Sci U S A 112:8608–8613

Zhang T et al (2018) FUS regulates activity of MicroRNA-mediated gene silencing. Mol Cell 69: 787–801.e8

Chapter 2
Basics and Recent Advances in Computational and Theoretical Methods for Understanding the Liquid–Liquid Phase Separation

Takefumi Yamashita

Abstract This chapter provides an overview of the basics and recent advances in computational and theoretical methods for understanding liquid–liquid phase separation (LLPS). This phenomenon plays a critical role in many biological processes, including the formation of membraneless organelles in cells. Computational and theoretical methods offer a complementary approach to experimental methods, allowing us to gain a deep understanding of the physical principles that govern LLPS in biology. This chapter discusses several theoretical and computational approaches, including the Flory–Huggins theory, theory of viscoelastic phase separation, all-atom molecular dynamics (MD) simulation, coarse-grained MD simulation, and informatics methods. These methods can help us to develop a comprehensive understanding of LLPS and its role in cellular function and disease.

Keywords Liquid–liquid phase separation · Thermodynamic properties · Membraneless organelles · Computational method · Phase separation theory · Flory-Huggins theory · Viscoelastic phase separation · Molecular dynamics simulation · All-atom model · Coarse-grained model · Force field · Informatics

2.1 Introduction

Liquid-liquid phase separation (LLPS) is a phenomenon in which a homogeneous mixture of two (or more liquids) separates into two distinct liquid phases with different compositions and properties. This process is driven by the thermodynamic properties of the components in the mixture, such as their tendency to form clusters

T. Yamashita (✉)
Laboratory for Systems Biology and Medicine, Research Center for Advanced Science and Technology, The University of Tokyo, Meguro-ku, Tokyo, Japan

Department of Physical Chemistry, School of Pharmacy and Pharmaceutical Sciences, Hoshi University, Shinagawa-ku, Tokyo, Japan
e-mail: yamashita@lsbm.org

R. Kurokawa (ed.), *Phase Separation in Living Cells*,
https://doi.org/10.1007/978-981-99-4886-4_2

or aggregates. In LLPS, the two phases that form are typically a dense liquid phase and a dilute liquid phase. The dense phase contains a higher concentration of some specific components (e.g., polymer), which have a strong tendency to cluster or aggregate, while the dilute phase contains a lower concentration of these components. This phase separation can occur spontaneously or be induced by changes in the temperature, pressure, or concentration of the components in the mixture. LLPS plays a critical role in material science, where it can be used to create complex structures and functional materials with unique properties.

LLPS is also important in many biological processes that underlie the formation of many membraneless organelles in cells (Hyman et al. 2014; Alberti et al. 2019; Alberti and Dormann 2019). These organelles are formed through the clustering and coalescence of specific biomolecules, such as proteins and RNA. One well-known example of LLPS in biology is the formation of the nucleolus, a large, membraneless organelle found within the nucleus of eukaryotic cells. The nucleolus is formed through the LLPS of specific RNA, DNA, and protein molecules, which cluster together to form a liquid-like droplet that serves as a site of ribosome biogenesis. Other examples of membraneless organelles formed through LLPS in cells include stress granules, P-bodies, and Cajal bodies. These organelles play important roles in processes such as cellular stress response, RNA metabolism, RNA processing, etc.

Although the study of LLPS in biology has become active, it is still complex and challenging. One of the primary difficulties is the inherent complexity of biomolecular systems. These systems contain a vast array of different biomolecules, each with unique properties and interactions. Another challenge is the heterogeneity of biomolecules involved in LLPS. They can be highly diverse and dynamic, making it difficult to isolate and study specific components of the system.

To overcome such difficulties associated with the study of LLPS, computational and theoretical methods can play crucial roles. One advantage of computational methods is that they can provide a more detailed and comprehensive view of the behavior of biomolecules involved in LLPS. For example, some computational methods can simulate the interactions between individual biomolecules at the molecular level, providing insights into the underlying mechanisms driving LLPS or experimental condition controlling LLPS. Computational methods can also help address the challenge of heterogeneity in biological systems. By modeling the behavior of individual biomolecules, researchers can gain insights into how the heterogeneity of these molecules contributes to LLPS behavior. This can help to identify specific biomolecules that are critical for LLPS and provide insights into how their interactions contribute to the formation and behavior of LLPS structures.

Theoretical studies are also important for understanding the LLPS phenomenon in biology. Theoretical models provide a framework for understanding the underlying physics that drive phase separation, allowing researchers to make predictions about the behavior of biomolecules and the formation of membraneless organelles. One example of a theoretical model used to study LLPS is the Flory-Huggins theory, which describes the thermodynamics of polymer solutions and can be used to predict the conditions under which phase separation will occur.

Computational and theoretical studies provide a complementary approach to experimental studies, allowing researchers to gain a deep understanding of the physical principles that govern LLPS in biology. By combining theoretical, computational, and experimental approaches, researchers can develop a comprehensive understanding of this important phenomenon and its role in cellular function and disease. In this chapter, we explain and discuss the basics and recent advances in computational and theoretical methods for understanding the LLPS phenomena.

2.2 Theoretical Approaches

2.2.1 Flory-Huggins Theory: Basic Theory for Understanding the LLPS Phenomena

The Flory-Huggins theory, which was first proposed independently by Paul Flory and Maurice Huggins in 1942, has been widely used to describe and predict LLPS and of course can be used to understand the biological LLPS (Flory 1953). The theory describes the free energy of mixing between two different types of molecules in a solution, such as proteins and water. The theory treats a polymer solution consisting of two components: the polymer itself and the solvent in which it is dissolved. To describe the LLPS simply, the Flory-Huggins theory introduces the assumption of a regular lattice structure for the molecular arrangement.

Following the idea of Flory-Huggins theory, we first introduce a lattice model (Fig. 2.1). Here, the number of cells is N, and the cell size is essentially the same as the solvent size and the size of the polymer segment. In this system, N_0 solvent molecules and N_1 polymers exist. Each polymer consists of P segments. (In the case of Fig. 2.1, $N = 36$, $N_0 = 18$, $N_1 = 3$, and $P = 6$.) Therefore, we obtain:

$$N = N_0 + P \times N_1$$

To evaluate the mixing entropy, let us count the patterns in which particles are placed in the mixed state (Fig. 2.1a). First, the number of patterns in which the first

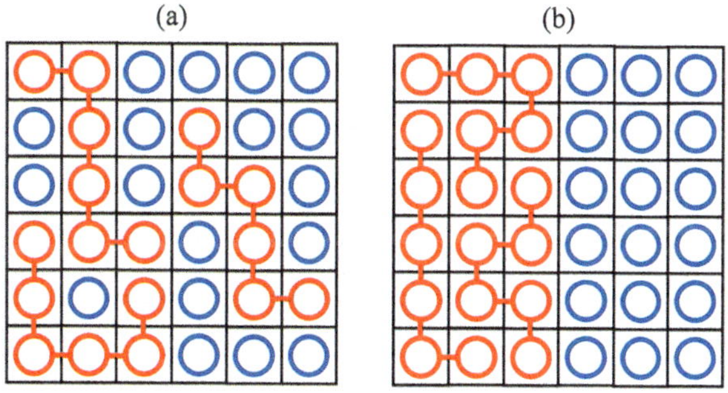

Fig. 2.1 Lattice model of a polymer solution in (**a**) the mixed state and (**b**) phase-separated state

segment of the first polymer is placed is N. Because the second segment is bonded to the first segment, the pattern number for the second segment is the same as the number of the nearest neighbor cells (Z). For the third segment, due to the bond to the second segment and the excluded volume effect from the first segment, the pattern number is $Z - 1$. If the pattern numbers for the other segments can be considered the same as that for the third segment, the number of patterns in which the first polymer is placed can be given by

$$\nu_1 = N \times Z(Z-1)^{P-2} = Nf_{\max}$$

Here, $f_{\max} = Z(Z-1)^{P-2}$ is the maximum flexibility, which represents the number of conformations or internal degree of freedom of the polymer. However, when the kth segment is placed, $(k - 1)$ cells were already occupied. Therefore, the probability that the cell in which the segment is placed is not occupied is

$$P_{\text{unocc}} = 1 - \frac{k-1}{N} = \frac{N-(k-1)}{N}$$

By taking this effect into account, for the kth segment ($k > 2$), the number of placement patterns can be approximated by the following equation:

$$(Z-1)P_{\text{unocc}} = (Z-1)\frac{N-(k-1)}{N}$$

This approximation can be considered as one kind of the mean field approximation. Consequently, the number of patterns in which the first polymer is placed can be also approximated by

$$\nu_1 = Z(Z-1)^{P-2} \times \frac{1}{N^{P-1}} \times \frac{N!}{(N-P)!}$$

By taking the presence of the first polymer into account, the pattern number of the second polymer placement is

$$\nu_2 = Z(Z-1)^{P-2} \times \frac{1}{N^{P-1}} \times \frac{(N-P)!}{(N-2P)!}$$

In the same way, the pattern number of the ith polymer placement is

$$\nu_i = Z(Z-1)^{P-2} \times \frac{1}{N^{P-1}} \times \frac{[N-P(i-1)]!}{[N-Pi]!}$$

Thus, the number of patterns in which N_1 polymers are placed can be given by

$$\prod_{i=1}^{N_1} \nu_i = f_{\max}^{N_1} \times \frac{1}{N^{(P-1)N_1}} \times \frac{N!}{N_0!}$$

The pattern number of the solvent placement is $N_0!$. Thus, the total pattern number for the placement can be given by

$$\Omega = \frac{1}{N_0! N_1! \delta^{N_1}} \left(\prod_{i=1}^{N} \nu_i \right) N_0! = \frac{1}{N_1! \delta^{N_1}} \times \prod_{i=1}^{N} \nu_i = \frac{1}{N_1! \delta^{N_1}} \times \left(\frac{f_{\max}}{N^{p-1}} \right)^{N_1} \times \frac{N!}{N_0!}$$

Here, the equation is divided by $N_0 ! N_1 !$, because we do not distinguish the same type of molecules one by one. Also, taking the polymer symmetry into account, we divide the number by δ^{N_1}, where the δ is the number of duplicates originated from the polymer symmetry. Thus, the entropy of this mixing state can be given by

$$S_{\text{mix}} = k_{\text{B}} \ln \Omega$$

where k_{B} is the Boltzmann constant. Then,

$$\begin{aligned} \frac{S_{\text{mix}}}{k_{\text{B}}} &= \ln \Omega = \ln \left[\frac{1}{N_1! \delta^{N_1}} \times \left(\frac{f_{\max}}{N^{p-1}} \right)^{N_1} \times \frac{N!}{N_0!} \right] \\ &= -\ln N_1! + N_1 \ln \left(\frac{f_{\max}}{\delta} \right) - N_1 (P-1) \ln N + \ln N! - \ln N_0! \end{aligned}$$

By using Stirling's formula, we obtain

$$\frac{S_{\text{mix}}}{k_{\text{B}}} = \ln \Omega = N_1 \ln \left(\frac{f_{\max}}{\delta} \right) - N_1 (P-1) - N_1 (\ln PN_1 - \ln P)$$

$$- N_0 \ln N_0 + N_0 \ln N + N_1 \ln N = N_1 \ln \left(\frac{P f_{\max}}{\delta e^{p-1}} \right) - N_1 \ln \left(\frac{PN_1}{N} \right) - N_0 \ln \left(\frac{N_0}{N} \right)$$

If we suppose the cell volume is ν, the volume fractions of solvent and polymer are given by

$$\varphi_0 = \frac{V_0}{V} = \frac{\nu N_0}{\nu N} = \frac{N_0}{N}$$

and

$$\varphi_1 = \frac{V_1}{V} = \frac{\nu N_1 P}{\nu N} = \frac{N_1 P}{N}$$

respectively. Thus, the entropy of the mixing state can be also given by

$$S_{\text{mix}} = k_B \ln \Omega(N_0, N_1, P) = k_B N_1 \ln\left(\frac{Pf_{\max}}{\delta e^{P-1}}\right) - k_B(N_1 \ln \varphi_1 + N_0 \ln \varphi_0)$$

In the phase separation state (Fig. 2.1b), N_1 polymers are placed in the N_1P cells and N_0 solvents are placed in N_0 cells. Using the same approach as for the mixing state, the number of placement patterns of polymer can be given by

$$\Omega_1 = \frac{1}{N_1!\delta^{N_1}} f_{\max}^{N_1} \frac{(N_1P)!}{(N_1P)^{(P-1)N_1}}$$

However, the placement pattern number is $\Omega_0 = 1$. Thus, the entropy of the phase separation state can be given by

$$S_{\text{sep}} = k_B \ln \Omega_0 \Omega_1 = k_B \ln \Omega_1$$

By using Stirling's formula, we obtain

$$S_{\text{sep}} = k_B N_1 \ln\left(\frac{Pf_{\max}}{\delta e^{P-1}}\right)$$

Now, we can evaluate the change in entropy for mixing the polymer phase and solvent phase using the following equation:

$$\Delta S_{\text{mix}} = S_{\text{mix}} - S_{\text{sep}} = -k_B(N_1 \ln \varphi_1 + N_0 \ln \varphi_0)$$

This is the so-called mixing entropy.

To obtain deeper insight, let us consider the case where the polymer segments are not bonded to each other. In this case, the mixing entropy can be given by

$$\Delta S'_{\text{mix}} = S'_{\text{mix}} - S'_{\text{sep}} = -k_B N_0 \ln \varphi_0 - k_B N_1 P \ln \varphi_1$$

By comparison of $\Delta S'_{\text{mix}}$ and ΔS_{mix}, we can realize that the polymerization decreases the mixing entropy by

$$| \, k_B N_1 (P-1) \ln \varphi_1 \, | \, .$$

Note that $\varphi_1 < 1$ and $\ln\varphi_1 < 0$.

To investigate the mixing enthalpy, let us consider that two solvent molecules are in contact with each other and that two polymer segments are in contact with each other. While the interaction energy between the solvent molecules is u_{00}, the interaction energy between the polymer segments is u_{11}. If these pairs are dissociated and then two solvent molecules are in contact with two polymer segments, respectively (the interaction energy between the solvent molecule and polymer segment is

u_{01}), the energetic change is $2u_{01} - (u_{00} + u_{11})$. For each new contact, the energy change can be written by

$$\Delta u = u_{01} - \frac{1}{2}(u_{00} + u_{11})$$

To simply evaluate the number of solvent–solvent contact pairs in the mixing state (N_{00}), a kind of the mean field approximation is used. Each solvent molecule is in contact with Z particles. The probability that the particle in the cell is a solvent is given by $\varphi_0 = N_0/N$. Thus, the expected number of the contact solvent molecules can be approximated by $Z\varphi_0$. Therefore, we obtain the following equation:

$$N_{00} = N_0 Z\varphi_0/2 = N\, Z{\varphi_0}^2/2$$

In the same way, the numbers of contact solvent molecules and contact polymer segments for a polymer segment can be given by $Z\varphi_0$ and $Z\varphi_1$, respectively. Thus, the number of solvent–segment contact pairs can be given by

$$N_{01} = PN_1 \times Z\varphi_0 = N\, Z\varphi_0\varphi_1$$

and the number of segment–segment contact pairs can be given by

$$N_{11} = PN_1 Z\varphi_1/2 = N\, Z{\varphi_1}^2/2$$

Then, the interaction energy of the mixing state can be written by

$$E_{\text{mix}} = N_{00}u_{00} + N_{11}u_{11} + N_{01}u_{01}$$

For the phase separation state, since the contact between the solvent molecule and polymer segment is negligible, the numbers of solvent–solvent contact pairs (N'_{00}), segment–segment contact pairs (N'_{11}), and solvent–segment contact pairs (N'_{01}) can be given by

$$\begin{aligned} N'_{00} &= ZN_0/2 = Z\varphi_0 N/2 \\ N'_{11} &= Z(N_1 P)/2 = Z\varphi_1 N/2 \\ N'_{01} &= 0 \end{aligned}$$

Then, the interaction energy of the phase separation state can be written by

$$E_{\text{sep}} = u_{00}N'_{00} + u_{11}N'_{11} + u_{01}N'_{01} = u_{00}N'_{00} + u_{11}N'_{11}$$

Consequently, the energy change associated with the mixing of the polymer phase and solvent phase can be given by

$$\Delta E_{\text{mix}} = E_{\text{mix}} - E_{\text{sep}} = u_{00}\left(N_{00} - N'_{00}\right) + u_{11}\left(N_{11} - N'_{11}\right) + u_{01}N_{01}$$
$$= -u_{00}N\,Z\varphi_0\varphi_1/2 - u_{11}NZ\varphi_0\varphi_1/2 + u_{01}N\,Z\varphi_0\varphi_1$$

Note that:

$$N_{00} - N'_{00} = N\,Z{\varphi_0}^2/2 - NZ\varphi_0/2 = -\frac{NZ\varphi_0}{2}(1-\varphi_0)$$
$$N_{11} - N'_{11} = N\,Z{\varphi_1}^2/2 - NZ\varphi_1/2 = -\frac{NZ\varphi_1}{2}(1-\varphi_1)$$

Then,

$$\Delta E_{\text{mix}} = NZ\varphi_0\varphi_1\left(u_{01} - \frac{u_{00}+u_{11}}{2}\right) = NZ\varphi_0\varphi_1\Delta u$$

This is the so-called mixing energy. As we consider the system volume is constant, this can be seen as the mixing enthalpy. Here, we introduce a dimensionless parameter

$$\chi = \frac{Z\Delta u}{k_BT}$$

which is called Flory's χ parameter. Thus, we obtain the following equation:

$$\Delta E_{\text{mix}} = N\,k_BT\chi\varphi_0\varphi_1$$

Consequently, the mixing free energy (the free energy change associated with the mixing) can be given by

$$\Delta F_{\text{mix}} = \Delta E_{\text{mix}} - T\Delta S_{\text{mix}} = k_BT\chi\varphi_0\varphi_1 + k_BT(N_1\ln\varphi_1 + N_0\ln\varphi_0)$$

Furthermore, the dimensionless mixing free energy per cell can be defined as

$$f(\phi,T) \equiv \frac{\Delta F_{\text{mix}}}{Nk_BT} = \chi\varphi_0\varphi_1 + \frac{N_0}{N}\ln\varphi_0 - \frac{N_1}{N}\ln\varphi_1$$

If the volume fraction of polymer (φ_1) is replaced by ϕ, we obtain the following equation:

$$f(\phi,T) = \frac{\phi}{P}\ln\phi + (1-\phi)\ln(1-\phi) + \chi\phi(1-\phi)$$

If the mixing free energy (ΔF_{mix} or f) is positive, the phase separation state will appear. Otherwise, the system will prefer the mixing state. While the first two terms are the entropic contribution, the last term is the enthalpic contribution. Because the

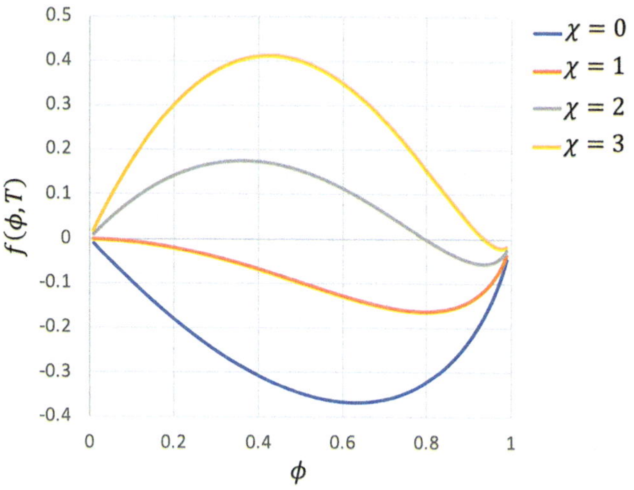

Fig. 2.2 Dimensionless mixing free energy as a function of volume fraction of polymer

entropic terms are always negative, the χ parameter, which determines the mixing energy, is the key factor that causes the phase separation (Fig. 2.2). Therefore, we can consider that the intermolecular interaction in the system is very important to understand or control the LLPS phenomena.

In summary, the Flory-Huggins theory provides a simple and useful framework for understanding the conditions under which LLPS occurs. As discussed above, this theory clearly shows that the intermolecular interaction is one key factor of LLPS. Although this theory enables us to understand the LLPS qualitatively, it has some serious limitations; the Flory-Huggins theory is based on the lattice model and thus cannot be used for the quantitative prediction. Furthermore, the mean field approximation, which is used in the derivation of the Flory-Huggins theory, is not accurate for the dilute solution. While the mean field approximation assumes that the polymer segments are uniformly distributed in the system, the polymer segments are localized in the dilute solution.

2.2.2 *Theory of Viscoelastic Phase Separation*

When LLPS occurs, the formation of the droplet of the minority phase is usually observed due to the minimization of the interface free energy. This means that membraneless organelles in the cell are usually droplet-like. (The minority phase is mainly composed of biological polymers, while the main component of the major phase is water.) However, it is known that membraneless organelles such as TIS granules, RNP granules, and centrosome aggregates have the mesh-like network structure. These membraneless organelles are formed by a different type of LLPS, the so-called viscoelastic phase separation, which is named after the fact that this phase separation is led by the viscoelasticity (Tanaka 2000, 2022).

Viscoelasticity is a material property that describes how a material deforms under the stress. Viscoelastic materials have both elastic (solid-like) and viscous (liquid-like) properties, which means that they exhibit a combination of elastic and viscous behavior under stress. For example, when you press a tube of toothpaste, the toothpaste flows out of the tube like a liquid. However, on the toothbrush, the toothpaste retains its shape and behaves as a solid.

The viscoelastic phase separation is basically attributed to the difference in the mobility between the minority phase and majority phase, which is called "dynamic asymmetry" (Tanaka 2022). Let us consider the mixture of the slow minor component (i.e., polymer) and fast major component (i.e., solvents). First, polymers aggregate and form small droplets. Solvents are excluded from the droplets and thus the concentration of the polymer is rapidly increased in the droplets, which results in the tight entanglement of the polymers. This suppresses the transfer of polymer between droplets in the droplet collision process, which consequently prohibits the coalescence and fusion of droplets. Finally, the droplets, which behave like gel balls, form the network structure. Again, the interaction between polymers is a dominant factor to lead to the tight entanglement and slowdown of polymers.

Ma and Mayr (2018) found that the RNA-binding protein TIS11B forms TIS granules, which enrich or exclude specific mRNAs and proteins. Although TIS granules have a mesh-like structure but not a droplet-like sphere, highly dynamic proteins were observed in TIS granules. More recent study (Ma et al. 2021) showed that the multivalent RNA–RNA interaction is the dominant factor to form the mesh-like structure. Furthermore, it was found that large unstructured regions of mRNAs highly tend to form mesh-like condensates, indicating that the unstructuredness of RNA is a key factor to form a broad RNA–RNA interaction network. We can consider that this is very important to induce not only the entanglement effect but also the dynamic asymmetry effect that causes the viscoelastic phase separation. Note that similar multivalent interaction was suggested to play an important role in the mammalian X-chromosome inactivation (Matsuno et al. 2019).

2.3 Computational Approaches

Computational science has emerged as a third science to complement experimental and theoretical science (Cao 2017). (Although information science is often considered a fourth science, here we include it in computational science.) By utilizing highly advanced computational technology, we are now able to obtain quantitative predictions and detailed understandings that could not be obtained by experiment or theory.

One area in which computational science has had a significant impact is the study of biological systems. Many biological systems involve multiple interacting components that are difficult to study experimentally or theoretically. However, using computational methods, one can simulate the behavior of these systems and gain insights into their underlying mechanisms. In addition to computational simulations, large databases on biomolecules have been developed, and informatics analysis including machine learning and AI is now applied to various biological problems.

In the study of LLPS, computational approaches have several advantages. One of the most important advantages is that they allow for a more detailed and systematic view of the behavior of the biomolecules involved in LLPS. As discussed above, the structure and interaction of biomolecules are the most important factors underlying the LLPS. By using computational techniques, one can analyze/predict the interactions between biomolecules at the molecular level, providing insights into the underlying mechanisms driving LLPS.

2.3.1 Molecular Dynamics Simulation

In computational studies of LLPS, a range of useful methods exist, and the selection of an appropriate method hinges upon the specific research question and system being analyzed. Among these methods, molecular dynamics (MD) simulations (Karplus and McCammon 2002; Yamashita 2016a, 2018, 2022) are frequently employed to investigate the temporal evolution of the biomolecular system. The MD simulation can provide detailed insights into the interactions between individual biomolecules and the mechanisms driving LLPS. However, these simulations can be computationally expensive and may require high-performance computing resources.

MD simulation is a computational method used to study the behavior of individual molecules over time. In this method, a computer program simulates the movement of molecules in a system, such as proteins, nucleic acids, or other biomolecules. The simulation begins with a set of initial conditions, such as the positions and velocities of each molecule in the system, and then calculates how the system evolves over time using Newton's laws of motion. To calculate the change in motion of an atom, the total force on this atom is necessary and is usually calculated based on their interactions, such as electrostatic forces and van der Waals forces. In such MD simulation, molecules are modeled as individual atoms, which means that each atom in a molecule is explicitly represented in the simulation. This model is called all-atom model (Fig. 2.3) and the MD simulation using the all-atom model is called all-atom MD (AA-MD) simulation. In an AA-MD simulation, the molecules in the system are typically described using a force field, which is a mathematical model that describes the potential energy of the system as a function of the positions of the atoms. The force field includes terms for bond stretching, angle bending, torsion energy, and nonbonded interactions such as van der Waals and electrostatic interactions. Note that forces can directly obtained from the electronic state calculation and the MD simulation using such a force calculation method is called ab initio MD simulation or first-principle MD simulation (Lippert et al. 1999) (see also Ref. (Yamashita and Takatsuka 2007)). Although forces obtained from electronic state calculations are generally more accurate than those from force fields, the

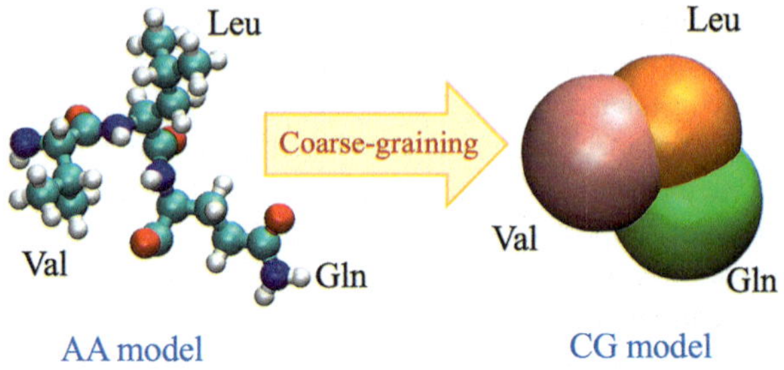

Fig. 2.3 Illustration of AA model and CG model of the Val-Leu-Gln peptide

computational cost associated with these calculations can be too high for studying the long-time behavior of molecular systems.

In MD simulations, the accuracy of a force field plays a crucial role in determining the reliability of the simulation outcomes. Hence, many force fields have been extensively developed, among which the AMBER, CHARMM, and OPLS-AA force fields are the most renowned (Wang et al. 2004; Zhu et al. 2012; Kony et al. 2002). These force fields are developed by fitting their parameters to experimental or ab initio calculation data, and thus, the accuracy of a force field is reliant on the quality of data used to develop it. Recently, force field parameters have been refined using highly precise ab initio calculations (Sasaki and Yamashita 2021; He et al. 2022). Notably, the FUJI force field is a modified version of the AMBER force field, of which main chain torsional parameters were based on the DF-LCCSD(T0)/aug-cc-pVTZ//DF-LMP2/aug-cc-pVTZ-level ab initio calculations (Fujitani et al. 2009, 2013; Yamashita et al. 2019a, 2022; Yamashita 2015).

One of the major challenges in simulating protein systems using AA-MD is the enormous number of atoms involved, which ranges from thousands to millions. However, with the recent advancements in computational hardware and software, it has become feasible to simulate such protein systems with the all-atom model in microsecond-order time. This has enabled us to investigate a wide range of protein issues, including membrane proteins, intrinsically disordered proteins, and protein–protein interactions. Furthermore, the application of advanced sampling techniques (e.g., the umbrella sampling and replica exchange methods) has enabled us to sample rare events and explore the free energy landscape of protein systems, which is important for understanding the molecular mechanisms underlying protein functions (Yamashita 2018, 2022; Bernardi et al. 2015; Yamashita and Fujitani 2014; Yamashita et al. 2014, 2015).

In summary, the AA-MD simulation has become an accurate tool for investigating various structures and interactions of biomolecules. As discussed above, the behavior of large unstructured region is critical to understanding and controlling LLPS. The AA-MD simulation is a promising tool for characterizing the structural features (Sasaki et al. 2018; Yamashita 2016b), clarifying the intermolecular interactions (Takamatsu et al. 2022; Nasrin et al. 2021; Mahmood and Yamashita 2021; Okajima et al. 2021; Koyama et al. 2019; Yamashita et al. 2019b; Miyanabe et al. 2018; Sakano et al. 2016; Nakayama et al. 2015; Mitsui et al. 2022), and ultimately understanding the molecular-level mechanisms of LLPS. In future, it will be challenging and important to computationally study the mechanisms of chemical reactions in the condensates generated by LLPS, as they can facilitate and enhance specific chemical reactions, such as the role of catalysts. For such studies, not only ab initio MD simulations but also MD simulations with reactive classical force fields (e.g., MS-EVB) will be helpful (Voth 2006; Xu et al. 2013; Yamashita and Voth 2010, 2012; Yamashita et al. 2012; Yamashita 2014).

2.3.2 Coarse-Grained Molecular Dynamics Simulation

AA-MD simulations are usually useful for studying molecular behavior that occurs on the microsecond time scale. However, the length of time that can be simulated depends on the size of the system, the level of detail of the simulation, and the computational resources available. For relatively small systems, such as small proteins or peptides, it is possible to perform AA-MD simulations that can capture molecular behavior on the microsecond time scale. However, for larger systems, such as membraneless organelles, the time scale that can be simulated is considerably limited to the nanosecond to microsecond range. As coarse-grained (CG) models (Tozzini 2005; Riniker et al. 2012; Clementi 2008; Dignon et al. 2018, 2019) allow for simulations to be performed over longer time scales with fewer computational resources, MD simulations with CG models (CG-MD simulations) can be advantageous for studying biological LLPS.

CG models simplify the molecular system by grouping several atoms together into a single "bead" or "particle," thereby reducing the number of degrees of freedom in the simulation (Fig. 2.3). This reduction in complexity can allow simulations to be performed on larger systems or for longer time scales than would be feasible with AA-MD simulations.

For example, to conduct CG-MD simulations of LLPS, Dignon et al. (2018) developed two CG models that describe one amino acid residue as a single particle. Some parameters of the CG models were based on previously determined, knowledge-based potentials, and others were parameterized to reproduce the data of experimentally measured radii of gyration of disordered proteins. Both the CG models generated similar intermolecular contact maps in the CG-MD simulations of the concentrate phase. By using the residue-level CG-MD simulation, they successfully determined the phase diagram and coexistence density of proteins in the two phases as a function of the total protein concentration and temperature. They also showed that the CG-MD simulations can capture qualitative changes in the phase diagram due to phosphorylation of FUS and the presence or absence of large folding domains in LAF-1. Note that the unstructured domain of the RNA-binding protein FUS and the DEAD-box helicase protein LAF-1 is well-known to undergo LLPS. Also, the effect of chain length on the phase diagram shown by CG-MD simulations was consistent with the Flory-Huggins theory.

Despite several successful CG-MD studies, CG models can also have limitations, as they sacrifice the detailed information about the structure and interaction which can be captured in the all-atom model (Khot et al. 2019). In some cases, the simplified representation of the system in CG-MD simulations may not accurately capture the physics of the system, and the results obtained may not be applicable to the real system. (For example, CG models may not be sufficiently accurate to investigate the influence of additives on LLPS (Bratek-Skicki et al. 2020), as the atomic-level information about the intermolecular interaction is essential in this case.) Therefore, it is important to carefully validate the CG models to ensure that the results obtained are meaningful and accurate.

There are two different strategies for developing CG models: top-down and bottom-up approaches. The top-down approach starts from experimental data, such as scattering or spectroscopic measurements, and builds a CG model to reproduce the observed properties of the system. One example is the Dignon's CG model mentioned above. However, the bottom-up approach involves starting from an all-atom (AA) model and simplifying it by grouping atoms together into larger particles or beads. Recently, we utilized this approach to develop a CG model of epoxy resin. By using this CG model, we unveiled the detailed mechanism of the density change associated with the conversion (Shoji et al. 2022).

2.3.3 Informatics: Database, Machine Learning, and AI

Recently, machine learning (ML) and artificial intelligence (AI) technologies have undergone significant advancements, fueled by the increasing availability of extensive databases and improvements in computing power. In the realm of biology, AlphaFold2, an AI developed by DeepMind, has emerged as a high-accuracy predictor of 3D protein structures from protein sequences (Jumper et al. 2021). (Figure 2.4 shows an application of AlphaFold2 to FUS.) The AlphaFold2 was trained on a dataset of approximately 170,000 protein structures. The performance of AlphaFold2 was exceptionally well in the 14th Critical Assessment of protein Structure Prediction (CASP14), which is a blind prediction experiment where protein structures are predicted without any prior knowledge of their actual structure. AlphaFold2 was able to accurately predict the structure of many proteins, outperforming other state-of-the-art methods. (See Ref. (Tarca et al. 2007; Chicco 2017; Yamashita and Shoji 2021; Yamashita et al. 2018) for another kinds of ML studies.)

To effectively apply such ML or AI technologies to LLPS problems, a large-scale database containing the LLPS-related information is essential. Fortunately, several LLPS-related databases have been made available recently, including LLPSDB, PhaSePro, PhaSepDB, DrLLPS, RNAgranuleDB, and HUMAN CELL MAP (Li et al. 2020). The first four databases contain LLPS-related proteins directly discovered from LLPS experiments, while latter two are databases containing data of proteome of organelles. With access to these databases, it becomes possible to investigate the features of LLPS and to apply ML or AI technologies to LLPS problems more efficiently.

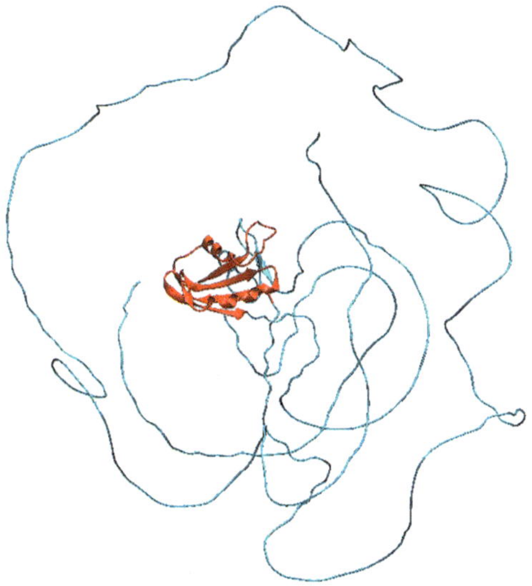

Fig. 2.4 The structure of FUS generated by AlphaFold2. Cyan color represents unstructured regions

2.4 Summary

In this chapter, we explained and discussed several theoretical and computational approaches that can be utilized to study the LLPS phenomena in cells. First, we explained the Flory-Huggins theory, which is the fundamental phase separation theory. Then, we discussed the viscoelastic LLPS to provide an understanding of the condensates' morphology. Furthermore, we explained several computational methods: AA-MD simulations, CG-MD simulations, and informatics approaches. These computational techniques are used to investigate the mechanisms underlying LLPS and to predict the properties of condensates.

Theoretical and computational studies provide a complementary approach to experimental studies, enabling researchers to gain a profound understanding of the physical principles that govern LLPS in biology. By combining theoretical, computational, and experimental approaches, researchers can develop a comprehensive understanding of LLPS and its role in cellular function and disease. The integration of these approaches allows us to address fundamental questions about LLPS, such as how it contributes to the regulation of cellular processes, how it is affected by environmental changes, and how it can cause pathological conditions. Therefore, theoretical and computational studies play a crucial role in advancing our knowledge of LLPS and in developing new therapeutic strategies for diseases caused by dysregulated phase separation.

Acknowledgments This research used computational resources of TSUBAME through the HPCI System Research Project (hp200068, hp210141, hp220134) and Research Center for Computational Science, Okazaki, Japan. We acknowledge support from Scientific Research (JSPS KAKENHI (C)21K03482, (C)18K05025, (B)16KT0050), GAP fund (UTokyo), and Program for Promoting Research on the Supercomputer Fugaku (Application of Molecular Dynamics

Simulation to Precision Medicine Using Big Data Integration System for Drug Discovery, JPMXP1020200201, hp200129, hp210172, hp220164).

References

Alberti S, Dormann D (2019) Liquid–liquid phase separation in disease. Annu Rev Genet 53:171–194

Alberti S, Gladfelter A, Mittag T (2019) Considerations and challenges in studying liquid-liquid phase separation and biomolecular condensates. Cell 176:419–434

Bernardi RC, Melo MCR, Schulten K (2015) Enhanced sampling techniques in molecular dynamics simulations of biological systems. Biochim Biophys Acta 1850:872–877

Bratek-Skicki A, Pancsa R, Meszaros B, Van Lindt J, Tompa P (2020) A guide to regulation of the formation of biomolecular condensates. FEBS J 287:1924–1935

Cao L (2017) Data science: a comprehensive overview. ACM Comput Surv 50:43

Chicco D (2017) Ten quick tips for machine learning in computational biology. BioData Mining 10:35

Clementi C (2008) Coarse-grained models of protein folding: toy models or predictive tools? Curr Opin Struct Biol 18:10–15

Dignon GL, Zheng W, Kim YC, Best RB, Mittal J (2018) Sequence determinants of protein phase behavior from a coarse-grained model. PLoS Comput Biol 14:e1005941

Dignon GL, Zheng W, Mittal J (2019) Simulation methods for liquid–liquid phase separation of disordered proteins. Curr Opin Chem Eng 23:92–98

Flory PJ (1953) Principles of polymer chemistry. Cornell University Press, Ithaca. ISBN 0-8014-0134-8

Fujitani H, Matsuura A, Sakai S, Sato H, Tanida Y (2009) High-level ab initio calculations to improve protein backbone dihedral parameters. J Chem Theory Comput 5:1155–1165

Fujitani H, Shinoda K, Yamashita T, Kodama T (2013) High performance computing for drug development on K computer. J Phys Conf Ser 454:012018

He X, Walker B, Man VH, Ren P, Wang J (2022) Recent progress in general force fields of small molecules. Curr Opin Struct Biol 72:187–193

Hyman AA, Weber CA, Jülicher F (2014) Liquid-liquid phase separation in biology. Annu Rev Cell Dev Biol 30:39–58

Jumper J, Evans R, Pritzel A et al (2021) Highly accurate protein structure prediction with AlphaFold. Nature 596:583–589

Karplus M, McCammon J (2002) Molecular dynamics simulations of biomolecules. Nat Struct Mol Biol 9:646–652

Khot A, Shiring SB, Savoie BM (2019) Evidence of information limitations in coarse-grained models. J Chem Phys 151:244105

Kony D, Damm W, Stoll S, Van Gunsteren WF (2002) An improved OPLS–AA force field for carbohydrates. J Comput Chem 23:1416–1429

Koyama T, Nakamoto M, Morishima K, Yamashita R, Yamashita T, Sasaki K, Kuruma Y, Mizuno N, Suzuki M, Okada K, Ieda R, Uchino T, Tasumi S, Hosoya S, Uno S, Koyama J, Toyoda A, Kikuchi K, Sakamoto T (2019) A SNP in a steroidogenic enzyme is associated with phenotypic sex in Seriola fishes. Curr Biol 29:1901–1909

Li Q, Wang X, Dou Z, Yang W, Huang B, Lou J, Zhang Z (2020) Protein databases related to liquid–liquid phase separation. Int J Mol Sci 21:6796

Lippert G, Hutter J, Parrinello M (1999) The Gaussian and augmented-plane-wave density functional method for ab initio molecular dynamics simulations. Theor Chem Accounts 103:124–140

Ma W, Mayr C (2018) A membraneless organelle associated with the endoplasmic reticulum enables 3 UTR-mediated protein-protein interactions. Cell 175:1492–1506
Ma W, Zheng G, Xie W, Mayr C (2021) In vivo reconstitution finds multivalent RNA–RNA interactions as drivers of mesh-like condensates. elife 10:e64252
Mahmood MI, Yamashita T (2021) Influence of lipid bilayer on the GPCR structure: comparison of all-atom lipid force fields. Bull Chem Soc Jpn 94:2569–2574
Matsuno Y, Yamashita T, Wagatsuma M, Yamakage H (2019) Convergence in LINE-1 nucleotide variations can benefit redundantly forming triplexes with lncRNA in mammalian X-chromosome inactivation. Mob DNA 10:33
Mitsui T, Wada M, Kamiya N, Matsuura A, Yamashita T (2022) Binding pose prediction of a drug candidate, cepharanthine, targeting the SARS-CoV-2 spike protein using large-scale MD simulations. AIP Conf Proc. 2611:020009
Miyanabe K, Yamashita T, Abe Y, Akiba H, Takamatsu Y, Nakakido M, Hamakubo T, Ueda T, Caaveiro J, Tsumoto K (2018) Tyrosine sulfation restricts the conformational ensemble of a flexible peptide, strengthening the binding affinity to an antibody. Biochemistry 57:4177–4185
Nakayama T, Mizohata E, Yamashita T, Nagatoishi S, Nakakido M, Iwanari H, Mochizuki Y, Kado Y, Yokota Y, Satoh R, Tsumoto K, Fujitani H, Kodama T, Hamakubo T, Inoue T (2015) Structural features of interfacial tyrosine residue in ROBO1 fibronectin domain-antibody complex: crystallographic, thermodynamic, and molecular dynamic analyses. Protein Sci 24: 328–340
Nasrin SR, Ganser C, Nishikawa S, Kabir AMR, Sada K, Yamashita T, Ikeguchi M, Uchihashi T, Hess H, Kakugo A (2021) Deformation of microtubules regulates translocation dynamics of kinesin. Sci Adv 7:eabf2211
Okajima R, Hiraoka S, Yamashita T (2021) Environmental effects on salt bridge stability in the protein-protein interface: the case of hen egg-white lysozyme and its antibody, HyHEL-10. J Phys Chem B 125:1542–1549
Riniker S, Allison JR, van Gunsteren WF (2012) On developing coarse-grained models for biomolecular simulation: a review. Phys Chem Chem Phys 14:12423–12430
Sakano T, Mahmood MI, Yamashita T, Fujitani H (2016) Molecular dynamics analysis to evaluate docking pose prediction. Biophys Physicobiol 13:181–194
Sasaki K, Yamashita T (2021) Modification and validation of the DREIDING force field for molecular liquid simulations (DREIDING-UT). J Chem Inf Model 61:1172–1179
Sasaki K, Okajima R, Yamashita T (2018) Liquid structures characterized by a combination of the persistent homology analysis and molecular dynamics simulation. AIP Conf Proc. 2040:020015
Shoji N, Sasaki K, Uedono A, Taniguchi Y, Hayashi K, Matsubara N, Kobayashi T, Yamashita T (2022) Effect of conversion on epoxy resin properties: combined molecular dynamics simulation and experimental study. Polymer 254:125041
Takamatsu Y, Hamakubo T, Yamashita T (2022) Molecular dynamics simulation of the antigen-antibody: complex formation process between hen egg-white lysozyme and HyHEL-10. Bull Chem Soc Jpn 95:1611–1619
Tanaka H (2000) Viscoelastic phase separation. J Phys Condens Matter 12:R207
Tanaka H (2022) Viscoelastic phase separation in biological cells. Commun Phys 5:167
Tarca AL, Carey VJ, Chen XW, Romero R, Drăghici S (2007) Machine learning and its applications to biology. PLoS Comp Biol 3:e116
Tozzini V (2005) Coarse-grained models for proteins. Curr Opin Struct Biol 15:144–150
Voth GA (2006) Computer simulation of proton solvation and transport in aqueous and biomolecular systems. Acc Chem Res 39:143–150
Wang J, Wolf RM, Caldwell JW, Kollman PA, Case DA (2004) Development and testing of a general amber force field. J Comput Chem. 25:1157–1174
Xu J, Yamashita T, Agmon N, Voth GA (2013) On the origin of proton mobility suppression in aqueous solutions of amphiphiles. J Phys Chem B 117:15426–15435
Yamashita T (2014) Properties of a hydrated excess proton near the cholesterol-containing phospholipid bilayer. JPS Conf Proc. 1:013086

Yamashita T (2015) Improvement in empirical potential functions for increasing the utility of molecular dynamics simulations. JPS Conf Proc 5:010003
Yamashita T (2016a) Towards physical understanding of molecular recognition in the cell: recent evolution of molecular dynamics techniques and free energy theories. Biomed Sci 2:34–47
Yamashita T (2016b) On the accurate molecular dynamics analysis of biological molecules. AIP Conf Proc. 1790:020026
Yamashita T (2018) Toward rational antibody design: recent advancements in molecular dynamics simulations. Int Immunol 30:133–140
Yamashita T (2022) Molecular dynamics simulation for investigating antigen–antibody interaction. In: Computer-aided antibody design. New York, Springer, pp 101–107
Yamashita T, Fujitani H (2014) On accurate calculation of the potential of mean force between antigen and antibody: a case of the HyHEL-10-hen egg white lysozyme system. Chem Phys Lett 609:50–53
Yamashita T, Shoji N (2021) Singular spectrum transformation for detecting molecular motion mode change of protein systems. AIP Conf Proc 2343:020010
Yamashita T, Takatsuka K (2007) Hydrogen-bond assisted enormous broadening of infrared spectra of phenol-water cationic cluster: an ab initio mixed quantum-classical study. J Chem Phys 126:074304
Yamashita T, Voth GA (2010) Properties of hydrated excess protons near phospholipid bilayers. J Phys Chem B 114:592–603
Yamashita T, Voth GA (2012) Insights into the mechanism of proton transport in cytochrome c oxidase. J Am Chem Soc 134:1147–1152
Yamashita T, Peng Y, Knight C, Voth GA (2012) Computationally efficient multiconfigurational reactive molecular dynamics. J Chem Theory Comput 8:4863–4875
Yamashita T, Ueda A, Mitsui T, Tomonaga A, Matsumoto S, Kodama T, Fujitani H (2014) Molecular dynamics simulation-based evaluation of the binding free energies of computationally designed drug candidates: importance of the dynamical effects. Chem Pharm Bull 62:661–667
Yamashita T, Ueda A, Mitsui T, Tomonaga A, Matsumoto S, Kodama T, Fujitani H (2015) The feasibility of an efficient drug design method with high-performance computers. Chem Pharm Bull 63:147–155
Yamashita T, Okajima R, Shoji N (2018) Efficiency strategy for peptide design: a comparative study on all-atom, coarse-grained, and machine learning approaches. AIP Conf Proc. 2040: 020014
Yamashita T, Okajima R, Miyanabe K, Tsumoto K (2019a) Modified AMBER force-field (FUJI) parameters for sulfated and phosphorylated tyrosine residues: development and application to CCR5-derived peptide systems. AIP Conf Proc. 2186:030013
Yamashita T, Mizohata E, Nagatoishi S, Watanabe T, Nakakido M, Iwanari H, Mochizuki Y, Nakayama T, Kado Y, Yokota Y, Matsumura H, Kawamura T, Kodama T, Hamakubo T, Inoue T, Fujitani H, Tsumoto K (2019b) Affinity improvement of a cancer-targeted antibody through alanine-induced adjustment of antigen-antibody Interface. Structure 27:519–527
Yamashita T, Mitsui T, Sasaki K, Wada M, Matsuura A, Kamiya N (2022) Effect of N343 glycosylation and N501Y mutation on the SARS-CoV-2 spike protein: modeling and MD simulations. AIP Conf Proc 2611:020008
Zhu X, Lopes PE, MacKerell AD Jr (2012) Recent developments and applications of the CHARMM force fields. Wiley Interdiscip Rev Comput Mol Sci 2:167–185

Chapter 3
Promotion of Liquid–Liquid Phase Separation by G-Quadruplex DNA and RNA

Al Amin and Takanori Oyoshi

Abstract G-quadruplex (G4), a non-canonical nucleic acid structure, formed by guanine-rich sequences in nucleic acids, can act as a scaffold to promote liquid–liquid phase separation (LLPS). The formation of G4-LLPS is driven by the unique structural properties of G4s, including their ability to stack and form a dense core of hydrogen-bonded guanine quartets. Moreover, G4 can promote the assembly of G4-binding proteins and small molecules that stabilize or destabilize G4 structures, and cluster together, leading to the formation of liquid droplets or condensates that act as membrane-less organelles and are separated from the surrounding medium. The function of G4-LLPS is multifaceted and context-dependent. Since G4 structures have been implicated in regulating gene expression, telomere maintenance, and DNA replication, among other processes, G4-LLPS can boost the processes. G4-mediated LLPS has also been linked to the development of neurodegenerative diseases and cancer, as it can promote the assembly of protein aggregates and oncogenic proteins into liquid droplets that drive disease progression. Additionally, hydrogels containing G4 have recently attracted attention as drug carriers and artificial enzymes. Stimuli-responsive hydrogels easily release drugs upon being excited with external and internal stimuli like pH and glucose. Biocompatible dynamic hydrogel, in which chemical reactions throughout the gelation process regulate drug entrapment and release, might be useful for a drug delivery system.

Keywords G-quadruplex · G4 binding protein · Hydrogel · RGG domain · Intrinsically disordered region (IDRs) · Liquid–liquid phase separation (LLPS)

A. Amin
Graduate School of Science and Technology, Shizuoka University, Shizuoka, Japan
e-mail: amin.md.al.20@shizuoka.ac.jp

T. Oyoshi (✉)
Graduate School of Science and Technology, Shizuoka University, Shizuoka, Japan

Graduate School of Integrated Science and Technology, Shizuoka University, Shizuoka, Japan

Research Institute of Green Science and Technology, Shizuoka University, Shizuoka, Japan
e-mail: oyoshi.takanori@shizuoka.ac.jp

R. Kurokawa (ed.), *Phase Separation in Living Cells*,
https://doi.org/10.1007/978-981-99-4886-4_3

3.1 Introduction

The smallest unit of life, a cell, is the structural and functional foundation of all living things and is typically undetectable to the human eyes, however, houses an ocean of mystery. Due to the development of modern science, these mysteries are being revealed day by day. In recent years, scientists have discovered that to control the spatiotemporal dynamics of complicated biochemical activities and carry out their regular function, cells arrange proteins, DNA, RNA, and other macromolecules into subcellular compartments. In addition to the organelles surrounded by the membrane, a cell contains several membranelles compartments formed by liquid–liquid phase separation (LLPS) (Banani et al. 2017; Boeynaems et al. 2018), like nucleoli (Hernandez-Verdun 2011), nuclear speckles (Spector and Lamond 2011), paraspeckles (Fox et al. 2002), Cajal bodies (Cioce and Lamond 2005), promyelocytic leukemia protein bodies (Lallemand-Breitenbach and de The 2010; Batty et al. 2012; Sawyer et al. 2019) and stress granules (SGs), processing bodies (P bodies) (Banani et al. 2017; Jain et al. 2016; Sachdev et al. 2019; Mitrea and Kriwacki 2016; Wheeler et al. 2016; Jiang et al. 2020).

Phase separation is also responsible for various diseases. When proteins aggregate and form condensation, it might cause several diseases such as amyotrophic lateral sclerosis (ALS), frontotemporal dementia (FTD), Alzheimer's disease (AD), and Parkinson's disease (PD) (Polymenidou and Cleveland 2011; Li et al. 2013). The shift from reversible dynamic LLPS to irreversible aggregation has been demonstrated for TAR DNA-binding protein 43 (TDP-43), translocated in liposarcoma/fused in sarcoma (TLS/FUS), Tau, and α-synuclein have been found as aggregations in afflicted neurons of patients with them (Patel et al. 2015; Ambadipudi et al. 2017; Schmidt et al. 2019; Ray et al. 2020; Zbinden et al. 2020; Wang et al. 2021). The proteins related to these diseases are nucleic acid-binding proteins and some of them have been reported as G-quadruplex (G4) binding proteins. G4 which is composed of two or more stacked guanine (G)-tetrad planes and a monovalent cation such as K^+ or Na^+ is formed at G-rich sequences in DNA and RNA (Oyoshi and Masuzawa 2020). G4 does not relate only to disease but to a lot of fundamental functions in a cell including transcription, RNA processing, mRNA localization and translation, transcription, telomere elongation, epigenetics, and replication, which has been extensively reviewed (Lipps and Rhodes 2009; Murat and Balasubramanian 2014; Rhodes and Lipps 2015; Varizhuk et al. 2019; Mukherjee et al. 2019; Spiegel et al. 2020). These functions of them are regulated by G4 binding proteins. While the component of hydrogel in a cell has been reviewed, little attention has been focused on how G4 and G4 binding protein affects the formation of hydrogels. Recent results revealed that G4 has the potential capabilities to form a hydrogel by itself (Sreenivasachary and Lehn 2008; Belda et al. 2017). Moreover, G4 binding proteins have similar potential and might affect the formation of a hydrogel with G4 which is related to diseases (Polymenidou and Cleveland 2011; Li et al. 2013; Patel et al. 2015; Ambadipudi et al. 2017). These studies also led to the development of therapeutic agents targeting these structures.

In this review, we focus on the LLPS of G4 and G4 binding proteins related to diseases. Furthermore, the hydrogels containing G4 have the capacity of adding the functions such as drug delivery and catalytic activity. These studies demonstrate the potential of G4 as a novel functional material. This paper also introduces these studies.

3.2 G4s to Promote the Formation of Hydrogels

G4, which has a lot of functions including regulation of gene expression in living cells, is gaining more and more interest for its potential to form membrane-less organelle or liquid droplets through the LLPS process. G-tetrad, which form G4, can self-assemble and extend further resulting in the formation of fibers also known as G wires (Marsh and Henderson 1994) which eventually form a gel. In the case of in vitro experiment, guanosine-5′-monophosphate (GMP) was seen to form gel (Belda et al. 2017; Gellert et al. 1962; Zimmerman 1976) (Fig. 3.1). The self-organization of four GMP molecules and a cation into a G-tetrad is the first step in the formation of gels from GMP. These G-tetrads are formed by Hoogsteen hydrogen bonding. Eight hydrogen bonds connect four guanines on the Watson-Crick and Hoogsteen sides of the base to form a G-tetrad. G-tetrads form a G4 when they are stacked on top of one another, stabilized by π-π stacking interactions and the coordination of cations, particularly Na^+, K^+, NH_4^+, and Pb^+ (Sreenivasachary and Lehn 2008; Kolesnikova and Curtis 2019). Later more G-tetrads stack with it through end stacking manner and a fibrous G wire is formed through the columnar assembly of G-tetrads. These fibrous columnar, G wires, culminate in the formation

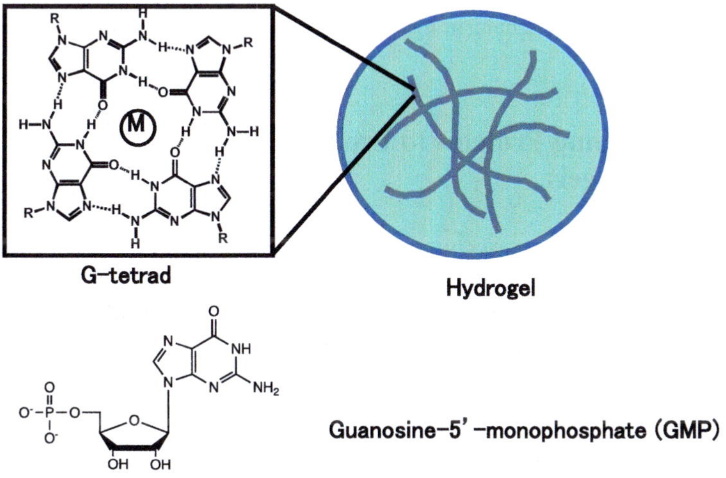

Fig. 3.1 An illustration of the hydrogel containing G-tetrad and the structure of guanosine-5′--monophosphate (GMP)

of gels in vitro (Sreenivasachary and Lehn 2008; Marsh and Henderson 1994; Kolesnikova and Curtis 2019). The addition of polyamines such as spermine and spermidine increases the gelation process several folds. The unique configuration of the G-tetrad results in the formation of an anionic outer side because of the presence of negatively charged phosphate in the periphery, which is stabilized by cations that are positioned in the heart of the G-tetrad. Thus G wires show an anionic outer side. This negatively charged phosphate-containing outer side becomes more stable when polycations make a reversible crosslink with it resulting in more gel formation (Belda et al. 2017; González-Rodríguez et al. 2009).

The mRNA forming G4 in cells might be involved in its storage and conservation of itself. SHORT ROOT RNA (SHR), which is one of the mRNAs transcribed in plant cells, contains G4-forming sequences and is capable of forming LLPS in the root of *Arabidopsis* (Zhang et al. 2019). This particular gene of SHR contributes to root development. A putative two-guanine-quarter containing G4 motif having the sequence $G_2L_3G_2L_3G_2L_2G_2$ (G; guanine, L; loop) present in the SHR mRNA is responsible for the G4 formation and it undergoes LLPS. When G4 is formed within the sequence, the structure becomes rigid which is the main driving force of this LLPS here. On the contrary, when the G4 structure is broken down, the hairpin structure made by the sequence becomes flexible again and the droplets are seen to dissolve. The LLPS rate depends on the concentration of K^+ ions. The concentration of K^+ within plant cells might change depending on the stage of development or the degree of stress the plant is experiencing (Greenway and Munns 1980; Zhang and Blumwald 2001). During drought, plants, for example, raise their cytosolic K^+ concentrations from 100 mM under normal cellular circumstances to 700 mM (Levi et al. 2011). The substantial elevation in intracellular K^+ is anticipated to stimulate the formation of additional G4, leading to an increase in the frequency of G4-induced, RNA-driven phase separation (Zhang et al. 2019). It is possible that in stressful conditions, such as drought, G4-triggered RNA-driven phase separation provides a regulatory mechanism that allows cells to preserve and/or store RNAs.

3.3 G4s Binding Proteins to Promote the Formation of Hydrogels

G4, which has the potential ability to form droplets by itself, is promoted to form droplets by G4-binding proteins, and G4 and G4-binding proteins with intrinsically disordered regions (IDRs) may form droplets in the cell. Proteins with IDRs can operate as scaffolding with characteristic short linear motifs to create a variety of interactions (Li et al. 2012; Lee et al. 2015; Lin et al. 2015) (Fig. 3.2). The multivalent interaction is the key driver behind LLPS. It is possible to classify all multivalent interactions that lead to LLPS into two broad groups. One is the weak, transitory, multivalent intermolecular interactions such as π–π interaction, cation–anion interactions, dipole–dipole interactions, and π–cation interactions among

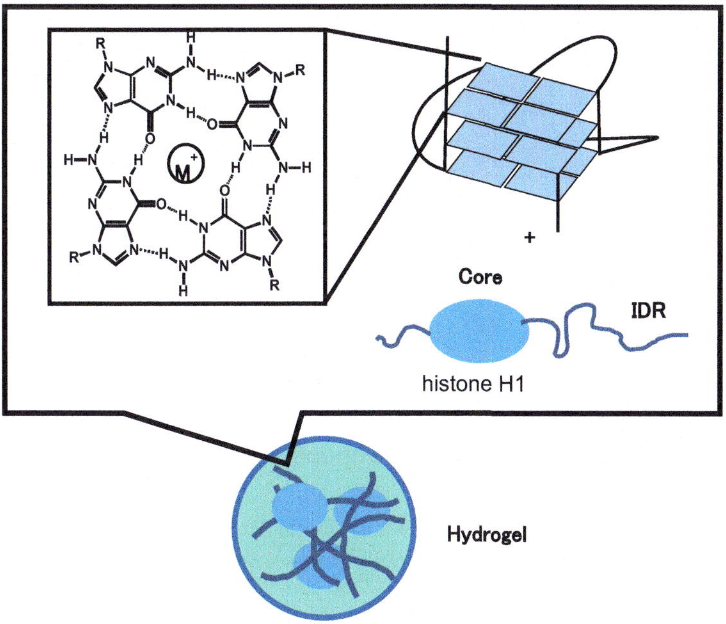

Fig. 3.2 A schematic illustration of DNA G-quadruplex and histone H1 in hydrogel

IDRs, and the other is the intracellular protein–protein, protein–RNA, and RNA–RNA interactions (Wang et al. 2018, 2021; Li et al. 2012; Zhang et al. 2020; Mimura et al. 2021). IDRs are richer in charged amino acids and are polar. G4 can interact with these IDRs and can form LLPS. The histone H1 (H1) C-terminal region contains positively charged IDR. This IDR becomes positively charged due to the presence of lysin. Through this IDR the C terminal region of H1 regulates nucleosome packing density. When H1 combines with the sequence from the *c-myc* promoter, which may fold into a G4, H1 goes through the LLPS process (Mimura et al. 2021; Shakya and King 2018; Ambrus et al. 2005). The droplet was found to be proportional to the amount of G4 supplied in the reaction media which ensures that the G4 structure was involved in H1-*c-myc* LLPS. Some of the Arg-Gly-Gly repeat (RGG) domain, which is found in many nucleic acid-binding proteins, has been reported to have G4-binding properties and droplet-forming ability. TLS/FUS, which is associated with ALS and the most characterized RNA binding protein (RBP), contains one QGSY-rich region and one RNA recognition motif (RRM), a Zn finger, three RGG domains, and a C-terminal region with a nuclear localization signal (Fig. 3.3a). TLS/FUS interacts with G4-generated human telomere and stabilizes G4 (Takahama et al. 2011a, b). The G4 recognition mechanism by TLS/FUS has been investigated in detail. RGG3, which is located in the C term area of TLS/FUS, has a specific affinity for the G4 structures of human telomeric DNA and telomeric repeat-containing RNA (TERRA), and it forms a ternary complex with both of these G4s in vitro (Takahama et al. 2013). Further study

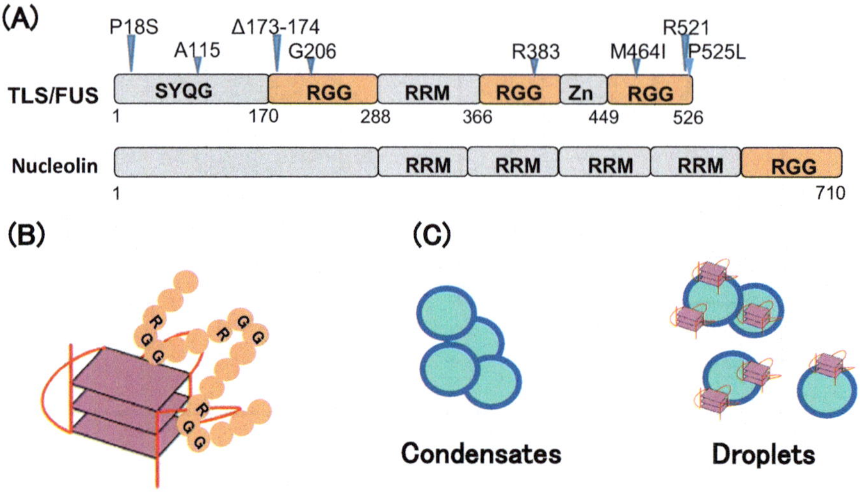

Fig. 3.3 G4 binding proteins forming RNA granules. (**a**) Schematic illustration of TLS/FUS and nucleolin. SYQG-rich; RGG 1, Arg-Gly-Gly-rich motif 1; RRM, RNA recognition motif; RGG 2, Arg-Gly-Gly-rich motif 2; ZnF, zinc finger; RGG 3, Arg-Gly-Gly-rich motif 3, finger; RGG, Arg-Gly-Gly-rich motif. ALS-linked TLS/FUS mutation points are shown at each site in the particular domains. (**b**) A proposed model describing the binding of Arg-Gly-Gly in the RGG domain to G4. (**c**) A model for condensates and droplets containing RNA G4 and TLS/FUS

revealed that Tyr in RGG3 of TLS/FUS can recognize 2'-OH groups on the ribose of loops in G4 structures (Takahama and Oyoshi 2013). NMR analysis revealed that Tyr and Phe in RGG3 recognize the G-tetrad plane in G4 (Kondo et al. 2018). Moreover, the ß-spiral structure of the RGG domain also might play a role in G4 binding (Yagi et al. 2018) (Fig. 3.3b). TLS/FUS binds to G4 by this recognition mechanism and forms droplets, the detailed mechanism of which is presented in the next paragraph.

3.4 RNA Granules G4 Related to Disease

RNA granules G4 are the aggregation of RNA molecules containing G4 structures, which are dynamic and form in response to cellular stress and have been implicated in the pathogenesis of various diseases. Examples of granules are the nucleus, SGs, P bodies, RNA foci, and paraspeckles (Dumas et al. 2021; Kharel et al. 2020). RNA granules themselves are membrane-less assembly and the G4 structure within these droplets attracts more G4 binding pathogenic protein and make a denser condensation which is responsible for various diseases. The C9orf72 gene has a hexanucleotide repeat expansion (HRE), $(GGGGCC)_n$, in its non-coding region that is associated with neurodegenerative disorders like FTD and ALS (Renton et al. 2011; DeJesus-Hernandez et al. 2011). The majority of normal human

C9orf72 alleles have less than eight repetitions and more than half have two repeats but the repetition number can be varied from 2 to 25 (Rutherford et al. 2012). Due to the disease, GGGGCC (G4C2) repeat number increases drastically. 10 to 1000 hexanucleotide repeats are capable to associate with ALS and FTD. G4C2 repeat DNA can create a four-stack antiparallel G4 motif, with the guanines on the outside of the stacks being more susceptible to chemical changes than the guanines buried in the inside of the sequence and the G4C2 repeat sequences of the C9orf72 HRE DNA or RNA are capable of adopting intermolecular, intramolecular, as well as parallel G4 structures and may also generate transcriptionally induced RNA–DNA hybrids known as R-loops (Haeusler et al. 2014). C9orf72HRE leads to a reduction in the processivity of RNA polymerase in the repeat region, which results in the accumulation of repeat-containing abortive transcripts and a loss of full-length transcripts. It is possible to say, in a word, that the presence of f G4s inside the C9orf72 HRE region inhibits the processivity of polymerase. Both full-length and abortive transcripts can form RNA G4. These RNA eventually develop into RNA foci (Gendron et al. 2013), which are a kind of RNA granule. The G4 structures of the C9orf72 HRE repeat are primarily responsible for the formation of G4C2 repeat RNA-mediated foci.

These foci are similar to paraspeckles in several ways, including the fact that they colocalize with G4 binding proteins such as nucleolin (NCL) and TLS/FUS (Česnik et al. 2019) and cause G4C2 repeat RNA condensation by interacting with RBPs including NCL and TLS/FUS and make membrane-less assemblies through phase separation (Dumas et al. 2021; Kharel et al. 2020) (Fig. 3.3a). These RNA foci can serve as a template to draw in and bind certain RBPs. NCL is a multifunctional, highly conserved, abundant RNA binding protein found in a variety of subcellular compartments, including a relatively stable major nucleolar pool as well as more dynamic nucleoplasmic, cytoplasmic, and plasma membrane complexes (Berger et al. 2015). In a normal cell, NCL is found condensed in the nucleolus. Yet, in the patient who possesses C9orf72HRE, NCL is observed distributed throughout the nucleus (Haeusler et al. 2014). NCL has central 4 RRM regions and one C-terminal RGG region (Masuzawa and Takanori Oyoshi 2020) (Fig. 3.3a). NCL is responsible for the transcription and translation of G4-containing regions, and it binds to DNA G4 via its RRM-RGG domains mainly by the RRM domain through the LLPS process. A Phe is located next to four Arg-Gly-Gly sequences in NCL's RGG domain. The RRM domain of NCL preferentially binds to the 5′- and 3′-terminal single strands containing guanine in the G4, and Phe in the RGG domain contributes to the folding of the G4 (Masuzawa and Takanori Oyoshi 2020). Thus, NCL selectively binds with the RNA G4s created by C9orf72-produced RNA foci, and denser LLPS is observed (Česnik et al. 2019). In a word, NCL represents a direct link between the characteristic G4s of C9orf72 HRE and the resulting cascade of pathological defects in ALS patients. Another G4 binding nucleoprotein TLS/FUS plays a significant role in neurodegenerative diseases like ALS. A missense mutation in TLS/FUS is related to this ALS disease (Fig. 3.3a). As we know TLS/FUS contains an N-terminal QGSY-rich region and one RNA recognition motif (RRM), a Zn finger, three RGG domains, and a C-terminal PY region, now the question is which region is responsible for the G4 recognition and droplet formation. An experiment was conducted using two mutants (P18S and A115N) in the N-terminal proximal QGSY-rich area, two mutants (Δ 173–174 and G206S) in the RGG1 domain, two mutants (R383C and M464I) in RGG2 and RGG3 domains, and two mutants (R521C and P525L) in exon 15 at the C-terminal of PY region to mimic ALS-linked TLS/FUS mutations. In RNA G4 free condition wild-type TLS/FUS immediately made protein condensation and in the presence of RNA G4, TLS/FUS reacted with the RNA G4 and made droplets (Fig. 3.3c). In a word, RNA G4 binds wild-type TLS/FUS resulting in the inhibition of condensation and acceleration of the formation of droplets in an LLPS manner. All the mutants of TLS/FUS showed a similar behavior of condensates and droplets formation. In the absence of RNA G4, all mutants made condensates; however, six mutants, P18S, A115N, Δ 173–174, G206S, M464I, and R521C showed fewer condensates formation than the wild type, whereas the other two mutants, R383C and P525L showed similar condensates formation like wild type. In the presence of RNA G4, all of the mutants made droplets, although the droplets formation level by all the mutants was less than the wild type. From this experiment, it was found that all ALS-linked TLS/FUS mutants whose binding ability was weaker than the wild type showed decreased binding ability with RNA G4 and lost their ability to control the LLPS and liquid-to-solid transition pathways (LST). This result shows that RNA G4 is an important part of this LLPS and LST process (Ishiguro et al. 2021).

3.5 Application of G4-Mediated Hydrogel

Hydrogels containing G4 have recently attracted attention not only as intracellular components but also as artificial materials and a variety of functional materials of them have been developed. Because of the hydrophilic nature of G4-mediated hydrogel, possible biocompatibility (Sreenivasachary and Lehn 2008), and programmability (Xie et al. 2021) hydrogel has attached attention as a drug carrier and artificial enzyme. Moreover, the excellent injectability (Thakur et al. 2019), high transparency, and rapid self-healing ability (Biswas et al. 2018) of hydrogels make it a suitable ingredient for 3D printing.

Hydrogels containing G-tetrads are three-dimensional hydrophilic polymeric networks that can absorb huge amounts of water and they are able to noncovalently hold a lot of drugs (Li et al. 2017; Biswas et al. 2020; Ghosh et al. 2020) (Fig. 3.4a). Conversely, a hydrogel formed by G-tetrad covalent bonding with drugs has the ability to be a drug carrier and protects them from environmental variations (Thakur et al. 2019). In the case of noncovalent bonding, drugs can be loaded with the G4 hydrogel during the gel formation. Moreover, scientists have become successful to make more efficient hydrogel by modifying guanosine. Due to the dynamic combinatorial chemistry and dynamic covalent bonds involved in making guanosine-based hydrogels, researchers can change, exchange, or rearrange the building blocks in hydrogels. This makes it possible to use a much wider range of materials to make

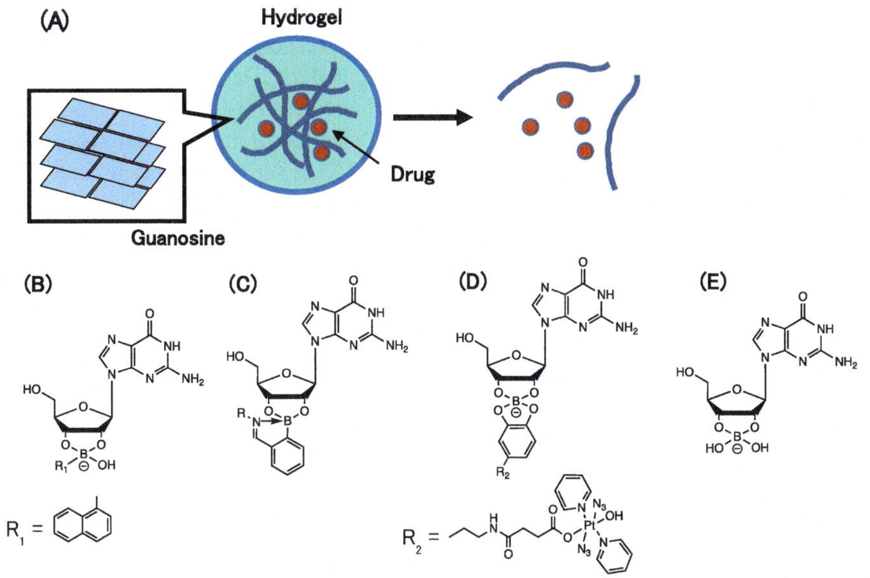

Fig. 3.4 Proposed applications of hydrogel containing G-tetrad. (**a**) Hydrogels formed with the help of G-tetrads hold the drug and release them as the gel decay. (**b**) Boronate ester bond between 1-naphthaleneboronic acid (1-NapBA) and guanosine. (**c**) Iminoboronate bond between guanosine and tris(2-aminoethyl) amine (TAEA) or guanosine and polyethylene glycol (PEG). (**d**) Borate ester bond between guanosine and dopamine-conjugated platinum IV anticancer complexes (Pt-DA) makes Pt-DA-borate ester-guanosine. (**e**) Borate ester bond between guanosine and boric acid

hydrogels (Lehn 2007). In guanosine, a ribose sugar (ribofuranose) moiety is attached to the guanine through a ß-glycosidic bond (Li et al. 2017). This ribose sugar moiety contains cis-1,2-diol which is very important for this guanosine modification. Different boronic acids like 1-naphthaleneboronic acid (1-NapBA) can bind with this cis-diol through a pH-sensitive, reversible cyclic boronate ester bond (Ghosh et al. 2020) (Fig. 3.4b). In the presence of K^+, 4 guanosine of G-rich single-stranded DNA and RNA sequence binds together through supramolecular Hoogsteen-type hydrogen-bonding interactions forming G-quartets. These cation-templated hydrogen-bonded G-quartets make G4 and through π–π interaction they form a long G4 fiber that contains modified guanines with it resulting in hydrogel formation. The π–π stacking of the boronic acid group present in the modified guanosine that makes G4 might give extra strength to the G4 wires resulting in more stable G4 hydrogel. Cyclic boronate ester (B-O-C) bond plays one of the key roles in hydrogel formation and because of this bond, the hydrogel becomes injectable (Ghosh et al. 2020). Drugs that are incorporated with the hydrogel through the B-O-C bond can be affected by pH. The cyclic boronate ester bond between the cis-diol of the sugar moiety of guanosine and the boronic acid or boronic acid derivatives which is the main driving force of hydrogel formation is broken down in low pH. For example, low pH (4.8) was responsible for the destruction of the boronate ester (B-O-C) bond between the sugar moiety of guanosine and the boronic acid group of 1-NapBA (G – NapBA) which was responsible for the formation of hydrogel in the presence of K^+ at physiological pH (7.4). Due to the breaking down of the B-O-C bond, loaded drugs were seen to be released (Ghosh et al. 2020). This injectable G4 hydrogel of G – NapBA has been reported to carry vitamin B_2 and B_{12} upon low pH (4.8) stimulus (Ghosh et al. 2020). Moreover, some other moieties can bind with these boronic acid derivatives and make more fibrous hydrogel. For example, a boronic acid derivative, 2-formylphenylboronic acid (2-FPBA) can bind with cis-diol present on the ribose group of guanosine through a boronate ester bond (Li et al. 2017; Biswas et al. 2020) (Fig. 3.4c). At the same time, the aldehyde group of 2-FPBA can bind with an amino group of polyamines like tris (2-aminoethyl) amine (TAEA) (Li et al. 2017) or polymers like polyethylene glycol (PEG) (Biswas et al. 2020) through imino bond. Due to this iminoboronate bond, G4, boronic acid derivative and polyamine make a fibrous hydrogel. Polyamine minimizes the repulsion among the negatively charged phosphate in the G wire and both the polyamines and polymers can make reversible crosslinks among G4 wires resulting in a complex hydrogel (Belda et al. 2017; Biswas et al. 2020). Through these types of hydrogels, drugs can be loaded in situ rather than of any other specific interaction during the gel formation. When the drug and gel-forming agents are mixed together and the gelation process is applied, drugs are encapsulated by the hydrogels by forming noncovalent bonds with them (Li et al. 2017; Biswas et al. 2020; Ghosh et al. 2020). Upon pH stimulation, the zero-order controlled release of loaded drug through slow diffusion is guaranteed by the dense, entangled, crosslinking fibrous network of the hydrogel. Acidic pH not only breaks down B-O-C bond between guanosine and 2-FPBA but also TAEA is also released due to the destruction of the imino bond between aldehyde and amino groups present in 2-FPBA and TAEA (Li et al. 2017). Acidic stimuli (pH 5.0) were responsible for the breakdown of iminoboronate bonding between guanosine, 2-FPBA, and TAEA resulting in the destruction of G4 hydrogel and release of methylene blue and FITC-lysozyme (Li et al. 2017). Another hydrogel named G4PEG hydrogel was made by using 2-FPBA, guanosine, and PEG in KOH solution at physiological pH conditions and was reported to carry and release doxorubicin, a model chemotherapeutic medication, in a zero-order way at low pH (4.8) environment. pH-sensitive iminoboronate bonds of G4PEG hydrogels were broken down by the low pH stimulus and doxorubicin was released from the G4 hydrogel (Biswas et al. 2020).

Conversely, due to a large number of binding sites for metal coordination and hydrogen bonding present in guanosine, the native nucleotides could form a hydrogel on their own by crosslinking the G4 fibers with metal ions (Ghosh et al. 2020; Lopez and Liu 2017; Pu et al. 2018; Xiao and Davis 2018; Sutyak et al. 2016). For instance, a hydrogel was made using a guanosine derivative, guanosine-5′-monophosphate (GMP) through double crosslinking by Fe^{3+} and Ca^{2+}. GMP can form a G4 structure in the presence of Na^+ and stack one after another resulting in G4 fiber formation (Thakur et al. 2019) (Fig. 3.1). The sugar moiety of this GMP contains a phosphate group that can bind with Fe^{3+} salt. In the water solution, Fe^{3+} forms $FeOH^{2+}$ species which are further polymerized to yield Fe^{3+}-

oxyhydroxide. The phosphate group in GMP can bond with Fe^{3+}-oxyhydroxide species either by forming an inner sphere complex through ligand exchange between phosphate oxygens and hydroxyl oxygens on the surface of the iron oxyhydroxide or by forming an outer sphere complex through electrostatic interactions between phosphate anions and $Fe\text{-}OH^{2+}$ species. Fe^{3+} salts make a lot of crosslinking between the nanofibrous structures of G-quadruplex, which causes hydrogelation and keeps the helically stacked G-quadruplex structure intact. Moreover, Ca^{2+} also acts as cross-linkers that connect the diol groups of the ribose moiety of GMP resulting in a more fibrous gel formation through dual crosslink (Thakur et al. 2019). Drugs are loaded with this type of gel during the gel formation and medicine binds with it through electrostatic interaction. By acidic pH stimulus, the drug is seen to be released from this hydrogel. Hydrogels that are constructed through the crosslinking of G4 wires by metal ions release drugs in different ways depending on how much the phosphate group is protonated or deprotonated at different pH levels. The hydrogel made by the dual crosslinking of Fe^{3+} and Ca^{2+} released drugs depending on the protonation/deprotonation of the phosphate group situated in the sugar moiety of guanosine and their binding affinity for the $FeOH^{2+}$ on Fe^{3+} oxyhydroxide surface. Under conditions of neutral pH, deprotonation of phosphate makes dianionic phosphate. This makes it easier for $FeOH^{2+}$ and the phosphate moiety to interact electrostatically. Conversely, when the pH is acidic, protonation of the phosphate moiety reduces negative charges. This weakens the electrostatic bonds between the metal surface and the phosphate groups. So, the compactness of the Ca-Fe-GMP hydrogel and the release of the drug from it depend on the interaction of the phosphate part of GMP and the Fe^{3+} metal ion at different pH levels. Doxorubicin was transported through a supramolecular hydrogel created by the self-assembly of GMP into a highly ordered G4 structure and then dual cross-linked by Fe^{3+} and Ca^{2+}. When it was stimulated by low pH, this hydrogel showed zero-order controlled release of doxorubicin (Thakur et al. 2019).

In addition to the noncovalent bond, the covalent bond also plays a significant role in hydrogel formation. A hydrogel was made where a photoactivable anticancer component, dopamine-conjugated platinum IV anticancer complex (Pt-DA) was attached with it via a noncovalent bond, borate ester bond (Fig. 3.4d). Light-sensitive hydrogel does not break the G4 hydrogels nor the borate ester bond that binds the medicine with the hydrogel. The application of laser light just activates the drug component by changing its oxidation state. Boric acid can modify guanosine. Boric acid binds with the –OH group of the sugar moiety of guanosine and makes guanosine mono-borate ester (Venkatesh et al. 2017). Conversely, boric acid can bind with the drug and make a mono-borate ester compound of the drug as well. These mono-borate esters can bind with the guanosine of the G4 sequence and can make guanosine-borate ester-guanosine and guanosine-borate ester-drugs through a diester bond. This modified guanosine of sequence then self-assembles and makes G4, then G4 fiber, and finally hydrogel which carries the drug with it. During the formation of hydrogel, at first, Pt-DA was reacted with potassium hydroxide and boric acid which results in the formation of Pt-DA-borate ester (Pt-DA-B), a mono-borate ester. This Pt-DA-B monoester made a bond with guanosine through a diester bond and made Pt-DA-borate ester-guanosine (Pt-DA-B-G). Conversely, in the presence of boric acid, some guanosine formed guanosine-borate diester (G-B-G) as well. All of these guanosine borates (one Pt-DA-B-G and three G-B-G) self-assembled into G4 which can carry Pt-DA with it. Notably, if four Pt-DA bound with the guanosine of the G4 sequence, it failed to form hydrogel due to the cause of steric disruption to the stacking of G-quartets made by Pt-DA situated in the arms of G-quartets. One Pt-DA-borate ester-guanosine diester and three guanosine-borate ester-guanosine diester combinations formed G4 hydrogel. Photoactivable diazol-Pt (IV) is a prodrug for the cancer cell. Upon stimulated by the light it can be oxidized to form Pt(II) which is cytotoxic. However, dopamine is a neurotransmitter as well as a constituent of adhesive protein in muscle fiber. Recently, a photoactivatable dopamine-conjugated platinum anticancer complex (Pt-DA) containing hydrogel has been prepared which showed photocytotoxicity against cisplatin-resistant A2780C is human ovarian cancer cells that were loaded and transported through G4 borate hydrogels to make this hydrogel (Venkatesh et al. 2017). This hydrogel was loaded with Pt(IV) which can be used as an anticancer complex after photoactivation.

G4 hydrogel can regulate the activity of the artificial enzyme by binding or releasing hemin in an ion-regulated manner. Hemin and G4s together show artificial enzyme-mimicking activity and due to its ion-triggerable programmability, it acts like a logic gate that is being studied frequently. Natural enzymes are exceptional in their activity, selectivity, and efficiency, and they play important roles in biological processes in living systems. However, they have several disadvantages such as difficult preparation, ease of denaturation, nominal thermal stability, expensive price, and low reusability (Zhao et al. 2015, 2016; Berglund et al. 2002; Nath et al. 2016; Zhong et al. 2018). Due to these problems, the large-scale production of natural enzymes is very difficult. Recently scientists are showing interest to make artificial enzymes using G4 hydrogels. The G4 structural motif is nuclease resistant (Alford et al. 2018) and it can load a large amount of cargo and can protect them from the external environment for a long time (Li et al. 2017) which allows the enzyme to perform its activity for a long time. It was reported that G4 and hemin together can act as an artificial enzyme mimicking the catalytic activities of horse-radish peroxide (HPR) (Bhattacharyya et al. 2017). In 2013, Lu reported the construction of hemin-G4 hydrogel that was capable to crosslink acrylamide resulting in the formation of acrylamide-hemin-G4 hydrogel which acts like an artificial enzyme. This hydrogel was made by the self-assembly of acrylamide chains. Acrylamide monomers and acrydite-modified oligonucleotides (acrydite-AA GGG) were polymerized to make a chain and upon addition of K^+, the gelation of copolymer chain occurred. The formation of gel was attributed to the formation of interchain G4 that crosslinks the hydrogel. This artificial enzyme hydrogel (AEH) was found to catalyze the oxidation of 2, 2′-azinobis-(3-ethylbenzthiazoline-6-sulfonic acid), ABTS2− to ABTS−, by the reduction of H_2O_2 to H_2O. By using 18-crown-6, which is a chelator of K^+, K^+ can be removed from the G4 resulting in the dissociation of hydrogel and losing the enzymatic activity of this artificial enzyme. In a word the removal of K^+ from the hydrogel was responsible for the loss of the enzymatic activity of AEH (Lu et al. 2013). Hemin can also bind at end stacking mode with the G4 structure of hydrogel that is made by using guanosine and boronic acid or boric acid. When boronic acid like phenylboronic acid is mixed with guanosine, it modifies the sugar moiety by attaching to it through the boronate ester bond (Bhattacharyya et al. 2017) (Fig. 3.4b). In addition, boric acid can also modify the sugar moiety of guanosine. When boric acid is mixed with guanosine, it binds with the sugar moiety of guanosine through the borate ester bond and makes guanosine-borate monoester (Zhong et al. 2018) (Fig. 3.4e). Then in the presence of cation, these modified sugar moieties containing guanosines make self-assembly resulting in G4 and hydrogel formation. Hemin binds to G4 in the presence of K^+. When a hydrogel's G4 structure, which has been stabilized by the K^+, is exposed to a Pb^{2+} solution, the K^+ from the G4 structure is replaced by Pb^{2+}. Due to the shorter M-O and O-O bonds, Pb^{2+} is more effective at stabilizing G4 than K^+. If the G4 is stabilized by Pb^{2+}, hemin is seen to be released from the G4 and G4 loses the enzyme-mimicking property. Due to the presence of Pb^{2+}, the parallel arrangement of G4 is induced which is not favorable for hemin to bind with. An HPR-mimicking enzyme made by hemin, G4, and boronic acid showed oxidation activity on 3,3′,5,5′-tetramethylbenzidine in the presence of H_2O_2. A similar phenomenon was observed by AEH made through hemin, boric acid, and G4. This AEH was also capable to oxidize 3,3′,5,5′-tetramethylbenzidine in the presence of H_2O_2 (Zhong et al. 2018; Bhattacharyya et al. 2017). In all cases, the activity and inactivity of AEH can be controlled by adding or removing K^+. AEH can be activated in the presence of K^+ ions. And the absence of this ion makes the AEH lose the enzymatic activity. Since the binding ability of hemin can be controlled by K^+, the activity of AEH can be controlled in a switch-on and switch-off manner. In conclusion, it can be said that K^+ -pH switchable AEH can be described as a two-input inhibit logic gate (Zhong et al. 2018; Bhattacharyya et al. 2017; Lu et al. 2013).

Stimuli-sensitive hydrogels respond to environmental changes by altering form or volume. They respond to physical stimuli like light, pressure, temperature, electric field, magnetic field, and ultrasound, chemical stimuli like pH, redox, ionic strength, CO_2, glucose, and biological stimuli like enzymes (Pillai and Panchagnula 2001). pH-sensitive stimuli-responsive hydrogels are the most investigated. Self-assembling supramolecular G4 hydrogels exhibit structural stability at physiological pH and temperature (Carducci et al. 2018; Bhattacharyya et al. 2018; Yoneda et al. 2021). As a result of pathology, alterations occur in the normal cellular pH environment. Cancer cells, for example, have a different metabolic profile and often live in a low pH (extracellular) environment (Persi et al. 2018). Cancer cells exhibit uncontrolled cell division and it requires a lot of energy. Otto Warburg and colleagues were the first to discover aberrant anaerobic glycolysis as well as metabolic changes in cancer cells (Warburg 1956). Despite the availability of oxygen, pyruvate, which is the last result of glycolysis, is transformed in cancer cells into lactate, which is the final product of anaerobic respiration and the production of lactate by cancer cells ultimately results in the acidification of the extracellular environment (Melkonian and Schury 2022; de la Cruz-López et al. 2019; Swietach 2019). In addition, cancer cells upregulate the expression and activation of transporters and pump such as the Na^+/H^+ exchanger (NHE-1), the H^+-lactate co-transporter, and the proton pump (H^+-ATPase). This results in an increase in the secretion of hydrogen ions into the extracellular environment of the cells, which further contributes to the acidification of the extracellular environment (Swietach 2019; Counillon et al. 2016). As a result, cancer cells have a pH gradient that is in the opposite direction, with a little higher pH level within the cell and a slightly lower pH level outside the cell (Koltai 2020; Swietach et al. 2014; Damaghi et al. 2013; Lee and Shanti 2021). Since hydrogels become activated and perform zero-order drug release at low pH, G4 hydrogels are quickly becoming increasingly popular as an agent for drug delivery systems in a variety of cancer treatments.

3.6 Conclusion

G4 is a secondary structure of nucleic acids that plays a vital role in LLPS. Multivalent interactions between G4, G4 binding proteins, RNA, etc. are the key driver of LLPS. Guanines of G-rich single-stranded DNA and RNA sequence bind together through supramolecular Hoogsteen-type hydrogen-bonding interactions, π–π stacking interactions, and the coordination of cations forming G-quartets and G4 structure which can self-assemble on top of one another and form G4 wires. These G4 wires are capable to produce LLPS by themselves. Moreover, polyamines can bind with it and enhance the gelation process. Polyamine minimizes the repulsion among the negatively charged phosphate in the G wire and can make reversible crosslinks among G4 wires. Conversely nucleic acids, proteins like TLS/FUS, RNA, RNA granules like nuclein, etc. with IDRs can interact with G4 to form LLPS. G4-mediated LLPS is responsible for numerous diseases in the human body like neurodegenerative diseases, cancer, etc. However, the application of G4-mediated LLPS has gained significant attention in recent years due to its potential implications in cellular processes, regulation of gene expression, and disease development. Nowadays G4 hydrogels are being used as a drug delivery system. G4 hydrogels can be made by modifying the guanosine that makes G4. Modified guanosine can be obtained through the interaction of cis-diol present on the ribose group of guanosine with boronic acid like 1-NapBA through a cyclic boronate ester bond (Fig. 3.4b). At the same time, the aldehyde group of some boronic acid derivatives like 2-FPBA can bind with the amino group of polyamines such as TAEA or with a polymer like PEG through imino bond and can make more complex hydrogel through iminoboronate bond (Fig. 3.4c). This modified guanosine later self-assembles and makes G4 wire resulting in the hydrogel. Cyclic boronate ester bond is the main driving force of hydrogel formation and due to this bond, the gel becomes injectable. These types of gels can carry drugs with them by binding the drugs through a noncovalent bond. Some guanosine derivatives like 5′ GMP can bind Fe^{3+} with it through the negatively charged phosphate (Fig. 3.1). Fe^{3+} salts make a lot of crosslinking between the nanofibrous structures of G-quadruplex, which causes hydrogelation and keeps the helically stacked G-quadruplex structure intact. Moreover, Ca^{2+} also acts as cross-linkers that connect the diol groups of the ribose moiety of GMP resulting in a more fibrous gel formation through dual crosslink. These types of hydrogels bind drugs with them by a covalent bond. All of these hydrogels are pH-sensitive and in acidic pH, the cyclic boronate ester bond breaks down resulting in the destruction of hydrogel and drug release. Moreover, pH breaks the G4 structure by breaking the hydrogen bond among the guanosines. Apart from a covalent bond, drugs can also be attached to the guanosine through a diester bond, a noncovalent bond when a drug, guanosine, and boric acid are mixed together for gel formation. Pt-DA can bind with guanosine through a Pt-DA-borate ester-guanosine diester bond. Upon stimulated by light Pt(IV) can be oxidized to form Pt(II) which is cytotoxic and can be used against cancer cells. Ion-triggered G4 hydrogel acts as an artificial enzyme as well. Acrylamide monomers and acrydite-modified oligonucleotides make acrylamide-hemin-G4 hydrogel which acts like an artificial enzyme. The activity of this enzyme can be switched off by removing the K^+ by using 18-crown-6, a chelator of K^+. Hemin may also bind at end stacking mode with the G4 hydrogel structure generated from guanosine and boronic or boric acid. Adding boronic acid like phenylboronic acid to guanosine modifies the sugar moiety through the boronate ester bond (Fig. 3.4b). Additionally, boric acid modifies the sugar moiety of guanosine by binding with it through a diester bond (Fig. 3.4e). Hemin binds with G4 in the presence of K^+ and hemin is released from G4 in the presence of Pb^{2+} since Pb^{2+} induces a parallel arrangement of G4. Parallel G4 structure is not favorable for hemin to bind, thus hemin is released and an artificial enzyme made by the combination of hemin and G4 loses its activity. Thus, it can be said that ion-triggered G4 acts as a two-input inhibit logic gate. Since these hydrogels are pH triggerable, they can be used in treating different physiological disorders like cancer. Cancer cells have a different metabolic profile and live in a low pH (extracellular) environment. This is due to aberrant anaerobic glycolysis, which transforms pyruvate into lactate. Additionally, cancer cells upregulate the expression and activation of transporters and pump such as the Na^+/H^+ exchanger (NHE-1), the H^+-lactate co-transporter, and the proton pump (H^+-ATPase). This increases the secretion of hydrogen ions into the extracellular environment, which further contributes to the acidification of the extracellular environment. This acidic environment can trigger G4 hydrogels to release drugs in a particular area in a controlled manner. Thus, G4 made drug delivery system is gaining interest in the field of medicine nowadays.

References

Alford A, Tucker B, Kozlovskaya V et al (2018) Encapsulation and ultrasound-triggered release of G-Quadruplex DNA in multilayer hydrogel microcapsules. Polymers (Basel) 10:1342

Ambadipudi S, Biernat J, Riedel D et al (2017) Liquid-liquid phase separation of the microtubule-binding repeats of the Alzheimer-related protein tau. Nat Commun 8:275

Ambrus A, Chen D, Dai J et al (2005) Solution structure of the biologically relevant G-quadruplex element in the human c-MYC promoter. Implications for G-quadruplex stabilization. Biochemistry 44:2048–2058

Banani SF, Lee HO, Hyman AA et al (2017) Biomolecular condensates: organizers of cellular biochemistry. Nat Rev Mol Cell Biol 18:285–298

Batty EC, Jensen K, Freemont PS (2012) PML nuclear bodies and other TRIM-defined subcellular compartments. Adv Exp Med Biol 770:39–58

Belda R, García-España E, Morris GA et al (2017) Guanosine-5'-monophosphate polyamine hybrid hydrogels: enhanced gel strength probed by z-spectroscopy. Chem Eur J 23:7755–7760

Berger CM, Gaume X, Bouvet P (2015) The roles of nucleolin subcellular localization in cancer. Biochimie 113:78–85

Berglund GI, Carlsson GH, Smith AT et al (2002) The catalytic pathway of horseradish peroxidase at high resolution. Nature 417:463–468

Bhattacharyya T, Kumar YP, Dash J (2017) Supramolecular hydrogel inspired from DNA structures mimics peroxidase activity. ACS Biomater Sci Eng 3:2358–2365

Bhattacharyya T, Saha P, Dash J (2018) Guanosine-derived supramolecular hydrogels: recent developments and future opportunities. ACS Omega 3:2230–2241

Biswas A, Malferrari S, Kalaskar DM et al (2018) Arylboronate esters mediated self-healable and biocompatible dynamic G-quadruplex hydrogels as promising 3D-bioinks. Chem Commun 54: 1778–1781

Biswas A, Ghosh T, Gavel PK et al (2020) PEG functionalized stimuli responsive self-healable injectable dynamic Imino-boronate G-quadruplex hydrogel for the delivery of doxorubicin. ACS Appl Bio Mater 3:1052–1060

Boeynaems S, Alberti S, Fawzi NL et al (2018) Protein phase separation: a new phase in cell biology. Trends Cell Biol 28:420–435

Carducci F, Yoneda JS, Itri R, Mariani P (2018) On the structural stability of guanosine-based supramolecular hydrogels. Soft Matter 14:2938–2948

Česnik AB, Simona Darovic S, Mihevc SP et al (2019) Nuclear RNA foci from C9ORF72 expansion mutation form paraspeckle-like bodies. J Cell Sci 132:jcs224303

Cioce M, Lamond AI (2005) Cajal bodies: a long history of discovery. Annu Rev Cell Dev Biol 21: 105–131

Counillon L, Bouret Y, Marchiq I et al (2016) Na(+)/H(+) antiporter (NHE1) and lactate/H(+) symporters (MCTs) in pH homeostasis and cancer metabolism. Biochim Biophys Acta 1863: 2465–2480

Damaghi M, Wojtkowiak JW, Gillies RJ (2013) pH sensing and regulation in cancer. Front Physiol 4:370

de la Cruz-López KG, Castro-Muñoz LJ, Reyes-Hernández DO et al (2019) Lactate in the regulation of tumor microenvironment and therapeutic approaches front. Oncologia 9:1143

DeJesus-Hernandez M, Mackenzie IR, Boeve BF et al (2011) Expanded GGGGCC hexanucleotide repeat in noncoding region of C9ORF72 causes chromosome 9p-linked FTD and ALS. Neuron 72:245–256

Dumas L, Herviou P, Dassi E et al (2021) G-Quadruplexes in RNA biology: recent advances and future directions. Trends Biochem Sci 46:270–283

Fox AH, Lam YW, Leung AK et al (2002) Paraspeckles: a novel nuclear domain. Curr Biol 12:13–25

Gellert M, Lipsett MN, Davies DR (1962) Helix formation by guanylic acid. Proc Natl Acad Sci U S A 48:2013–2018

Gendron TF, Bieniek KF, Zhang YJ et al (2013) Antisense transcripts of the expanded C9ORF72 hexanucleotide repeat form nuclear RNA foci and undergo repeat-associated non-ATG translation in c9FTD/ALS. Acta Neuropathol 126:829–844

Ghosh T, Biswas A, Gavel PK et al (2020) Engineered dynamic Boronate Ester-mediated self-healable biocompatible G-Quadruplex hydrogels for sustained release of vitamins. Langmuir 36:1574–1584

González-Rodríguez D, van Dongen JL, Lutz M, Spek AL et al (2009) G-quadruplex self-assembly regulated by coulombic interactions. Nat Chem 1:151–155

Greenway H, Munns R (1980) Mechanisms of salt tolerance in nonhalophytes. Ann Rev Plant Physiol 31:149–190
Haeusler AR, Donnelly CJ, Periz G et al (2014) C9orf72 nucleotide repeat structures initiate molecular cascades of disease. Nature 507:195–200
Hernandez-Verdun D (2011) Assembly and disassembly of the nucleolus during the cell cycle. Nucleus 2:189–194
Ishiguro A, Lu J, Ozawa D et al (2021) ALS-linked FUS mutations dysregulate G-quadruplex-dependent liquid-liquid phase separation and liquid-to-solid transition. J Biol Chem 29:101284
Jain S, Wheeler JR, Walters RW et al (2016) ATPase-modulated stress granules contain a diverse proteome and substructure. Cell 164:487–498
Jiang S, Fagman JB, Chen C et al (2020) Protein phase separation and its role in tumorigenesis. elife 9:e60264
Kharel P, Becker G, Tsvetkov V et al (2020) Properties and biological impact of RNA G-quadruplexes: from order to turmoil and back. Nucleic Acids Res 48:12534–12555
Kolesnikova S, Curtis EA (2019) Structure and function of multimeric G-Quadruplexes. Molecules 24:3074
Koltai T (2020) Targeting the pH paradigm at the bedside: a practical approach. Int J Mol Sci 21:9221
Kondo K, Mashima T, Oyoshi T (2018) Plastic roles of phenylalanine and tyrosine residues of TLS/FUS in complex formation with the G-quadruplexes of telomeric DNA and TERRA. Sci Rep 8:2864
Lallemand-Breitenbach V, de The H (2010) PML nuclear bodies. Cold Spring Harb Perspect Biol 2:a000661
Lee S, Shanti A (2021) Effect of exogenous pH on cell growth of breast cancer cells. Int J Mol Sci 22:9910
Lee C, Occhipinti P, Gladfelter AS (2015) PolyQ-dependent RNA–protein assemblies control symmetry breaking. J Cell Biol 208:533–544
Lehn JM (2007) From supramolecular chemistry towards constitutional dynamic chemistry and adaptive chemistry. Chem Soc Rev 36:151–160
Levi A, Paterson AH, Cakmak I, Saranga Y (2011) Metabolite and mineral analyses of cotton near-isogenic lines introgressed with QTLs for productivity and drought-related traits. Physiol Plant 141:265–275
Li P, Banjade S, Cheng HC et al (2012) Phase transitions in the assembly of multivalent signalling proteins. Nature 483:336–340
Li YR, King OD, Shorter J, Gitler AD (2013) Stress granules as crucibles of ALS pathogenesis. J Cell Biol 201:361–372
Li Y, Liu Y, Ma R et al (2017) A G-Quadruplex hydrogel via multicomponent self-assembly: formation and zero-order controlled release. ACS Appl Mater Interfaces 9:13056–13067
Lin Y, Protter DSW, Rosen M, Parker R (2015) Formation and maturation of phase-separated liquid droplets by RNA-binding proteins. Mol Cell 60:208–219
Lipps HJ, Rhodes D (2009) G-quadruplex structures: in vivo evidence and function. Trends Cell Biol 19:414–422
Lopez A, Liu J (2017) Self-assembly of nucleobase, nucleoside and nucleotide coordination polymers: from synthesis to applications. Chem Nano Mater 3:670–684
Lu CH, Qi XJ, Orbach R et al (2013) Switchable catalytic acrylamide hydrogels cross-linked by hemin/G-quadruplexes. Nano Lett 13:1298–1302
Marsh TC, Henderson E (1994) G-wires: self-assembly of a telomeric oligonucleotide, d (GGGGTTGGGG), into large superstructures. Biochemistry 33:10718–10724
Masuzawa T, Takanori Oyoshi T (2020) Roles of the RGG domain and RNA recognition motif of Nucleolin in G-Quadruplex stabilization. ACS Omega 5:5202–5208
Melkonian EA, Schury MP (2022) Biochemistry, anaerobic glycolysis. StatPearls Publishing, Treasure Island, FL

Mimura M, Tomita S, Shinkai Y et al (2021) Quadruplex folding promotes the condensation of linker histones and DNAs via liquid-liquid phase separation. J Am Chem Soc 143:9849–9857
Mitrea DM, Kriwacki RW (2016) Phase separation in biology; functional organization of a higher order. Cell Commun Signal 14:1
Mukherjee AK, Sharma S, Chowdhury S (2019) Non-duplex G-Quadruplex structures emerge as mediators of epigenetic modifications. Trends Genet 35:129–144
Murat P, Balasubramanian S (2014) Existence and consequences of G-quadruplex structures in DNA. Curr Opin Genet Dev 25:22–29
Nath I, Chakraborty J, Verpoort F (2016) Metal organic frameworks mimicking natural enzymes: a structural and functional analogy. Chem Soc Rev 45:4127–4170
Oyoshi T, Masuzawa T (2020) Modulation of histone modifications and G-quadruplex structures by G-quadruplex-binding proteins. Biochem Biophys Res Commun 53:39–44
Patel A, Lee HO, Jawerth L et al (2015) A liquid-to-solid phase transition of the ALS protein FUS accelerated by disease mutation. Cell 162:1066–1077
Persi E, Duran-Frigola M, Damaghi M et al (2018) Systems analysis of intracellular pH vulnerabilities for cancer therapy. Nat Commun 9:2997
Pillai O, Panchagnula R (2001) Polymers in drug delivery. Curr Opin Chem Biol 5:447–451
Polymenidou M, Cleveland DW (2011) The seeds of neurodegeneration: prion-like spreading in ALS. Cell 147:498–508
Pu F, Ren J, Qu X (2018) Nucleobases, nucleosides, and nucleotides: versatile biomolecules for generating functional nanomaterials. Chem Soc Rev 47:1285–1306
Ray S, Singh N, Kumar R et al (2020) α-Synuclein aggregation nucleates through liquid-liquid phase separation. Nat Chem 12:705–716
Renton AE, Majounie E, Waite A et al (2011) A hexanucleotide repeat expansion in C9ORF72 is the cause of chromosome 9p21-linked ALS-FTD. Neuron 72:257–268
Rhodes D, Lipps HJ (2015) G-quadruplexes and their regulatory roles in biology. Nucleic Acids Res 43:8627–8637
Rutherford NJ, Heckman MG, Dejesus-Hernandez M et al (2012) Length of normal alleles of C9ORF72 GGGGCC repeat do not influence disease phenotype. Neurobiol Aging 33:2950. e5–2950.e7
Sachdev R, Hondele M, Linsenmeier M et al (2019) Pat1 promotes processing body assembly by enhancing the phase separation of the DEAD-box ATPase Dhh1 and RNA. elife 8:e41415
Sawyer IA, Sturgill D, Dundr M (2019) Membraneless nuclear organelles and the search for phases within phases. Wiley Interdiscip Rev RNA 10:e1514
Schmidt HB, Barreau A, Rohatgi R (2019) Phase separation-deficient TDP43 remains functional in splicing. Nat Commun 10:4890
Shakya A, King JT (2018) Non-Fickian molecular transport in protein-DNA droplets. ACS Macro Lett 7:1220–1225
Spector DL, Lamond AI (2011) Nuclear speckles. Cold Spring Harb Perspect Biol 1(3):a000646
Spiegel J, Adhikari S, Balasubramanian S (2020) The structure and function of DNA G-Quadruplexes. Trends Chem 2:123–136
Sreenivasachary N, Lehn JM (2008) Structural selection in G-quartet-based hydrogels and controlled release of bioactive molecules. Chem Asian J 3:134–139
Sutyak KB, Zavalij PY, Robinson ML et al (2016) Controlling molecularity and stability of hydrogen bonded G-quadruplexes by modulating the structure's periphery. Chem Commun (Camb) 52:11112–11115
Swietach P (2019) What is pH regulation, and why do cancer cells need it? Cancer Metastasis Rev 38:5–15
Swietach P, Vaughan-Jones RD, Harris AL et al (2014) The chemistry, physiology and pathology of pH in cancer. Philos Trans R Soc Lond Ser B Biol Sci 369:20130099
Takahama K, Oyoshi T (2013) Specific binding of modified RGG domain in TLS/FUS to G-quadruplex RNA: tyrosines in RGG domain recognize 2'-OH of the ribosomes of loops in G-quadruplex. J Am Chem Soc 135:18016–18019

Takahama K, Kino K, Arai S et al (2011a) Identification of Ewing's sarcoma protein as a G-quadruplex DNA- and RNA-binding protein. FEBS J 278:988–998

Takahama K, Sugimoto C, Arai S et al (2011b) Loop lengths of G-quadruplex structures affect the G-quadruplex DNA binding selectivity of the RGG motif in Ewing's sarcoma. Biochemistry 50: 5369–5378

Takahama K, Takada A, Tada S et al (2013) Regulation of telomere length by G-quadruplex telomere DNA- and TERRA-binding protein TLS/FUS. Chem Biol 20:341–350

Thakur N, Sharma BS, Bishnoi S, Jain S, Nayak D, Sarma TK (2019) Biocompatible Fe^{3+} and Ca^{2+} dual cross-linked G-quadruplex hydrogels as effective drug delivery system for pH-responsive sustained zero-order release of doxorubicin. ACS Appl Bio Mater 2:3300–3311

Varizhuk A, Isaakova E, Pozmogova G (2019) DNA G-Quadruplexes (G4s) modulate epigenetic (re) programming and chromatin remodeling: transient genomic G4s assist in the establishment and maintenance of epigenetic marks, while persistent G4s may erase epigenetic marks. BioEssays 41:e1900091

Venkatesh V, Mishra NK, Romero-Canelón I et al (2017) Supramolecular photoactivatable anti-cancer hydrogels. J Am Chem Soc 139:5656–5659

Wang J, Choi JM, Holehouse AS et al (2018) A molecular grammar governing the driving forces for phase separation of prion-like RNA binding proteins. Cell 174:688–699

Wang B, Zhang L, Dai T et al (2021) Liquid–liquid phase separation in human health and diseases. Sig Transduct Target Ther 6:290

Warburg O (1956) On respiratory impairment in cancer cells. Science 124:269–270

Wheeler JR, Matheny T, Jain S et al (2016) Distinct stages in stress granule assembly and disassembly. elife 5:e18413

Xiao S, Davis JT (2018) G4-quartet hydrogels from 5′-hydrazino-guanosine for the non-covalent and covalent remediation of contaminants from water. Faraday Discuss 209:97–112

Xie XQ, Zhang Y, Wang M et al (2021) Programmable transient supramolecular chiral G-quadruplex hydrogels via a chemically fueled non-equilibrium self-assembly strategy. Angew Chem Int Ed Engl 61:e202114471

Yagi R, Miyazaki T, Oyoshi T (2018) G-quadruplex binding ability of TLS/FUS depends on the β-spiral structure of the RGG domain. Nucleic Acids Res 46:5894–5901

Yoneda JS, de Araujo DR, Sella F et al (2021) Self-assembled guanosine-hydrogels for drug-delivery application: structural and mechanical characterization, methylene blue loading and controlled release. Mater Sci Eng C Mater Biol Appl 121:111834

Zbinden A, Pérez-Berlanga M, De Rossi P et al (2020) Phase separation and neurodegenerative diseases: a disturbance in the force. Dev Cell 55:45–68

Zhang H-X, Blumwald E (2001) Transgenic salt-tolerant tomato plants accumulate salt in foliage but not in fruit. Nat Biotechnol 19:765–768

Zhang Y, Yang M, Duncan S et al (2019) G-quadruplex structures trigger RNA phase separation. Nucleic Acids Res 47:11746–11754

Zhang H, Ji X, Li P et al (2020) Liquid-liquid phase separation in biology: mechanisms, physiological functions and human diseases. Sci China Life Sci 63:953–985

Zhao YN, Yuan QP, Li C et al (2015) Dynamic layer-by-layer films: a platform for zero-order release. Biomacromolecules 16:2032–2039

Zhao YN, Gu JJ, Jia SY et al (2016) Zero order release of polyphenolic drugs from dynamic, hydrogen bonded LBL films. Soft Matter 12:1085–1092

Zhong R, Xiao M, Zhu C et al (2018) Logic catalytic interconversion of G-molecular hydrogel. ACS Appl Mater Interfaces 10:4512–4518

Zimmerman SB (1976) X-ray study by fiber diffraction methods of a self-aggregate of guanosine-5'-phosphate with the same helical parameters as poly (rG). J Mol Biol 106:663–672

Chapter 4
Chaperons Against Self-Association for Phase-Separating RNA-Binding Proteins

Takuya Yoshizawa

Abstract Droplets formed through liquid–liquid phase separation (LLPS) are fluid-like liquids, and this fluidity is based on weak and transient interactions between internal molecules. In biology, the flexible formation and dispersion of droplets are considered important for the functioning of membrane-less organelles. Such liquid droplets, composed of biological macromolecules, are based on an exquisite balance of multiple interactions, and misregulation leads to irreversible aggregation. Abnormal aggregation of phase-separating RNA-binding proteins, including fused-in-sarcoma (FUS) and TAR DNA binding protein 43 (TDP-43), has been observed in severe neurodegenerative diseases, suggesting that the dysregulation of LLPS and diseases are closely related. Therefore, the droplet state must be properly maintained to achieve its function as a membrane-less organelle. To maintain the LLPS state properly, chaperone proteins that maintain fluidity and inhibit the aggregation of phase-separating proteins have been identified. Nuclear transport receptor Karyopherin-β family proteins (Kapβs) and small heat shock proteins (sHSP) have been unveiled playing a key role as chaperones for LLPS. Kapβs and sHSP prevent aberrant phase separation of RNA-binding proteins including FUS and TDP-43.

Keywords RNA-binding proteins · FUS · TDP-43 · Karyopherinβ · Small heat shock proteins · Chaperone · Intrinsically disordered region

4.1 Introduction

Over 1500 human proteins have been identified as RNA-binding proteins, many of which partially or mainly consist of intrinsically disordered regions that do not have a specific ternary structure; the specific functions of intrinsically disordered regions are unclear (Hentze et al. 2018; Corley et al. 2020). In the past decade, numerous RNA granules composed of RNA and RNA-binding proteins have been found to function in cells, and the intrinsically disordered regions of RNA-binding proteins

T. Yoshizawa (✉)
College of Life Sciences, Ritsumeikan University, Kusatsu, Shiga, Japan

R. Kurokawa (ed.), *Phase Separation in Living Cells*,
https://doi.org/10.1007/978-981-99-4886-4_4

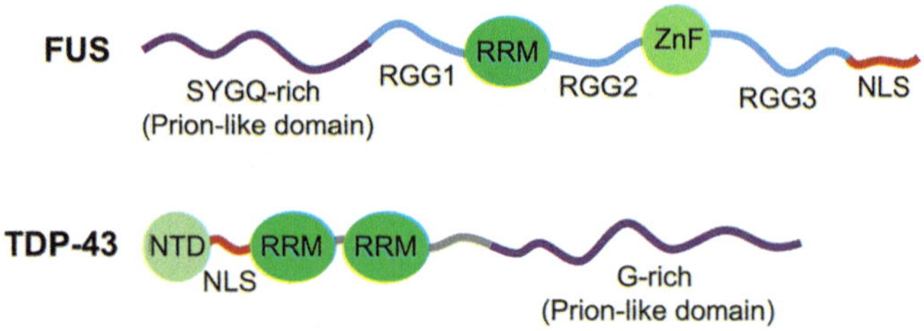

Fig. 4.1 Domain structure of FUS and TDP-43

contribute to droplet formation via liquid–liquid phase separation (LLPS) (Hyman et al. 2014; Banani et al. 2017).

The RNA-binding protein fused-in-sarcoma (FUS), a commonly used target of phase separation studies, is a 526 amino acids protein that plays a role in transcription, splicing, DNA repair, and mRNA transport. Approximately 70% of the FUS region, except for the RNA recognition motif (RRM) and zinc finger (ZnF), is intrinsically disordered (Fig. 4.1). The intrinsically disordered regions of FUS can be divided into two types: an N-terminus low-complexity region and an arginine-glycine-glycine (RGG) repeat. The N-terminus low-complexity region, also known as the SYGQ-rich region, is enriched in serine, tyrosine, glycine, and glutamate. Three RGG repeats were distributed in the C-terminal region and flanked by SYGQ-rich, RRM, and ZnF. Regarding self-association, multiple cation–π interactions between arginine residues in the RGG-repeats and tyrosine residues in the SYGQ-rich region contribute to LLPS. The SYGQ-rich region forms an amyloid-like cross-β polymer that is thought to drive droplet formation and aggregation (Kato et al. 2012; Burke et al. 2015; Murray et al. 2017). FUS has a proline-tyrosine nuclear localisation signal (PY-NLS) at the C-terminal end and is recognised and transported by the nuclear transport receptor Karyopherinβ2 (Kapβ2) (Zhang and Chook 2012). FUS aggregates have been observed in amyotrophic lateral sclerosis (ALS) and frontotemporal lobar degeneration (FTLD), suggesting that aggregation due to aberrant phase transition of FUS is related to these diseases. Several FUS mutations have been reported in patients with familial ALS. Mutations were concentrated in the nuclear localisation signal (NLS) (Kwiatkowski et al. 2009). Thus, the regulation by Kapβ2 is important to prevent aggregation.

Mutations in the RNA-binding protein, TAR DNA binding protein 43 (TDP-43) have also been reported to be associated with ALS and FTLD (Ling et al. 2013). TDP-43 plays a role in RNA metabolism and consists of an RNA-binding domain and an intrinsically disordered region, such as FUS. There are several differences between FUS and TDP-43: (1) TDP-43 has a folded N-terminal domain (NTD) that is known to oligomerise and the oligomerisation is required for TDP-43 function, (2) canonical-NLS that follows the NTD and it is recognised by Kapα/β1 heterodimer, (3) tandem RNA-binding domains without Zn finger, and (4) C-terminal intrinsically disordered region enriched in glycine (Fig. 4.1). In addition, most disease-related mutations are found in the C-terminal intrinsically disordered regions (Harrison and Shorter 2017).

Aggregation of FUS and TDP-43 is known to be a hallmark of ALS; therefore, proper regulation is required. In this section, chaperone proteins focus on maintaining the phase state of phase-separating RNA-binding proteins.

4.2 Kapβ Family

The nucleus is a membrane-bound organelle surrounded by a lipid bilayer, and the exchange of biological macromolecules between the nucleus and the cytoplasm occurs mainly through the nuclear pore (Fernandez-Martinez and Rout 2021). The nuclear pore is a large ring complex composed of approximately 30 proteins called nucleoporins, and the inside is filled with intrinsically disordered regions called phenylalanine-glycine (FG) repeats. A molecular weight of 40 kDa or less can pass through the nuclear pore, whereas those above 40 kDa require nuclear import receptors namely the Kapβ family proteins (Wing et al. 2022).

The Kapβs consist of a solenoid-like structure of stacked α-helices called HEAT repeats (Fig. 4.2). Kapβs recognise their binding partners at the central cleft. Proteins transported to the nucleus have an NLS, which is recognised by and bound to Kapβs, forming a complex that transports the protein to the nucleus. Twenty family proteins have been identified in humans. Each Kapβ recognises a specific NLS. Structural studies revealed that Kapβ2 recognises the proline-tyrosine motif at the C-terminal end of NLS and basic residues of its upstream (Lee et al. 2006). Kapβ1 recognises cNLS via Kapα. Kapα binds to a lysine-rich sequence, and Kapβ1 recognises the region of Kapα exposed upon binding, forming a ternary complex that enters the nucleus (Xu et al. 2010; Wing et al. 2022). In the nucleus, the GTP-binding small GTPase Ran binds to Kapβ, and transport is completed by releasing substrates. The Kapβ-RanGTP complex is ejected from the nucleus and leaves Kapβ when the GTP hydrolyses GDP through the GTPase activity of Ran. Kapβ families perform nuclear transport of various nuclear proteins in this way. FUS and TDP-43 are imported to

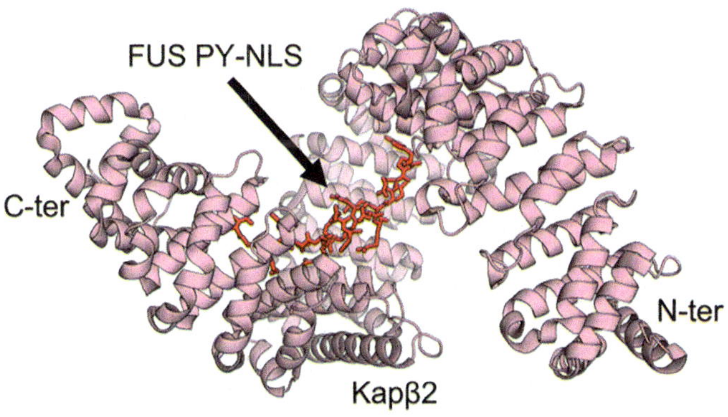

Fig. 4.2 Structure of Kapβ2 in complex with FUS PY-NLS

nucleus by Kapβ2 and Kapα/β1, respectively in this manner. Crm1, a member of the Kapβ family, is known to be responsible for nuclear export, recognising the sequence called nuclear export signal and transporting (Fu et al. 2018).

4.3 Kapβ Inhibits Self-Association of Substrate

Kapβ family not only transports transport substrates into the nucleus but also suppresses their phase separation (Guo et al. 2018; Hofweber et al. 2018; Qamar et al. 2018; Yoshizawa et al. 2018). Purified FUS protein forms droplets owing to its self-association propensity, but in the presence of Kapβ2, FUS loses its droplet-forming ability. Furthermore, FUS droplets vanish upon the addition of Kapβ2, and FUS and Kapβ2 become stable in a dispersed state as a 1:1 complex. RanGTP binding to Kapβ2 resulted in the loss of Kapβ2 phase separation inhibitory activity. The phase separation inhibitory activity of Kapβ2 was also lost by M9M, a PY-NLS binding inhibitor peptide that binds more tightly than the native PY-NLS. Thus, Kapβ2 suppresses phase separation through its interaction with PY-NLS (Yoshizawa et al. 2018).

Kapα/β also inhibited TDP-43 self-association similarly (Guo et al. 2018). Using chimera in which the PY-NLS of FUS was replaced by cNLS, Kapα alone did not inhibit phase separation, but the addition of Kapβ with Kapα inhibited the phase separation of FUS chimera. While, replacing PY-NLS of FUS with nuclear export signal were not inhibited phase separation by the nuclear export receptor Crm1. Thus, the inhibitory activity of self-association for RNA-binding proteins is common in nuclear import receptors but not in nuclear export receptors. Nuclear import receptors protect and transport aggregation-prone RNA-binding proteins into the nucleus. The phase separation inhibitory ability of the Kapβ family is also thought to play a role in permeabilising the phase-separated state inside the nuclear pore composed by its FG repeats (Yoshizawa and Matsumura 2020; Yoshizawa and Guo 2021).

4.4 Mechanism for LLPS Inhibition

Kapβ binding to the substrate via NLS is important for inhibiting phase separation. While the SYGQ-rich region and the RGG repeats where are regions outside of NLS drive phase separation. How does Kapβ suppress client phase separation? Nuclear magnetic resonance (NMR) analysis of Kapβ2 interactions with FUS revealed that Kapβ2 weakly interacts with various regions of FUS, including SYGQ-rich region and RGG repeats, suggesting that Kapβ2 anchors substrates through strong NLS-mediated interactions and suppresses phase separation by interacting with regions that promote phase separation of FUS (Yoshizawa et al. 2018). Kapβ2 shows a high ability to inhibit phase separation of FUS in the liquid phase, but less activity in aggregated phase. This may be due to a change in the balance of interactions between FUS proteins during the phase transition from the liquid state to the aggregate state. In the highly fluid liquid phase of FUS, there are many electrostatic interactions such as cation–π interactions, while with aggregation, there is an increase in hydrophobic interactions and elongation of amyloid-like cross-β polymers, which can no longer compete with the interaction between Kapβ2 and FUS (Wang et al. 2018; Kato and McKnight 2021).

There is also diversity in the droplet state. Molecular dynamics simulations indicated that the hydrophobic interactions were more stable under high-pressure conditions than under atmospheric pressure. It was reported that the ability of Kapβ2 to inhibit phase separation is weakened for droplets formed under high-pressure conditions above 2000 bar (Kitahara et al. 2021; Li et al. 2021, 2022a), suggesting that differences in the interactions inside the droplet affect the ability of Kapβ2 to inhibit phase separation.

Experiments using a construct lacking the NLS of FUS (FUSΔNLS) showed that Kapβ2 can also inhibit the phase separation of FUSΔNLS (Gonzalez et al. 2021). The cleft on which Kapβ2 recognises PY-NLS is acidic, which shows high affinity with RGG repeats. Thus, anchoring to Kapβ2 via the RGG repeat also inhibits phase separation. Acidic clefts that recognise NLSs play a pivotal role as the starting point for phase separation inhibition. HEAT repeats (H1-H20) comprising Kapβ2 recognise PY-NLSs using the broad region from H8 to the C-terminal region(H20). Deletion of part of this region causes the loss of the inhibiting phase separation activity of Kapβ2. While the construct lacks the N-terminal region, Kapβ2(H8-H20) inhibited the phase separation as the full-length form (Fare et al. 2023). This indicates that the C-terminal region, including the PY-NLS-binding site, is important for phase separation inhibition.

4.5 Dysfunction of Kapβ2 and Disease

C9orf72 is known as a causative gene of familial ALS (fALS). In fALS, a hexanucleotide repeat (GGGGCC) is expanded in the intron of *C9orf72*. Repeat expansion results in repeat-associated non-AUG translation, which produces a repeating peptide of two amino acids (DeJesus-Hernandez et al. 2011; Renton et al. 2011). Five types of dipeptide repeats are produced according to their reading frames: glycine-alanine (GA), proline-alanine (PA), glycine-arginine (GR), proline-arginine (PR), and glycine-proline (GP). PR and GR repeats are known to be highly toxic. These arginine-rich dipeptide repeats have been found to co-aggregate with TDP-43 and FUS and induce aggregation by enhancing self-aggregation. Accumulation in the nucleoli has also been reported, and is thought to affect various phase separation states (Kwon et al. 2014).

The arginine-rich dipeptide repeats have been also reported to disturb Kapβ2 function, i.e., Kapβ2 lacks inhibition activity of FUS phase separation in the presence of arginine-rich dipeptide repeats (Hayes et al. 2020; Hutten et al. 2020;

Nanaura et al. 2021). Biochemical analysis revealed that PR repeat peptides with eight repeats did not bind to Kapβ2, whereas those with 18 repeats bound to Kapβ2, indicating that the binding is repeat-dependent. Comparisons of the interaction analysis of PY-NLS with Kapβ2 and PR repeat peptide with Kapβ2 were performed using NMR that showed that the PR binding region on Kapβ2 partially overlaps with the binding region of PY-NLS. The arginine-rich dipeptide repeats also bind to the Kapα/β complex in a repeat frequency-dependent manner (Nanaura et al. 2021). Although various possible mechanisms of arginine-rich dipeptide repeat toxicity have been reported and its mechanisms remain unclear, it has been suggested that dysregulation of the Kapβ family may lead to aberrant phase separation related to neurodegenerative diseases (Zhang et al. 2018).

In addition, as mentioned earlier, missense mutations in FUS have been reported in many cases, particularly in fALS. Although the mutations are concentrated in the PY-NLS, Kapβ2 shows phase separation inhibitory activity against FUS lacking the PY-NLS in vitro. This inhibition was due to the binding of the RGG repeats instead of PY-NLS (Gonzalez et al. 2021). RGG repeats have been shown to interact not only with Kapβ2 but also with other members of the Kapβs, including Kapβ1, transportin-3, and importin7, suggesting crosstalk in phase separation inhibition. It should be noted that many arginines of RGG repeats are methylated in the cell (Dormann et al. 2012; Hofweber et al. 2018; Qamar et al. 2018), and the methylation alters the charge balance, thus the binding capability to Kapβs may differ, leading to mislocalisation and aberrant phase separation.

4.6 HSP as Phase Regulator

Small heat shock proteins (sHSPs) are molecular chaperones found in many organisms, and their expression is induced by various environmental stresses to protect cells. Unlike major heat shock proteins, sHSPs are ATP-independent chaperones that are approximately 20 kDa in size (Acunzo et al. 2012; Garrido et al. 2012; Bakthisaran et al. 2015). The α-crystallin domain of sHSP is known to be oligomerised, and the oligomerisation is presumably important for the chaperone functions. Hsp27 (HspB1) has been reported to prevent aggregation and suppress huntingtin-induced toxicity, and its oligomerisation is important for chaperone function. In contrast, Hsp22 (HspB8) exerts its chaperone function without self-oligomerisation.

sHSPs regulate LLPS state. Hsp27 interacts with the SYGQ-rich region of FUS and suppresses FUS phase separation (Liu et al. 2020). The NTD of Hsp27, rather than the α-crystallin domain conserved among sHSPs, is important for LLPS suppression. NTD contains stress-induced phosphorylation sites, phosphorylated Hsp27 localises in droplets with FUS, and suppresses amyloid-like fibrosis, suggesting that NTD is important for both the suppression of droplet formation and amyloid-like fibrosis. The α-crystallin domain contributes to inhibiting amyloid fibrilisation although it does not inhibit LLPS.

It was also reported that Hsp22 suppresses aggregation in LLPS droplets through the interaction of the α-crystallin domain with the RNA-binding domain of FUS (Boczek et al. 2021; Lu et al. 2022). Misfolding and fibrilisation of the RNA-binding domain of FUS may occur during the aggregation process, suggesting that stabilisation of the RNA-binding domain structure plays a role in the inhibition of aggregation.

In addition to sHSP, HSP70 and HSP40 may also contribute to the regulation of LLPS within droplets (Gu et al. 2020; Yu et al. 2021; Li et al. 2022b), revealing phase-separating chaperones as a new role for HSPs.

4.7 Perspective

In this section, the Kapβ and HSP families were discussed as phase separation regulators: the Kapβ family acts to dissolve, while HSP co-localises in droplets to maintain fluidity (Fig. 4.3). The HEAT repeats structured proteins like the Kapβ family are widely present, suggesting a conserved role for the HEAT repeats, and membrane-less organelles appear in various biological events (Yoshimura and Hirano 2016). As mentioned in above, many proteins in the HSP family may act as phase separation chaperones. It has also been reported that intrinsically disordered proteins, such as heat-resistant obscure proteins, regulate phase separation (Tsuboyama et al. 2020). The chaperonin subunit CCT2 functions as a sensor of aberrant phase separation and induces autophagy (Ma et al. 2022). Many candidates can be categorised as phase separation chaperones.

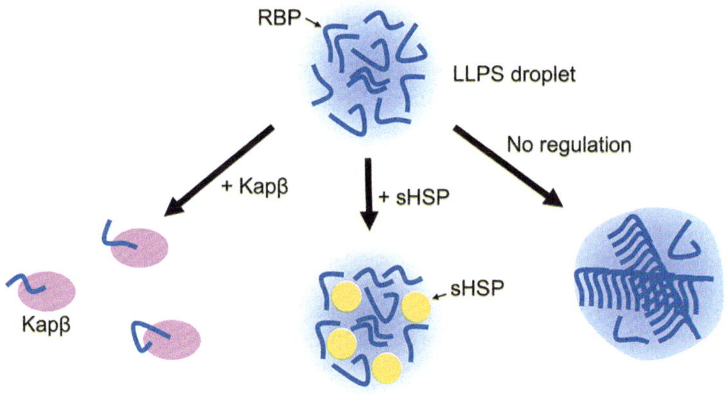

Fig. 4.3 Schematic diagram of phase regulation by Kapβ and sHSP

References

Acunzo J, Katsogiannou M, Rocchi P (2012) Small heat shock proteins HSP27 (HspB1), αb-crystallin (HspB5) and HSP22 (HspB8) as regulators of cell death. Int J Biochem Cell Biol 44:1622–1631. https://doi.org/10.1016/j.biocel.2012.04.002

Bakthisaran R, Tangirala R, Rao CM (2015) Small heat shock proteins: role in cellular functions and pathology. Biochim Biophys Acta 1854:291–319

Banani SF, Lee HO, Hyman AA, Rosen MK (2017) Biomolecular condensates: organizers of cellular biochemistry. Nat Rev Mol Cell Biol 18:285–298. https://doi.org/10.1038/nrm.2017.7

Boczek EE, Fürsch J, Niedermeier ML et al (2021) HspB8 prevents aberrant phase transitions of FUS by chaperoning its folded RNA binding domain. elife 10:e69377. https://doi.org/10.7554/eLife.69377

Burke KA, Janke AM, Rhine CL, Fawzi NL (2015) Residue-by-residue view of in vitro FUS granules that bind the C-terminal domain of RNA polymerase II. Mol Cell 60:231–241. https://doi.org/10.1016/j.molcel.2015.09.006

Corley M, Burns MC, Yeo GW (2020) How RNA-binding proteins interact with RNA: molecules and mechanisms. Mol Cell 78:9–29. https://doi.org/10.1016/j.molcel.2020.03.011

DeJesus-Hernandez M, Mackenzie IR, Boeve BF et al (2011) Expanded GGGGCC Hexanucleotide repeat in noncoding region of C9ORF72 causes chromosome 9p-linked FTD and ALS. Neuron 72:245–256. https://doi.org/10.1016/j.neuron.2011.09.011

Dormann D, Madl T, Valori CF et al (2012) Arginine methylation next to the PY-NLS modulates Transportin binding and nuclear import of FUS. EMBO J 31:4258–4275. https://doi.org/10.1038/emboj.2012.261

Fare CM, Rhine K, Lam A et al (2023) A minimal construct of nuclear-import receptor Karyopherin-β2 defines the regions critical for chaperone and disaggregation activity. J Biol Chem 299:102806. https://doi.org/10.1016/j.jbc.2022.102806

Fernandez-Martinez J, Rout MP (2021) One ring to rule them all? Structural and functional diversity in the nuclear pore complex. Trends Biochem Sci 1–13:595. https://doi.org/10.1016/j.tibs.2021.01.003

Fu SC, Fung HYJ, Cağatay T et al (2018) Correlation of CRM1-NES affinity with nuclear export activity. Mol Biol Cell 29:2037–2044. https://doi.org/10.1091/mbc.E18-02-0096

Garrido C, Paul C, Seigneuric R, Kampinga HH (2012) The small heat shock proteins family: the long forgotten chaperones. Int J Biochem Cell Biol 44:1588–1592. https://doi.org/10.1016/j.biocel.2012.02.022

Gonzalez A, Mannen T, Çağatay T et al (2021) Mechanism of karyopherin-β2 binding and nuclear import of ALS variants FUS(P525L) and FUS(R495X). Sci Rep 11:3754. https://doi.org/10.1038/s41598-021-83196-y

Gu J, Liu ZZ, Zhang S et al (2020) Hsp40 proteins phase separate to chaperone the assembly and maintenance of membraneless organelles. Proc Natl Acad Sci 117:202002437. https://doi.org/10.1073/pnas.2002437117

Guo L, Kim HJ, Wang H et al (2018) Nuclear-import receptors reverse aberrant phase transitions of RNA-binding proteins with prion-like domains. Cell 173:677–692.e20. https://doi.org/10.1016/j.cell.2018.03.002

Harrison AF, Shorter J (2017) RNA-binding proteins with prion-like domains in health and disease. Biochem J 474:1417–1438. https://doi.org/10.1042/BCJ20160499

Hayes LR, Duan L, Bowen K et al (2020) C9orf72 arginine-rich dipeptide repeat proteins disrupt karyopherin-mediated nuclear import. elife 9:1–29. https://doi.org/10.7554/eLife.51685

Hentze MW, Castello A, Schwarzl T, Preiss T (2018) A brave new world of RNA-binding proteins. Nat Rev Mol Cell Biol 19:327–341. https://doi.org/10.1038/nrm.2017.130

Hofweber M, Hutten S, Bourgeois B et al (2018) Phase separation of FUS is suppressed by its nuclear import receptor and arginine methylation. Cell 173:706–719.e13. https://doi.org/10.1016/j.cell.2018.03.004

Hutten S, Usluer S, Bourgeois B et al (2020) Nuclear import receptors directly bind to arginine-rich dipeptide repeat proteins and suppress their pathological interactions. Cell Rep 33:108538. https://doi.org/10.1016/j.celrep.2020.108538
Hyman AA, Weber CA, Jülicher F (2014) Liquid-liquid phase separation in biology. Annu Rev Cell Dev Biol 30:39–58. https://doi.org/10.1146/annurev-cellbio-100913-013325
Kato M, McKnight SL (2021) The low-complexity domain of the FUS RNA binding protein self-assembles via the mutually exclusive use of two distinct cross-β cores. Proc Natl Acad Sci U S A 118:1–10. https://doi.org/10.1073/pnas.2114412118
Kato M, Han TW, Xie S et al (2012) Cell-free formation of RNA granules: low complexity sequence domains form dynamic fibers within hydrogels. Cell 149:753–767. https://doi.org/10.1016/j.cell.2012.04.017
Kitahara R, Yamazaki R, Ide F et al (2021) Pressure-jump kinetics of liquid-liquid phase separation: comparison of two different condensed phases of the RNA-binding protein, fused in sarcoma. J Am Chem Soc 143:19697–19702. https://doi.org/10.1021/jacs.1c07571
Kwiatkowski TJ, Bosco DA, LeClerc AL et al (2009) Mutations in the FUS/TLS gene on chromosome 16 cause familial amyotrophic lateral sclerosis. Science 323:1205–1208. https://doi.org/10.1126/science.1166066
Kwon I, Xiang S, Kato M et al (2014) Poly-dipeptides encoded by the C9orf72 repeats bind nucleoli, impede RNA biogenesis, and kill cells. Science 345:1139–1145. https://doi.org/10.1126/science.1254917
Lee BJ, Cansizoglu AE, Süel KE et al (2006) Rules for nuclear localization sequence recognition by Karyopherinβ2. Cell 126:543–558. https://doi.org/10.1016/j.cell.2006.05.049
Li S, Yoshizawa T, Yamazaki R et al (2021) Pressure and temperature phase diagram for liquid-liquid phase separation of the RNA-binding protein fused in sarcoma. J Phys Chem B 125:6821–6829. https://doi.org/10.1021/acs.jpcb.1c01451
Li S, Yoshizawa T, Shiramasa Y et al (2022a) Mechanism underlying liquid-to-solid phase transition in fused in sarcoma liquid droplets. Phys Chem Chem Phys 24:19346–19353. https://doi.org/10.1039/D2CP02171D
Li Y, Gu J, Wang C et al (2022b) Hsp70 exhibits a liquid-liquid phase separation ability and chaperones condensed FUS against amyloid aggregation. iScience 25:25. https://doi.org/10.1016/j.isci.2022.104356
Ling SC, Polymenidou M, Cleveland DW (2013) Converging mechanisms in also and FTD: disrupted RNA and protein homeostasis. Neuron 79:416–438. https://doi.org/10.1016/j.neuron.2013.07.033
Liu ZZ, Zhang S, Gu J et al (2020) Hsp27 chaperones FUS phase separation under the modulation of stress-induced phosphorylation. Nat Struct Mol Biol 27:363–372. https://doi.org/10.1038/s41594-020-0399-3
Lu S, Hu J, Arogundade OA et al (2022) Heat-shock chaperone HSPB1 regulates cytoplasmic TDP-43 phase separation and liquid-to-gel transition. Nat Cell Biol 24:1378. https://doi.org/10.1038/s41556-022-00988-8
Ma X, Lu C, Chen Y et al (2022) CCT2 is an aggrephagy receptor for clearance of solid protein aggregates. Cell 185:1325–1345.e22. https://doi.org/10.1016/j.cell.2022.03.005
Murray DT, Kato M, Lin Y et al (2017) Structure of FUS protein fibrils and its relevance to self-assembly and phase separation of low-complexity domains. Cell 171:615–627.e16. https://doi.org/10.1016/j.cell.2017.08.048
Nanaura H, Kawamukai H, Fujiwara A et al (2021) C9orf72-derived arginine-rich poly-dipeptides impede phase modifiers. Nat Commun 12:1–12. https://doi.org/10.1038/s41467-021-25560-0
Qamar S, Wang GZ, Randle SJ et al (2018) FUS phase separation is modulated by a molecular chaperone and methylation of arginine cation-π interactions. Cell 173:720–734. https://doi.org/10.1016/j.cell.2018.03.056
Renton AE, Majounie E, Waite A et al (2011) A hexanucleotide repeat expansion in C9ORF72 is the cause of chromosome 9p21-linked ALS-FTD. Neuron 72:257–268. https://doi.org/10.1016/j.neuron.2011.09.010

Tsuboyama K, Osaki T, Matsuura-Suzuki E et al (2020) A widespread family of heat-resistant obscure (hero) proteins protect against protein instability and aggregation. PLoS Biol 18: e3000632. https://doi.org/10.1371/journal.pbio.3000632

Wang J, Choi JM, Holehouse AS et al (2018) A molecular grammar governing the driving forces for phase separation of prion-like RNA binding proteins. Cell 174:688–699.e16. https://doi.org/10.1016/j.cell.2018.06.006

Wing CE, Fung HYJ, Chook YM (2022) Karyopherin-mediated nucleocytoplasmic transport. Nat Rev Mol Cell Biol 0123456789:307. https://doi.org/10.1038/s41580-021-00446-7

Xu D, Farmer A, Chook YM (2010) Recognition of nuclear targeting signals by Karyopherin-β proteins. Curr Opin Struct Biol 20:782–790. https://doi.org/10.1016/j.sbi.2010.09.008

Yoshimura SH, Hirano T (2016) HEAT repeats—versatile arrays of amphiphilic helices working in crowded environments? J Cell Sci 129:3963–3970. https://doi.org/10.1242/jcs.185710

Yoshizawa T, Guo L (2021) Karyopherin- β s play a key role as a phase separation regulator. J Biochem 1:1–9. https://doi.org/10.1093/jb/mvab072

Yoshizawa T, Matsumura H (2020) Effect of nuclear import receptors on liquid–liquid phase separation. Biophys Physicobiol 12:103–117. https://doi.org/10.2142/biophysico.BSJ-2019052

Yoshizawa T, Ali R, Jiou J et al (2018) Nuclear import receptor inhibits phase separation of FUS through binding to multiple sites. Cell 173:693–705. https://doi.org/10.1016/j.cell.2018.03.003

Yu H, Lu S, Gasior K et al (2021) HSP70 chaperones RNA-free TDP-43 into anisotropic intranuclear liquid spherical shells. Science 371:eabb4309. https://doi.org/10.1126/science.abb4309

Zhang ZC, Chook YM (2012) Structural and energetic basis of ALS-causing mutations in the atypical proline–tyrosine nuclear localization signal of the fused in sarcoma protein (FUS). Proc Natl Acad Sci 109:12017–12021. https://doi.org/10.1073/pnas.1207247109

Zhang K, Daigle JG, Cunningham KM et al (2018) Stress granule assembly disrupts nucleocytoplasmic transport. Cell 173:958–971.e17. https://doi.org/10.1016/j.cell.2018.03.025

Part II
Molecular Biology

Chapter 5
Positive and Negative Aspects of Protein Aggregation Induced by Phase Separation

Riki Kurokawa

Abstract Our experiments have demonstrated that RNA-binding protein (RBP) FUS/TLS exerts distinctive inhibition against histone acetyltransferase (HAT) activities from CBP, p300, but not PCAF. With the HAT inhibitory effect, TLS has been shown to repress transcription of the cyclin D1 gene that is one of the target genes of TLS. These data indicate that TLS works as a transcriptional repressor with its inhibitory activity against HAT. The HAT inhibitory activity requires its binding to specific RNA sequences including RNAs with GGUG consensus, and also forming of a protein complex consisting of CBP, CREB, and probably other RNA-binding proteins. Therefore, our TLS project has been started to examine its transcriptional roles. Just after our study, it has been reported that TLS should work as a causative gene for amyotrophic lateral sclerosis (ALS). Subsequent reports showed that TLS undergoes phase separation that generates biomolecular condensates, droplets, or membrane-less organelles for divergent cellular activities. It is occasionally happened that phase-separated TLS forms aggregation or precipitation, which hampers its biological functions in living cell. In particular, in the case of aggregation formation in neuronal cells neurodegenerative diseases occur. This aggregation formation should be a possible cause for neurodegenerative diseases including ALS, frontotemporal lobar degeneration (FTLD), Parkinson's disease, Alzheimer's disease. Recent days, there have been significant impacts on biological functions and dysfunctions of TLS. Then, a demand has emerged to elucidate molecular mechanisms of actions of TLS regarding basic science and also clinical applications. In this chapter, I would review biological outcomes of the phase separation of TLS and related RBPs, and also pay attentions its relation to diseases of mostly neurodegenerative diseases. Although most data have been published regarding harmful aspects of the aggregation, functional and beneficial aggregates have been reported as a regulator for neuronal functions including memory formation. The author would overview these divergent topics of the phase separation and

R. Kurokawa (✉)
Division of Biomedical Sciences, School of Medicine, Saitama Medical University, Hidaka, Saitama, Japan
e-mail: rkurokaw@saitama-med.ac.jp

R. Kurokawa (ed.), *Phase Separation in Living Cells*,
https://doi.org/10.1007/978-981-99-4886-4_5

aggregation of RBPs. These discussions would support comprehensive understanding of roles of the phase separation and aggregation in biological programs.

Keywords FUS · TLS · RNA-binding protein · HAT · CBP · ALS · Functional amyloid · Phase separation · Aggregation

5.1 Introduction

In very beginning of experiments with RNA-binding protein (RBP) TLS/FUS, we had been working on the identification of an interaction factor of CREB-binding protein (CBP) that is a general coactivator regulates many transcription factors including cAMP Response Element-Binding Protein (CREB), nuclear receptor, and AP1 (Chrivia et al. 1993; Arany et al. 1994; Kwok et al. 1994; Kamei et al. 1996; Kurokawa et al. 1998). Having started with an experiment of TLS/FUS, searching interacting protein to CREB-binding protein (CBP) was executed in order to identify inhibitory activity against histone acetyltransferase (HAT)of CBP (Fig. 5.1). Mass spectrometric analysis of binding fractions of FLAG-tagged

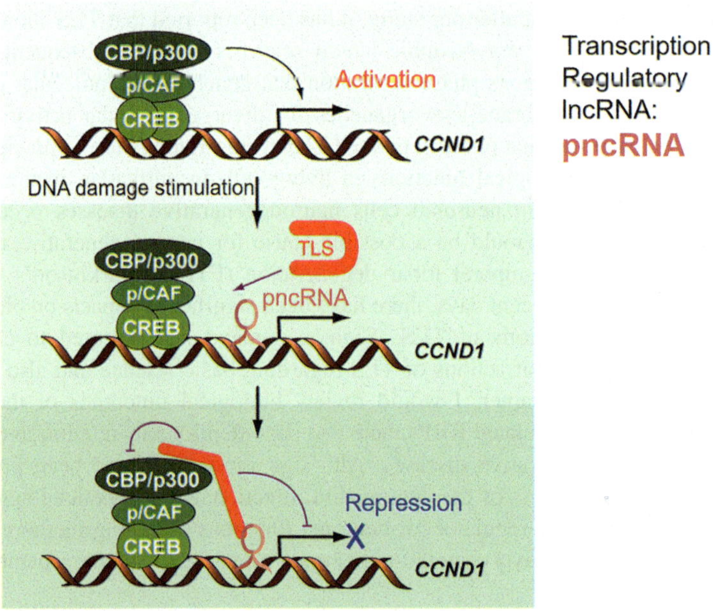

Fig. 5.1 TLS/FUS works as a transcriptional repressor. Our experiments are initiated with searching for binding molecules with coactivator CREB-binding protein CBP with criteria of inhibitory activity against histone acetyltransferase (HAT) activity. In this project, TLS has been identified as a potent HAT inhibitory factor (Wang et al. 2008). Actually, TLS binds CBP on the promoter of the cyclin D1 (CCND1) gene that is a target gene of TLS, and represses transcription of CCND1 through inhibitory effect on the HAT of CBP

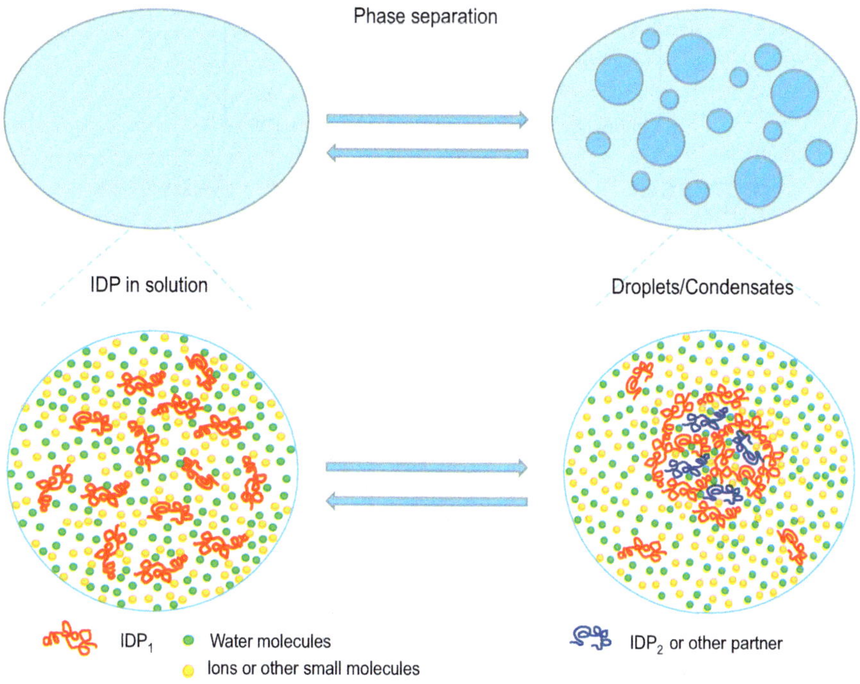

Fig. 5.2 Phase separation. Solute and solvent form droplets or condensates

baculovirus-expressed CBP with nuclear extract of a human cervical tumor HeLa cell line provided a few protein bands with coomassie brilliant blue staining of an SDS-PAGE gel. One of the detected bands was human TLS protein that possesses inhibitory activity against the CBP HAT activity (Wang et al. 2008). Subsequent series of experiments demonstrated that TLS exerts distinctive inhibition against HAT activities of CBP, p300, but not PCAF. As a result of the HAT inhibition, TLS has been shown to repress transcription of the cyclin D1 gene that is one of target genes of TLS. For HAT inhibitory activity, TLS requires binding to specific RNA sequences including GGUG consensus (Lerga et al. 2001), and forms protein complexes consisting of CBP, CREB, and other RNA-binding proteins including hnRNPH and hnRNPAB. Therefore, our TLS project has started to examine its transcriptional roles (Glass and Rosenfeld 2000; Kurokawa et al. 2009). Later, phase separation turned out to be involved in transcriptional regulations (Hnisz et al. 2017; Boija et al. 2018; Lu et al. 2018; Nair et al. 2019; Shao et al. 2022).

Just after our study, it has been reported that TLS should work as a causative gene for amyotrophic lateral sclerosis (ALS) (Kwiatkowski Jr. et al. 2009; Lagier-Tourenne and Cleveland 2009; Vance et al. 2009; Da Cruz and Cleveland 2011; Taylor et al. 2016). Subsequent reports showed that TLS undergoes phase separation that generates biomolecular condensates, droplets, or membrane-less organelles for divergent cellular activities (Fig. 5.2) (Kwon et al. 2013; Castello et al. 2016; Liepelt

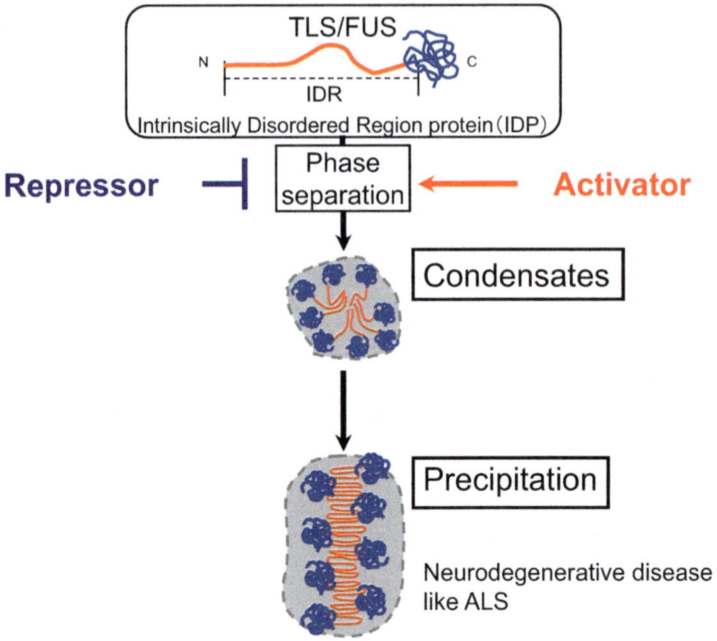

Fig. 5.3 Phase separation and aggregation of TLS/intrinsically disordered region protein (IDP). At higher concentration (more than 10 μg/μL), IDP goes phase separation to form condensates, and occasionally forms precipitation or aggregation with cytotoxicity. The toxic aggregates cause neurodegenerative diseases. Biotinylated isoxazole (BISOX) (activator) activates the phase separation, while long noncoding RNAs (repressor) repress the phase separation

et al. 2016). It is occasionally happened that phase-separated TLS forms aggregation or precipitation, which hampers its biological functions in living cell (Mackenzie et al. 2010; Hofweber et al. 2018; Wegmann et al. 2018; Portz et al. 2021; Korobeynikov et al. 2022). In particular, the formation of aggregation in neuronal cells should be linked to diseases. This aggregation should be a possible cause for neurodegenerative diseases comprising ALS, frontotemporal lobar degeneration (FTLD), Parkinson's disease, and Alzheimer's disease (Fig. 5.3) (Broustal et al. 2010; Van Langenhove et al. 2010; Irwin et al. 2013; Guo and Lee 2014; Osterberg Valerie et al. 2015; Martinelli et al. 2022). Recent days, there have been significant impacts on biological functions and dysfunctions of TLS. Then, emerging demand has been arisen to elucidate molecular mechanisms of actions of TLS regarding basic science and also clinical applications. In this chapter, I would review biological outcomes of the phase separation of TLS and related RNA-binding proteins, and also pay attentions its relation to diseases of mostly neurodegenerative diseases. Therefore, it is crucial to suppress the unscheduled phase separation and aggregation in living cells. Then, I also outline the regulatory apparatus of the phase separation and resultant aggregation in biological environments.

Although the majority of data published regarding harmful aspects of the aggregation, good and beneficial aggregates have been reported in divergent organisms regarding functions of biofilm formation of bacteria, nitrogen catabolism of yeast, melanin formation and also memory formation of human (Richter 2007; Pham et al. 2014; Jackson and Hewitt 2017; Shanmugam et al. 2019). Unexpectedly, the remarkable function of amyloid is one of the most refined and complexed missions in biological systems, memory formation (Si et al. 2003; Majumdar et al. 2012; Hervas et al. 2020; Sheng et al. 2020; Ulamec and Radford 2020; Soria et al. 2022; Kozlov et al. 2023). We keep this point in our mind. The author would overview these divergent topics of the phase separation and aggregation of IDP/TLS.

5.2 Intrinsically Disordered Protein Induces Phase Separation

Proteins form three-dimensional structures and exert specific functions based upon their structures (Sela et al. 1957; Bryngelson et al. 1995; Arai and Kuwajima 2000). However, it has been reported that some groups of proteins do not form a specific 3-D structure and consist of intrinsically disordered region (IDR), and intrinsically disordered proteins (IDP) undergo phase separation and form aggregates (Fig. 5.4) (Shin et al. 2017; Kulkarni et al. 2022). Recent decade, the phase separation has been paid attentions for its divergent biological programs including biomolecular condensates, droplets, or membrane-less organelles, and also relation to human diseases especially neurodegenerative diseases (Uversky 2017; Shen et al. 2020; Tauber et al. 2020; Yoshizawa et al. 2020; Zbinden et al. 2020).

Historically, Nuclear Magnetic Resonance (NMR) analysis of solution structures of proteins presents concept of naturally unfolded protein or IDP (Wright and Dyson 1999; Hu and Tycko 2010; Keul et al. 2018). There has been shown that IDP undergoes phase separation in vitro experimental system (Li et al. 2012). Just mixing

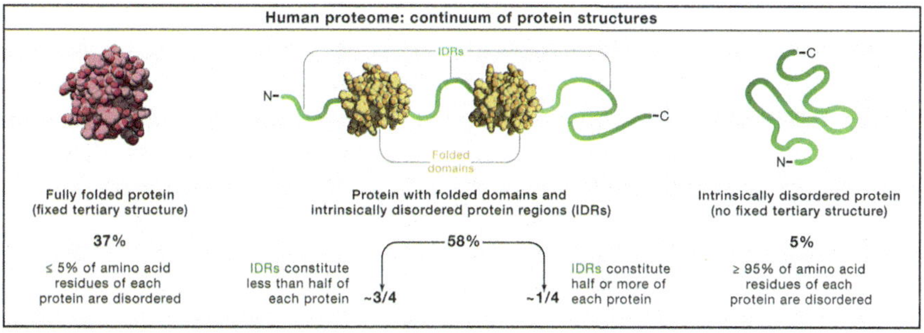

Fig. 5.4 Human proteome indicates continuum of protein structures. Description of the details is at the text. (This figure is reproduced from the reference Tsang et al. (2020) upon permission of Elsevier inc (Tsang et al. 2020). License Number:5497511289290)

two kinds of solutions of distinctive IDPs, src homology 3 (SH3), and proline rich motifs (PRM) generates droplets through phase separation. One more report has been published that TLS/FUS undergoes phase separation and resultant hydrogels (Kato et al. 2012). It has been known that RNA granules contain RNA-binding proteins which have repetitive sequence of low complexity (LC) domains with RNA-binding domains. FUS and also hnPNPA2 containing LC domains are mixed to be examined with a microscope. After incubation of these RBPs, these proteins were migrated to form β sheets and consequently dynamic fibers within hydrogels. The N-terminus of TLS has low complexity domain or prion domain consisting of 27 repeats of glycine, serine, proline, and tyrosine. Mutations of these amino acid repeat cause reduction of the formation of hydrogels (Murray et al. 2017).

Droplet formations are soft and dynamic process unlike precipitation of solid materials from solid–liquid phase separation and phase transition between water and ice. Therefore, droplet formation is able to respond subtle alterations in pH, ionic strength, temperature, concentrations of small molecules from metabolic reactions (Yu et al. 2016). Actually, it has been shown that IDR does not form higher structures less than random sequence structures. The simple repetitive sequences, LC domains, have been designated through the evolution of organisms to conduct biological phase separations (Dunker et al. 2015; Kulkarni 2020; Pajkos and Dosztányi 2021).

There is preference of biological species in expressions of IDP. Eukaryotic cells have been shown to express IDP more than prokaryotic cells. Extensive analysis of data bases by 3D structure prediction program have demonstrated that percentages of IDP are 4.2% in eubacteria, 2.0% in archaea, and 33% in eukaryote (Ward et al. 2004). There are also preferences of IDP for functions of proteins. IDR proteins are involved in gene regulations containing transcription, translation, and signal transductions containing transcription factors, nucleic acid binding proteins, RNA polymerases. Contrarily, IDR has not seen frequently in metabolic enzymes. In human proteome, 58% of human proteins contain both folded protein domains and IDRs (Tsang et al. 2020). The proteins just consisting of fully folded region is 37%, while proteins composing only from IDRs are 5% (Fig. 5.4) (Tsang et al. 2020).

These data indicate that emerging of IDP might be linked to biological evolution (Lichtinger et al. 2021; Ho and Huang 2022). In primitive stage of the earth primordial single cell organisms have protein molecules with folded and ordered three-dimensional structures and each specific function based upon its proper structure. During evolution of these creatures into multicellular systems, they have developed molecular systems of operating genetic information processes and also cellular signaling networks. They need to develop IDPs for fine regulations of these multicellular systems in living cells.

5.3 Harmful Aggregates Generated by the Phase Separation

The solution of two components of IDP undergoes phase separation to droplets or condensates, and occasionally form aggregates of these proteins. These aggregates frequently exert harmful effect or cytotoxicity (Blokhuis et al. 2013; Popper et al. 2021).

Representative proteins forming aggregation in pathological lesions of neurodegenerative disease are Transactive response DNA-binding protein 43 (TDP43) and TLS (Mackenzie et al. 2010; Patel et al. 2015b; Prasad et al. 2019). Both of these proteins are prone to precipitate in neuronal cells affected. However, mechanisms of induced neuronal cell damaging remain uncovered. Observation of the brains of the ALS patients showed frequent accumulations of digested fragments of TDP43 (Rohn 2008; Hasegawa et al. 2011). Then, series of overexpression experiment with TDP43 molecules were performed. Overexpression of the full-length TDP43 in cultured mammalian cells induced remarkable apoptosis there, although no aggregation of TDP43 that are observed in the patient's brains (Yamashita et al. 2014). On the other hand, overexpression of the truncated TDP43 in the cells caused cytosolic aggregation containing the truncated TDP43 similar to the aggregate observed in the ALS patients. These cells with truncated TDP43 aggregation did not have apoptosis-induced cell death, while they exhibited arrest of cellular growth. Analyzing these aggregates of the truncated TDP43 indicated accumulated transcription factors and related cofactors including AP1, CBP, and RNA polymerase II. These aggregation formations turned out to repress transcriptional activity of cellular growth-related genes, and later cause cell death. These experiments postulate two distinctive pathways to exert cytotoxicity from TDP43 (Hasegawa et al. 2011; Yamashita et al. 2014). The initial experiment showed that overexpression of full-length TDP43 causes apoptotic cell death, but does not induce aggregation. The second experiment demonstrates overexpression of truncated TDP43 induces aggregation that sequestrate transcriptional factors, leading to repression of expression of genes related cellular growth and to non-apoptotic cell death. Most probably, the truncated TDP43 aggregation causing cellular toxicity might be a major cause for ALS, because up to 95% of the patients have aggregation of TDP43 in motor neurons (Yang et al. 2014; Porta et al. 2021).

More focusing study regarding axonal TDP43 aggregation in motor neuron and neuromuscular junction, their disruption has been shown (Altman et al. 2021). Using induced pluripotent stem cells (iPS cells) from ALS patient bearing C9ORF72 mutation, motor neurons induced from the iPS cells had a significant increment in the amount of TDP43 compared to the control cell line. Then, mislocalization of TDP43 occurs from the nuclei of the neural cell body to the axon of the motor neurons. The axonal localization of TDP43 with phosphorylation assembles ribonucleoprotein (RNA and protein complex: RNP) consisting of phosphorylated TDP43, mRNA, and GTPase-activating protein binding protein 1(G3BP1) (Yang et al. 2020). The G3BP1 with phosphorylated TDP43 forms the RNP condensates

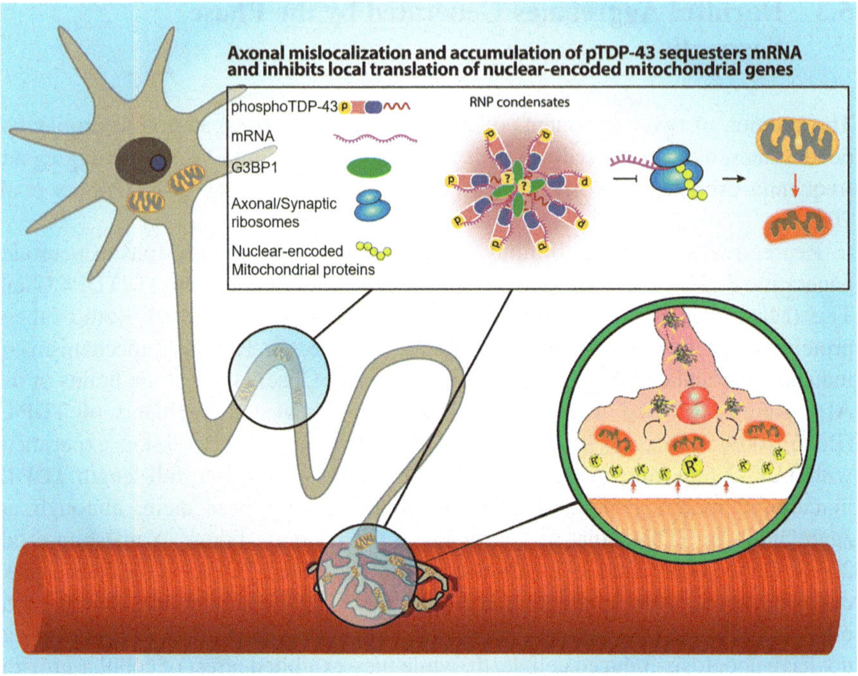

Fig. 5.5 Axonal mislocalization and accumulation of TDP-43 sequesters mRNA and inhibits local translation of nuclear-encoded mitochondrial genes. Detailed description is at the text. (This figure is reproduced from the reference Altman et al. (2021) upon permission of Springer (Altman et al. 2021))

(Sanchez et al. 2021). The G3BP1 positive RNP condensates ihibit translation of nuclear-encoded mitochondrial proteins by sequestration of their mRNA around the motor neuron axon and neuromuscular junction (NMJ)s. Inhibition of translation of mitochondrial proteins induces mitochondrial toxicity and finally NMJ degeneration (Fig. 5.5) (Altman et al. 2021). Notably, the clearance of G3BP1 and TDR43 RNP complex is able to restore damaged NMJ and also motor neuron (MN) (Yang et al. 2020). This restoration could be a hopeful target for therapeutics of ALS.

TLS has been shown to be precipitated in neuronal cells upon its mutation and in knock-in mice (Guerrero et al. 2016). The mutation in TLS caused aggregation and neuronal death in vivo. Strikingly, these toxic events were recovered with administration of anti-sense oligonucleotides against expression of TLS. The administration of anti-sense oligonucleotides to TLS should also be effective strategy for therapeutics to ALS (Korobeynikov et al. 2022).

Phase separation has been reported to exert various biological events in living cells. Accidentally, it generates unexpected and unwanted outcomes of aggregation of RBPs which frequently bother regular activities of biological events (Ratti and Buratti 2016; Aksoy et al. 2020). Therefore, it has been developed regulatory apparatus of the phase separation and aggregation formation in biological evolution.

5.4 Regulation of Phase Separation and Resultant Aggregation

Phase separation plays pivotal roles in divergent biological events including formation of membrane-less organelles, translation, transcription, embryonic development, and cellular growth; (Hnisz et al. 2017; Shin et al. 2017; Kim et al. 2019; Ong and Torres 2020; Ramat and Simonelig 2022; Shao et al. 2022). It generates fruitful benefits on cellular functions, but also causes aggregation having harmful adverse influences. Therefore, the regulation of the phase separation should be essential to achieve safety outcomes (Fig. 5.6)) (Gibson et al. 2019; Hondele et al. 2019; Lu et al. 2022). There are macromolecules, proteins and RNA with potency to regulate phase separation to repress risk of precipitation of the proteins (Hofweber et al. 2018; Maharana et al. 2018; Qamar et al. 2018; Liu et al. 2020; Ma et al. 2022a).

We initiate discussion from a main partner of RBP susceptible to phase separation, ribonucleic acid, RNA. RNA binds and might regulate functions of RBPs. Actually, it has been reported that RNA regulates phase separation and resultant aggregation, suggesting the possibility of therapeutic outcomes (Maharana et al. 2018; Ries et al. 2019; Elguindy and Mendell 2021). Most of RBPs including TLS, TDP43, and hnRNPA1/B1 contain low complexity domain or IDR which is tightly associated with phase separation and aggregation. The aggregation is linked to neurodegenerative diseases like ALS (Molliex et al. 2015; Patel et al. 2015a; Mateju et al. 2017). In physiological conditions, RBPs are associated with each other through their IDRs to form droplets, and accidentally more tightly migrated into aggregates (Brangwynne Clifford et al. 2015; Shin et al. 2017). Somehow to block these tight association trough IDRs of RBPs leads to dissociate the aggregates and to inspire an insight to therapeutics against neurodegenerative diseases (Fig. 5.3).

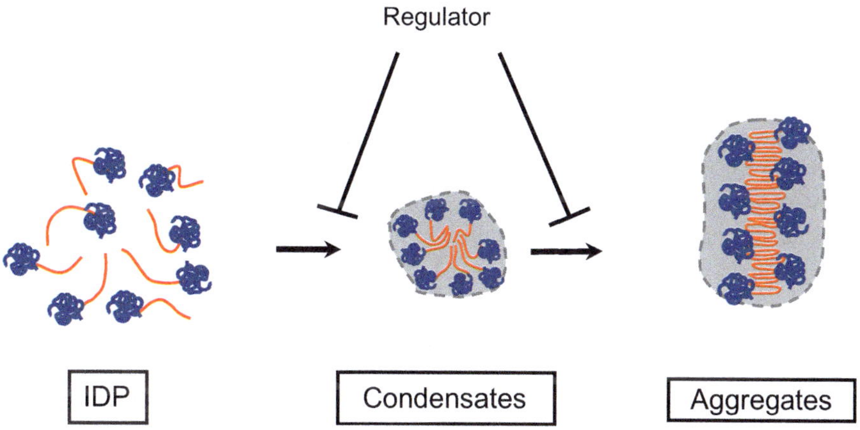

Fig. 5.6 Regulation of phase separation and aggregation. Regulatory factors repress steps of phase separation and formation of aggregation

TLS in cellular nucleus is usually soluble states, while TLS is mislocalized to cytoplasm and aggregated (Patel et al. 2015b; Hofweber et al. 2018; Mann and Donnelly 2021; Reber et al. 2021). Investigating the molecular mechanism of the relocation related aggregation in cytoplasm, RNA at higher concentrations in nucleus is found to block the TLS aggregation there (Maharana et al. 2018). Accidental leak of TLS into cytoplasm in which the RNA concentration is lower than in nucleus, promotes the droplets of TLS into aggregates, suggesting onset of ALS (Kwiatkowski Jr. et al. 2009; Vance et al. 2009). In this context, RNA works as a nonspecific molecular mass rather than specific binding molecule to TLS. This study also showed that lncRNA Neat1 stimulates formation of droplet of TLS even in the nucleus, mentioning that highly structured lncRNA like NEAT1 behaves differently to activate droplet formation.

Contrarily to this observation of RNA in nucleus, our experiments indicate that lncRNA pncRNAD1 blocks the aggregation of TLS in experimental conditions (Hamad et al. 2021). This aggregation is supposed to be mediated by phase separation, because it was sensitive to 1.6-hexanediol. Each lncRNA has its specific action on phase separation and aggregation.

Next, we take a look at adenosine triphosphate (ATP) for regulatory functions in phase separation. ATP is often referred to as the "molecular unit of currency" of intracellular energy transfer, and also a precursor to DNA and RNA. Recent data shows that ATP is an amphiphilic molecule and possesses two parts in the molecule. One portion of ATP has hydrophilic phosphorus, while other portion contains hydrophobic residue of nucleic acid salt (Hodgdon and Kaler 2007; Patel et al. 2017). This sort of molecules is identified as hydrotrope which is capable to dissolve water-insoluble chemicals in water (Sarkar and Mondal 2021; Tian and Qian 2021; Pandey et al. 2022). It has been shown that ATP interacts with fluorescent molecule like fluorescein and anilino naphthalene-sulfonic acid, and that ATP prevents heated egg white from aggregation formation. It has been also shown that ATP represses the formation of droplets of TLS (Rice and Rosen 2017). Biochemical experiments indicated that ATP at 1 mM promotes formation of TLS droplets, and at 8 mM blocks droplet formation of TLS (Patel et al. 2017). The physiological concentration of ATP in living cells is between 5 and 10 mM. In this range of concentration, ATP has been found to keep TLS and other RBP proteins soluble. This is one of reasons why cellular concentration of ATP is much higher than at a micromolar order that is sufficient for ATP-dependent enzymes in energy transfers functions in living cells.

Molecular chaperone assists the conformational folding or unfolding of high molecular weight proteins or protein complexes (Hartl et al. 2011; Saibil 2013; Balchin et al. 2016; Bobori et al. 2017; Yoo et al. 2022). One of major functions of chaperones is to prevent aggregation of misfolded proteins. Then, chaperone HSP 70 inhibits the liquid-to-solid phase transition of TLS. HSP70 predominantly interacts with IDR of TLS (Li et al. 2022). HSP70 under goes phase separation and is associated with TLS in stress granules (SGs). Then, HSP70 prevents aggregation of TLS. Knock down experiments of HSP70 indicated that depletion of HSP70 has no effect on the SG assembly but resulted in the liquid-to-solid transition in SGs. NMR analyses shown that HSP70 binds the IDR of N-terminus of TLS. These data indicate that molecular chaperone is one of the specific regulators of phase separations and resultant aggregations.

Regulation of phase separation is a first step of repression of the aggregation formation. RNA and small molecules like ATP do not always perfectly suppress the protein aggregation. After aggregation is initiated, it requires next barrier for the aggregation risk. Removing the aggregation is a next layer of risk management of the aggregation formation (Noda et al. 2020; Sun et al. 2020; Zhang et al. 2021; Ma et al. 2022a, b). Recently, Chaperonin Containing TCP1 Subunit 2CCT2 of a subunit of chaperonin has been reported to work as an autophagy specific for aggregated proteins, aggrephagy receptor for removing accidental protein aggregation (Ma et al. 2022a; Zhang and Klionsky 2022). Chaperonin is a call of molecular chaperones specific functions to prevent protein aggregations. CCT2 works as a chaperonin subunit to prevent proteins susceptible to aggregation from toxic aggregation formation in mouse neuronal cells. Then, actually excessive aggregations are caused by cellular stress or other stimuli, CCT2 switch its function from a part of chaperonin into aggrephagy receptor for recruitments of aggregated proteins into autophagosomes using LC3 as adaptor to hold them for transferring to autophagic degradation or clearance (Fig. 5.7) (Ma et al. 2022a) This is a next step of safety machinery for avoiding the risk coming from aggregation of proteins.

5.5 Good Aggregates, Functional Amyloid for Memory Formation

Amyloid aggregate is just a symbolic state of a bad aggregate, a pathogenic agent of the amyloid with its strong cytotoxicity on neuronal cells. The aggregations cause neuronal disorders including Alzheimer's disease (Chiti and Dobson 2006; Greenwald and Riek 2010; Riek and Eisenberg 2016). However, some sorts of amyloid aggregates have been shown to be functional and beneficial activities (Chapman et al. 2002; Maddelein et al. 2002; Gebbink et al. 2005; Maji et al. 2009). This section focuses on one of most notable functions of amyloids, memory formation. The molecular mechanism of memory formation still remains highly elusive. The Si laboratory elucidated the structure of neuronal amyloid proteins from Drosophila. These amyloid molecules consist of self-aggregated oo18 RNA-binding protein (Orb2), Drosophila homolog of RNA-binding protein cytoplasmic polyadenylation element-binding (CPEB) and have been shown to work on the long-term potentiation (Keleman et al. 2007; Khan et al. 2015; Hervás et al. 2016). It has been revealed that amyloid could form in time-and cell-type-specific manners to induce memory formation, although amyloid was thought to be induced by molecular damaging with cellular stresses and by misfolding with accidents. Mostly, amyloid has been recognized as a pathological agent of denatured precipitations for Alzheimer's disease and more. The proteins are aberrantly aggregated and generate troubles on neuronal systems. The Shi group has found a functional

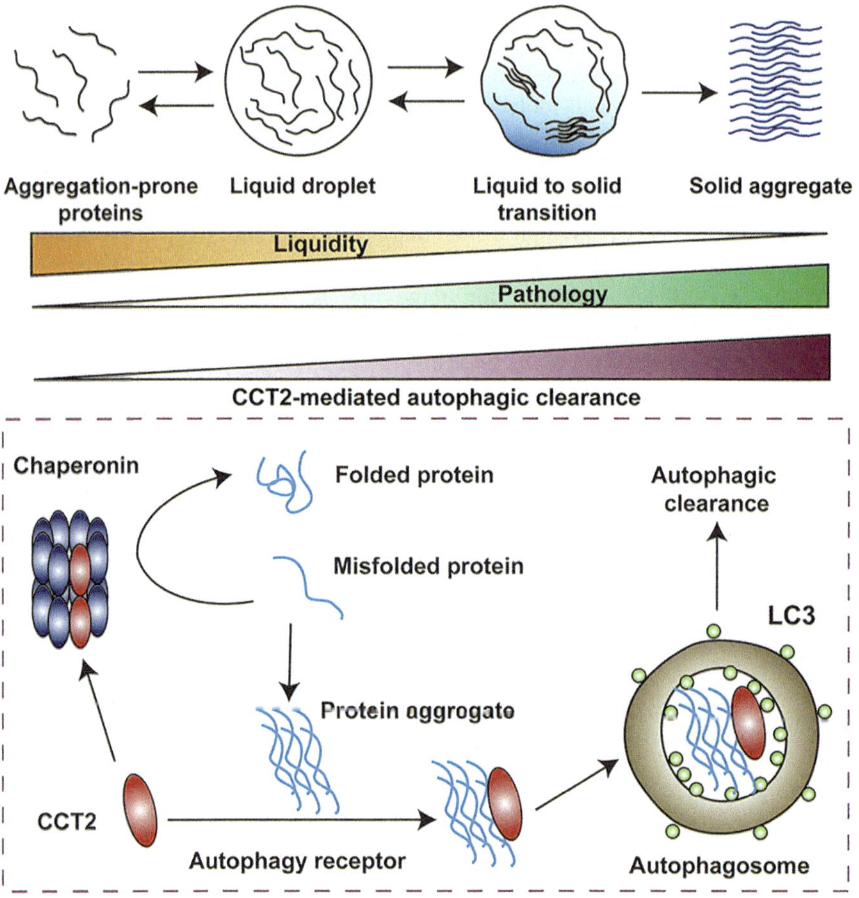

Fig. 5.7 Aggrephagy receptor CCT2 works for clearance of solid protein aggregates CCT2 regulates aggrephagy via associating with ATG8s and aggregation-prone proteins. The detailed description appears at the text. (This figure is reproduced from the reference Ma et al. (2022) upon permission of Elsevier Inc. (Ma et al. 2022a) License no.5482760384831)

amyloid in neuronal cells during the investigation of CPEB in Aplysia californica. Afterword, it has been demonstrated that aggregation of Orb2 in Drosophila is essential for its synaptic functions (Si et al. 2003, 2010).

The series of experiments indicate that aggregation of Orb2 exerts distinctive functions and localize specifically in the brain. Major format of Orb2 is supposed to be the monomers that repress synaptic translation while once memory formed these monomers are migrated into biochemically active form of self-aggregates that promote synaptic translation. The formation of the self-aggregation should be essential for memory persistence. These data indicate that Orb2 works as a good aggregate of amyloid in memory formation (Hervás et al. 2016).

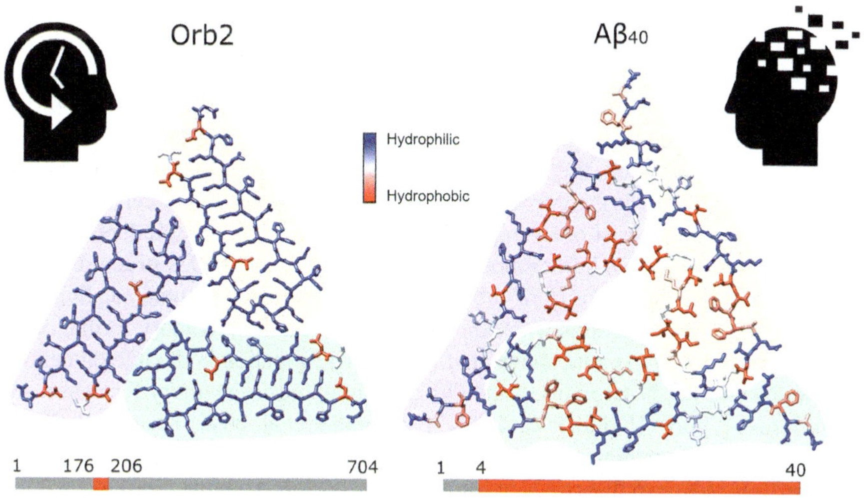

Fig. 5.8 Comparison between the functional amyloid fold of Orb2 from Drosophila brain (Left) versus the toxic fibril architecture of Aβ40 from human brain (Right) (Ulamec and Radford 2020). Orb2 is involved in long-term memory formation, while Aβ40 is involved in the neurodegenerative disorder Alzheimer's disease. These two structures possess three-fold symmetry and with three protofilaments marked with yellow, green, and pink, that form the mature fibril structures. The core of Orb2 is mainly hydrophilic (blue), while the Aβ40 core is hydrophobic (red). Hydrophobic interaction typically promotes aggregation of proteins. At the lower panel: Aβ40 forms almost structured conformation in the fibril core (red bar). The majority of the protein sequence of Orb2 is highly disordered. This should allow functionally important interactions with other molecules to attain memory formation. (This figure is reproduced from the reference Ulamec and Radford (2020) upon permission (no. 5500090793075) of Cell press)

Profound understanding of memory formation in Orb2 amyloid genesis requires more structural analysis of Orb2. The recombinant Orb2 expressed in vitro is not biochemically active. Therefore, they decided to purify endogenous Orb2 from three million of the Drosophila heads and achieved 97% purity of the purified Orb2 sample. The purified Orb2 turns out to be active and successfully forms functional self-aggregated filaments. Cryo-electron microscopy using the purified samples has demonstrated the atomic structure of Orb2 at the resolution 2.6 Å (Hervas et al. 2020) (Fig. 5.8). The Orb2 has been found to form the stacked three-fold helical symmetry of the filament core at 75 nanometers long.

These data confirm that the cross-β structures of the Orb2 amyloid form aggregates, and the hairpin-like fold of the core of protofilaments consisting of 31 amino acids is packed through hydrophilic interfaces (Fig. 5.8). The hydrophilic core of the Orb2 amyloid might allow it to be amenable to alteration upon the cellular environment in case of lowering pH. This flexibility of the Orb2 core is remarkable difference from the hydrophobic core of β-amyloid1–40 (Aβ40) filaments seeded from the brain of Alzheimer's disease (Lu et al. 2013; Yang et al. 2022) (Fig. 5.8). This features of Orb2 core make it possible to transform into flexible structures to promote formation of memory in the brain, suggesting that this regulatable structures of Orb2 practically should function in memory formation in the brain. Most of the amyloid has been recognized as pathological molecules that bother biological functions of living cell, while relatively small numbers of good or beneficial aggregations have been identified. One of intriguing question is what makes difference between functional aggregate and non-functional one. In the context of the Orb2 memory formation, the amenability of the hydrophilic core of glutamine and histidine might make it flexible and dynamic activities in neuronal cells, while the Aβ40 core mainly consist of hydrophobic amino acid residues that induce aggregation of protein (Fig. 5.8). Furthermore, the whole protein of Aβ40 is structured in the fibril core, while the majority of the Orb2 protein sequence is dynamically disordered permitting interactions with other molecules regarding memory formation.

Another piece of the data shows that Orb2 fibril formation is flexible on cellular environments. The N-terminus of the Orb2A is required for memory formation and forms cross-β fibrils (Krüttner et al. 2012; Majumdar et al. 2012; Khan et al. 2015). However, this N-terminus of Orb2A is not part of the core in the fibrils. Previous experiments indicated that the N-terminus of Orb2A binds anionic lipid membranes as an amphipathic helix besides formation of cross-beta fibrils. Recent solid-state NMR and electron paramagnetic resonance (EPR) analysis demonstrated that the Orb2A N-terminus could similarly interact with calcium activated calmodulin and that this interaction prevents fibril formation. This interaction possibly tells the regulatory role of Orb2A in memory formation. (Soria et al. 2022), and also presents that Orb2 forms flexible conformation to generate memory formation.

5.6 Future Prospects of Phase Separation and Aggregation

Structural alterations of Orb2 in neuronal cells resulted in switching of its functions of RBP into one of the memory formations. The environment in brain might allow a protein to transform a different conformation to get another function. This alteration makes it possible for a single protein to switch its function into another one. It should be useful for memory formation in neuronal cells to generate multiple functions from a single protein, because the process in memory formation should require a lot of labor.

Phase separation is efficient system to utilize a protein for three distinctive forms, liquid, gel, and aggregate. One solution of protein could be transformed into three distinctive modes of the states. Each one has each specific conformation and might have each specific function. In this context, the phase separation permits a protein with a function to transform into another function (Fig. 5.9).

In traditional view of protein, a protein should have a specific structure and its distinctive function (Liu et al. 2013; Evans and Mangelsdorf 2014; Girbig et al. 2022). The discussion postulates a bit modified model of protein functions. In particular, memory formation functions require a protein to transform one state to another conformation in order to switch its function to another one.

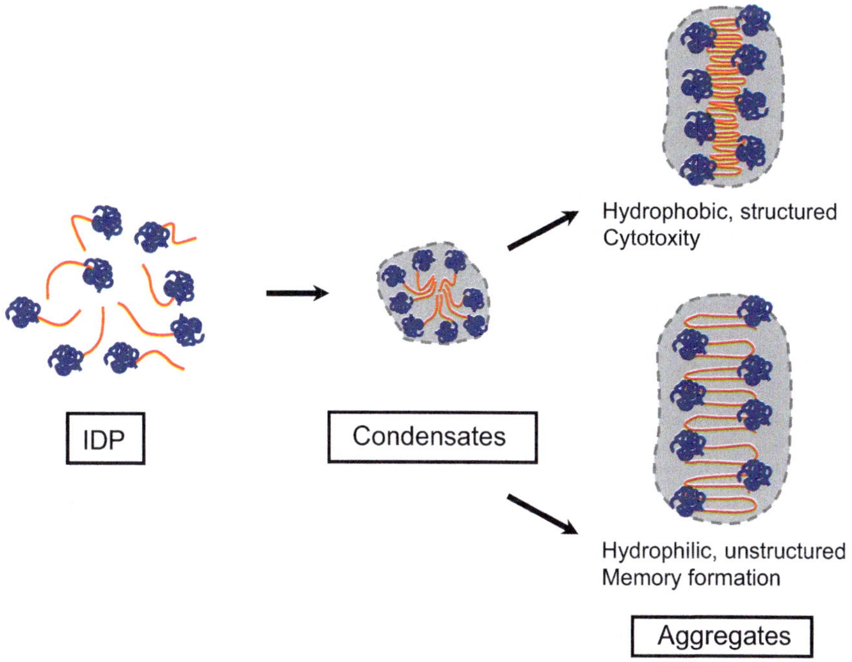

Fig. 5.9 Phase separation creates a novel function of protein. Phase separation alters conformation of a protein and generates its novel functions. At proteins' higher concentration, intrinsically disordered protein (IDP) undergoes phase separation into condensates or membrane-less organelles that have a distinctive function from the IDP in solution. The condensates occasionally transform to aggregates. In the structured and hydrophobic aggregates, they have cytotoxicity and induce neurodegenerative disease. On the other hand, the flexible and hydrophilic aggregates could have beneficial functions like memory formation. Then, the phase separation could expand functions of proteins

The conformational flexibility of some proteins opens the door to another possibility that protein might have more than one function depending upon conformational settings. The central nervous systems have one of the most complex functions in biological systems, but need to use just limited numbers of the genes to perform its higher order functions. The memory in the brain consists of huge numbers of fragments of information. The brain needs to process these pieces of memory. This job should require tremendous numbers of functional units and needs many proteins with them. There are, however, limited numbers of proteins in neuronal cells. Therefore, the brain needs to save numbers of protein and have multiple functions of proteins.

Then, the central nervous system has adopted itself to employ a strategy to alter structure of a protein to gain another function on its altered structure. Instead of generating new proteins, just generating a different shape of a protein could create a novel function. Actually, Orb2 in the brain of Drosophila has been reported to undergo phase separation into droplets that function as RBP to regulate translation or processing, and form aggregates, functional amyloid contribute to memory formation (Hervas et al. 2020; Ashami et al. 2021; Soria et al. 2022). The phase separation is excellent system to endow a protein with two or more functions. It is possible for the phase separation to make a protein solution into a droplet and aggregate that could have distinctive functions. In this phase separation, the aggregates should exert functions. Then, the phase separation could generate one more function on it (Fig. 5.9).

The phase separation allows a protein to expand its function upon its conformational alterations. This should be a strategy during the evolution to utilize limited numbers of proteins to conduct almost infinite mission of biological events, especially higher brain functions.

Acknowledgments The author would thank for fruitful and constructive discussion by all members of the Kurokawa laboratory. The author also thank Christopher K. Glass and Michael Geoffrey Rosenfeld for profound insight into development of projects. This study was supported by Grant-in-Aid for Scientific Research (C: 18 K06939; C: 22 K06939) and the joint usage/research programs of the Institute of Advanced Energy, Kyoto University (ZE2022A-10).

References

Aksoy YA, Deng W, Stoddart J, Chung R, Guillemin G, Cole NJ, Neely GG, Hesselson D (2020) "STRESSED OUT": the role of FUS and TDP-43 in amyotrophic lateral sclerosis. Int J Biochem Cell Biol 126:105821

Altman T, Ionescu A, Ibraheem A, Priesmann D, Gradus-Pery T, Farberov L, Alexandra G, Shelestovich N, Dafinca R, Shomron N et al (2021) Axonal TDP-43 condensates drive neuromuscular junction disruption through inhibition of local synthesis of nuclear encoded mitochondrial proteins. Nat Commun 12:6914

Arai M, Kuwajima K (2000) Role of the molten globule state in protein folding. Adv Protein Chem 53:209–282

Arany Z, Sellers WR, Livingston DM, Eckner R (1994) E1A-associated p300 and CREB-associated CBP belong to a conserved family of coactivators. Cell 77:799–800

Ashami K, Falk AS, Hurd C, Garg S, Cervantes SA, Rawat A, Siemer AB (2021) Droplet and fibril formation of the functional amyloid Orb2. J Biol Chem 297:100804

Balchin D, Hayer-Hartl M, Hartl FU (2016) In vivo aspects of protein folding and quality control. Science 353:aac4354

Blokhuis AM, Groen EJ, Koppers M, van den Berg LH, Pasterkamp RJ (2013) Protein aggregation in amyotrophic lateral sclerosis. Acta Neuropathol 125:777–794

Bobori C, Theocharopoulou G, Vlamos P (2017) Molecular chaperones in neurodegenerative diseases: a short review. Adv Exp Med Biol 987:219–231

Boija A, Klein IA, Sabari BR, Dall'Agnese A, Coffey EL, Zamudio AV, Li CH, Shrinivas K, Manteiga JC, Hannett NM et al (2018) Transcription factors activate genes through the phase-separation capacity of their activation domains. Cell 175:1842–1855.e1816

Brangwynne Clifford P, Tompa P, Pappu RV (2015) Polymer physics of intracellular phase transitions. Nat Phys 11:899–904

Broustal O, Camuzat A, Guillot-Noel L, Guy N, Millecamps S, Deffond D, Lacomblez L, Golfier V, Hannequin D, Salachas F et al (2010) FUS mutations in frontotemporal lobar degeneration with amyotrophic lateral sclerosis. J Alzheimers Dis 22:765–769

Bryngelson JD, Onuchic JN, Socci ND, Wolynes PG (1995) Funnels, pathways, and the energy landscape of protein folding: a synthesis. Proteins 21:167–195

Castello A, Fischer B, Frese CK, Horos R, Alleaume AM, Foehr S, Curk T, Krijgsveld J, Hentze MW (2016) Comprehensive identification of RNA-binding domains in human cells. Mol Cell 63:696–710
Chapman MR, Robinson LS, Pinkner JS, Roth R, Heuser J, Hammar M, Normark S, Hultgren SJ (2002) Role of Escherichia coli curli operons in directing amyloid fiber formation. Science 295: 851–855
Chiti F, Dobson CM (2006) Protein misfolding, functional amyloid, and human disease. Annu Rev Biochem 75:333–366
Chrivia JC, Kwok RP, Lamb N, Hagiwara M, Montminy MR, Goodman RH (1993) Phosphorylated CREB binds specifically to the nuclear protein CBP. Nature 365:855–859
Da Cruz S, Cleveland DW (2011) Understanding the role of TDP-43 and FUS/TLS in ALS and beyond. Curr Opin Neurobiol 21:904–919
Dunker AK, Bondos SE, Huang F, Oldfield CJ (2015) Intrinsically disordered proteins and multicellular organisms. Semin Cell Dev Biol 37:44–55
Elguindy MM, Mendell JT (2021) NORAD-induced Pumilio phase separation is required for genome stability. Nature 595:303–308
Evans RM, Mangelsdorf DJ (2014) Nuclear receptors, RXR, and the big bang. Cell 157:255–266
Gebbink MF, Claessen D, Bouma B, Dijkhuizen L, Wosten HA (2005) Amyloids—a functional coat for microorganisms. Nat Rev Microbiol 3:333–341
Gibson BA, Doolittle LK, Schneider MWG, Jensen LE, Gamarra N, Henry L, Gerlich DW, Redding S, Rosen MK (2019) Organization of chromatin by intrinsic and regulated phase separation. Cell 179:470–484.e421
Girbig M, Misiaszek AD, Müller CW (2022) Structural insights into nuclear transcription by eukaryotic DNA-dependent RNA polymerases. Nat Rev Mol Cell Biol 23:603–622
Glass CK, Rosenfeld MG (2000) The coregulator exchange in transcriptional functions of nuclear receptors. Genes Dev 14:121–141
Greenwald J, Riek R (2010) Biology of amyloid: structure, function, and regulation. Structure 18: 1244–1260
Guerrero EN, Wang H, Mitra J, Hegde PM, Stowell SE, Liachko NF, Kraemer BC, Garruto RM, Rao KS, Hegde ML (2016) TDP-43/FUS in motor neuron disease: complexity and challenges. Prog Neurobiol 145-146:78–97
Guo JL, Lee VMY (2014) Cell-to-cell transmission of pathogenic proteins in neurodegenerative diseases. Nat Med 20:130–138
Hamad N, Yoneda R, So M, Kurokawa R, Nagata T, Katahira M (2021) Non-coding RNA suppresses FUS aggregation caused by mechanistic shear stress on pipetting in a sequence-dependent manner. Sci Rep 11:9523
Hartl FU, Bracher A, Hayer-Hartl M (2011) Molecular chaperones in protein folding and proteostasis. Nature 475:324–332
Hasegawa M, Nonaka T, Tsuji H, Tamaoka A, Yamashita M, Kametani F, Yoshida M, Arai T, Akiyama H (2011) Molecular dissection of TDP-43 proteinopathies. J Mol Neurosci 45:480–485
Hervás R, Li L, Majumdar A, Fernández-Ramírez Mdel C, Unruh JR, Slaughter BD, Galera-Prat A, Santana E, Suzuki M, Nagai Y et al (2016) Molecular basis of Orb2 amyloidogenesis and blockade of memory consolidation. PLoS Biol 14:e1002361
Hervas R, Rau MJ, Park Y, Zhang W, Murzin AG, Fitzpatrick JAJ, Scheres SHW, Si K (2020) Cryo-EM structure of a neuronal functional amyloid implicated in memory persistence in drosophila. Science 367:1230–1234
Hnisz D, Shrinivas K, Young RA, Chakraborty AK, Sharp PA (2017) A phase separation model for transcriptional control. Cell 169:13–23
Ho WL, Huang JR (2022) The return of the rings: evolutionary convergence of aromatic residues in the intrinsically disordered regions of RNA-binding proteins for liquid-liquid phase separation. Protein Sci 31:e4317
Hodgdon TK, Kaler EW (2007) Hydrotropic solutions. Curr Opin Colloid Interface Sci 12:121–128

Hofweber M, Hutten S, Bourgeois B, Spreitzer E, Niedner-Boblenz A, Schifferer M, Ruepp MD, Simons M, Niessing D, Madl T et al (2018) Phase separation of FUS is suppressed by its nuclear import receptor and arginine methylation. Cell 173:706–719.e713

Hondele M, Sachdev R, Heinrich S, Wang J, Vallotton P, Fontoura BMA, Weis K (2019) DEAD-box ATPases are global regulators of phase-separated organelles. Nature 573:144–148

Hu KN, Tycko R (2010) What can solid state NMR contribute to our understanding of protein folding? Biophys Chem 151:10–21

Irwin DJ, Lee VMY, Trojanowski JQ (2013) Parkinson's disease dementia: convergence of α-synuclein, tau and amyloid-β pathologies. Nat Rev Neurosci 14:626–636

Jackson MP, Hewitt EW (2017) Why are functional amyloids non-toxic in humans? Biomol Ther 7: 71

Kamei Y, Xu L, Heinzel T, Torchia J, Kurokawa R, Gloss B, Lin SC, Heyman RA, Rose DW, Glass CK et al (1996) A CBP integrator complex mediates transcriptional activation and AP-1 inhibition by nuclear receptors. Cell 85:403–414

Kato M, Han TW, Xie S, Shi K, Du X, Wu LC, Mirzaei H, Goldsmith EJ, Longgood J, Pei J et al (2012) Cell-free formation of RNA granules: low complexity sequence domains form dynamic fibers within hydrogels. Cell 149:753–767

Keleman K, Krüttner S, Alenius M, Dickson BJ (2007) Function of the drosophila CPEB protein Orb2 in long-term courtship memory. Nat Neurosci 10:1587–1593

Keul ND, Oruganty K, Schaper Bergman ET, Beattie NR, McDonald WE, Kadirvelraj R, Gross ML, Phillips RS, Harvey SC, Wood ZA (2018) The entropic force generated by intrinsically disordered segments tunes protein function. Nature 563:584–588

Khan MR, Li L, Pérez-Sánchez C, Saraf A, Florens L, Slaughter BD, Unruh JR, Si K (2015) Amyloidogenic oligomerization transforms drosophila Orb2 from a translation repressor to an activator. Cell 163:1468–1483

Kim TH, Tsang B, Vernon RM, Sonenberg N, Kay LE, Forman-Kay JD (2019) Phospho-dependent phase separation of FMRP and CAPRIN1 recapitulates regulation of translation and deadenylation. Science 365:825–829

Korobeynikov VA, Lyashchenko AK, Blanco-Redondo B, Jafar-Nejad P, Shneider NA (2022) Antisense oligonucleotide silencing of FUS expression as a therapeutic approach in amyotrophic lateral sclerosis. Nat Med 28:104–116

Kozlov EN, Tokmatcheva EV, Khrustaleva AM, Grebenshchikov ES, Deev RV, Gilmutdinov RA, Lebedeva LA, Zhukova M, Savvateeva-Popova EV, Schedl P et al (2023) Long-term memory formation in drosophila depends on the 3'UTR of CPEB gene orb2. Cell 12:318

Krüttner S, Stepien B, Noordermeer JN, Mommaas MA, Mechtler K, Dickson BJ, Keleman K (2012) Drosophila CPEB Orb2A mediates memory independent of its RNA-binding domain. Neuron 76:383–395

Kulkarni P (2020) Intrinsically disordered proteins: insights from Poincaré, Waddington, and Lamarck. Biomol Ther 10:1490

Kulkarni P, Bhattacharya S, Achuthan S, Behal A, Jolly MK, Kotnala S, Mohanty A, Rangarajan G, Salgia R, Uversky V (2022) Intrinsically disordered proteins: critical components of the wetware. Chem Rev 122:6614–6633

Kurokawa R, Kalafus D, Ogliastro MH, Kioussi C, Xu L, Torchia J, Rosenfeld MG, Glass CK (1998) Differential use of CREB binding protein-coactivator complexes. Science 279:700–703

Kurokawa R, Rosenfeld MG, Glass CK (2009) Transcriptional regulation through noncoding RNAs and epigenetic modifications. RNA Biol 6:233–236

Kwiatkowski TJ Jr, Bosco DA, Leclerc AL, Tamrazian E, Vanderburg CR, Russ C, Davis A, Gilchrist J, Kasarskis EJ, Munsat T et al (2009) Mutations in the FUS/TLS gene on chromosome 16 cause familial amyotrophic lateral sclerosis. Science 323:1205–1208

Kwok RP, Lundblad JR, Chrivia JC, Richards JP, Bächinger HP, Brennan RG, Roberts SG, Green MR, Goodman RH (1994) Nuclear protein CBP is a coactivator for the transcription factor CREB. Nature 370:223–226

Kwon SC, Yi H, Eichelbaum K, Fohr S, Fischer B, You KT, Castello A, Krijgsveld J, Hentze MW, Kim VN (2013) The RNA-binding protein repertoire of embryonic stem cells. Nat Struct Mol Biol 20:1122–1130

Lagier-Tourenne C, Cleveland DW (2009) Rethinking ALS: the FUS about TDP-43. Cell 136: 1001–1004

Lerga A, Hallier M, Delva L, Orvain C, Gallais I, Marie J, Moreau-Gachelin F (2001) Identification of an RNA binding specificity for the potential splicing factor TLS. J Biol Chem 276:6807–6816

Li P, Banjade S, Cheng H-C, Kim S, Chen B, Guo L, Llaguno M, Hollingsworth JV, King DS, Banani SF et al (2012) Phase transitions in the assembly of multivalent signalling proteins. Nature 483:336–340

Li Y, Gu J, Wang C, Hu J, Zhang S, Liu C, Zhang S, Fang Y, Li D (2022) Hsp70 exhibits a liquid-liquid phase separation ability and chaperones condensed FUS against amyloid aggregation. iScience 25:104356

Lichtinger SM, Garaizar A, Collepardo-Guevara R, Reinhardt A (2021) Targeted modulation of protein liquid-liquid phase separation by evolution of amino-acid sequence. PLoS Comput Biol 17:e1009328

Liepelt A, Naarmann-de Vries IS, Simons N, Eichelbaum K, Fohr S, Archer SK, Castello A, Usadel B, Krijgsveld J, Preiss T et al (2016) Identification of RNA-binding proteins in macrophages by Interactome capture. Mol Cell Proteomics 15:2699–2714

Liu X, Bushnell DA, Kornberg RD (2013) RNA polymerase II transcription: structure and mechanism. Biochim Biophys Acta 1829:2–8

Liu Z, Zhang S, Gu J, Tong Y, Li Y, Gui X, Long H, Wang C, Zhao C, Lu J et al (2020) Hsp27 chaperones FUS phase separation under the modulation of stress-induced phosphorylation. Nat Struct Mol Biol 27:363–372

Lu JX, Qiang W, Yau WM, Schwieters CD, Meredith SC, Tycko R (2013) Molecular structure of β-amyloid fibrils in Alzheimer's disease brain tissue. Cell 154:1257–1268

Lu H, Yu D, Hansen AS, Ganguly S, Liu R, Heckert A, Darzacq X, Zhou Q (2018) Phase-separation mechanism for C-terminal hyperphosphorylation of RNA polymerase II. Nature 558:318–323

Lu S, Hu J, Arogundade OA, Goginashvili A, Vazquez-Sanchez S, Diedrich JK, Gu J, Blum J, Oung S, Ye Q et al (2022) Heat-shock chaperone HSPB1 regulates cytoplasmic TDP-43 phase separation and liquid-to-gel transition. Nat Cell Biol 24:1378–1393

Ma X, Lu C, Chen Y, Li S, Ma N, Tao X, Li Y, Wang J, Zhou M, Yan YB et al (2022a) CCT2 is an aggrephagy receptor for clearance of solid protein aggregates. Cell 185:1325–1345.e1322

Ma X, Zhang M, Ge L (2022b) A switch of chaperonin function regulates the clearance of solid protein aggregates. Autophagy 18:2746–2748

Mackenzie IR, Rademakers R, Neumann M (2010) TDP-43 and FUS in amyotrophic lateral sclerosis and frontotemporal dementia. Lancet Neurol 9:995–1007

Maddelein ML, Dos Reis S, Duvezin-Caubet S, Coulary-Salin B, Saupe SJ (2002) Amyloid aggregates of the HET-s prion protein are infectious. Proc Natl Acad Sci U S A 99:7402–7407

Maharana S, Wang J, Papadopoulos DK, Richter D, Pozniakovsky A, Poser I, Bickle M, Rizk S, Guillén-Boixet J, Franzmann TM et al (2018) RNA buffers the phase separation behavior of prion-like RNA binding proteins. Science 360:918–921

Maji SK, Perrin MH, Sawaya MR, Jessberger S, Vadodaria K, Rissman RA, Singru PS, Nilsson KP, Simon R, Schubert D et al (2009) Functional amyloids as natural storage of peptide hormones in pituitary secretory granules. Science 325:328–332

Majumdar A, Cesario WC, White-Grindley E, Jiang H, Ren F, Khan MR, Li L, Choi EM, Kannan K, Guo F et al (2012) Critical role of amyloid-like oligomers of drosophila Orb2 in the persistence of memory. Cell 148:515–529

Mann JR, Donnelly CJ (2021) RNA modulates physiological and neuropathological protein phase transitions. Neuron 109:2663–2681

Martinelli I, Zucchi E, Pensato V, Gellera C, Traynor BJ, Gianferrari G, Chio A, Mandrioli J (2022) G507D mutation in FUS gene causes familial amyotrophic lateral sclerosis with a specific genotype-phenotype correlation. Neurobiol Aging 118:124–128

Mateju D, Franzmann TM, Patel A, Kopach A, Boczek EE, Maharana S, Lee HO, Carra S, Hyman AA, Alberti S (2017) An aberrant phase transition of stress granules triggered by misfolded protein and prevented by chaperone function. EMBO J 36:1669–1687

Molliex A, Temirov J, Lee J, Coughlin M, Kanagaraj AP, Kim HJ, Mittag T, Taylor JP (2015) Phase separation by low complexity domains promotes stress granule assembly and drives pathological fibrillization. Cell 163:123–133

Murray DT, Kato M, Lin Y, Thurber KR, Hung I, McKnight SL, Tycko R (2017) Structure of FUS protein fibrils and its relevance to self-assembly and phase separation of low-complexity domains. Cell 171:615–627.e616

Nair SJ, Yang L, Meluzzi D, Oh S, Yang F, Friedman MJ, Wang S, Suter T, Alshareedah I, Gamliel A et al (2019) Phase separation of ligand-activated enhancers licenses cooperative chromosomal enhancer assembly. Nat Struct Mol Biol 26:193–203

Noda NN, Wang Z, Zhang H (2020) Liquid-liquid phase separation in autophagy. J Cell Biol 219: e202004062

Ong JY, Torres JZ (2020) Phase separation in cell division. Mol Cell 80:9–20

Osterberg Valerie R, Spinelli Kateri J, Weston Leah J, Luk Kelvin C, Woltjer Randall L, Unni VK (2015) Progressive aggregation of alpha-Synuclein and selective degeneration of Lewy inclusion-bearing neurons in a mouse model of parkinsonism. Cell Rep 10:1252–1260

Pajkos M, Dosztányi Z (2021) Functions of intrinsically disordered proteins through evolutionary lenses. Prog Mol Biol Transl Sci 183:45–74

Pandey MP, Sasidharan S, Raghunathan VA, Khandelia H (2022) Molecular mechanism of hydrotropic properties of GTP and ATP. J Phys Chem B 126:8486–8494

Patel A, Lee Hyun O, Jawerth L, Maharana S, Jahnel M, Hein Marco Y, Stoynov S, Mahamid J, Saha S, Franzmann Titus M et al (2015a) A liquid-to-solid phase transition of the ALS protein FUS accelerated by disease mutation. Cell 162:1066–1077

Patel A, Lee HO, Jawerth L, Maharana S, Jahnel M, Hein MY, Stoynov S, Mahamid J, Saha S, Franzmann TM et al (2015b) A liquid-to-solid phase transition of the ALS protein FUS accelerated by disease mutation. Cell 162:1066–1077

Patel A, Malinovska L, Saha S, Wang J, Alberti S, Krishnan Y, Hyman AA (2017) ATP as a biological hydrotrope. Science 356:753–756

Pham CLL, Kwan AH, Sunde M (2014) Functional amyloid: widespread in nature, diverse in purpose. Essays Biochem 56:207–219

Popper B, Scheidt T, Schieweck R (2021) RNA-binding protein dysfunction in neurodegeneration. Essays Biochem 65:975–986

Porta S, Xu Y, Lehr T, Zhang B, Meymand E, Olufemi M, Stieber A, Lee EB, Trojanowski JQ, Lee VM (2021) Distinct brain-derived TDP-43 strains from FTLD-TDP subtypes induce diverse morphological TDP-43 aggregates and spreading patterns in vitro and in vivo. Neuropathol Appl Neurobiol 47:1033–1049

Portz B, Lee BL, Shorter J (2021) FUS and TDP-43 phases in health and disease. Trends Biochem Sci 46:550–563

Prasad A, Bharathi V, Sivalingam V, Girdhar A, Patel BK (2019) Molecular mechanisms of TDP-43 Misfolding and pathology in amyotrophic lateral sclerosis. Front Mol Neurosci 12:25

Qamar S, Wang G, Randle SJ, Ruggeri FS, Varela JA, Lin JQ, Phillips EC, Miyashita A, Williams D, Ströhl F et al (2018) FUS phase separation is modulated by a molecular chaperone and methylation of arginine cation-π interactions. Cell 173:720–734.e715

Ramat A, Simonelig M (2022) Activating translation with phase separation. Science 377:712–713

Ratti A, Buratti E (2016) Physiological functions and pathobiology of TDP-43 and FUS/TLS proteins. J Neurochem 138(Suppl 1):95–111

Reber S, Jutzi D, Lindsay H, Devoy A, Mechtersheimer J, Levone BR, Domanski M, Bentmann E, Dormann D, Mühlemann O et al (2021) The phase separation-dependent FUS interactome

reveals nuclear and cytoplasmic function of liquid-liquid phase separation. Nucleic Acids Res 49:7713–7731
Rice AM, Rosen MK (2017) ATP controls the crowd. Science 356:701–702
Richter JD (2007) CPEB: a life in translation. Trends Biochem Sci 32:279–285
Riek R, Eisenberg DS (2016) The activities of amyloids from a structural perspective. Nature 539: 227–235
Ries RJ, Zaccara S, Klein P, Olarerin-George A, Namkoong S, Pickering BF, Patil DP, Kwak H, Lee JH, Jaffrey SR (2019) M(6)a enhances the phase separation potential of mRNA. Nature 571: 424–428
Rohn TT (2008) Caspase-cleaved TAR DNA-binding protein-43 is a major pathological finding in Alzheimer's disease. Brain Res 1228:189–198
Saibil H (2013) Chaperone machines for protein folding, unfolding and disaggregation. Nat Rev Mol Cell Biol 14:630–642
Sanchez II, Nguyen TB, England WE, Lim RG, Vu AQ, Miramontes R, Byrne LM, Markmiller S, Lau AL, Orellana I et al (2021) Huntington's disease mice and human brain tissue exhibit increased G3BP1 granules and TDP43 mislocalization. J Clin Invest 131:e140723
Sarkar S, Mondal J (2021) Mechanistic insights on ATP's role as a hydrotrope. J Phys Chem B 125: 7717–7731
Sela M, White FH Jr, Anfinsen CB (1957) Reductive cleavage of disulfide bridges in ribonuclease. Science 125:691–692
Shanmugam N, Baker MODG, Ball SR, Steain M, Pham CLL, Sunde M (2019) Microbial functional amyloids serve diverse purposes for structure, adhesion and defence. Biophys Rev 11:287–302
Shao W, Bi X, Pan Y, Gao B, Wu J, Yin Y, Liu Z, Peng M, Zhang W, Jiang X et al (2022) Phase separation of RNA-binding protein promotes polymerase binding and transcription. Nat Chem Biol 18:70–80
Shen Y, Ruggeri FS, Vigolo D, Kamada A, Qamar S, Levin A, Iserman C, Alberti S, George-Hyslop PS, Knowles TPJ (2020) Biomolecular condensates undergo a generic shear-mediated liquid-to-solid transition. Nat Nanotechnol 15:841–847
Sheng J, Olrichs NK, Gadella BM, Kaloyanova DV, Helms JB (2020) Regulation of functional protein aggregation by multiple factors: implications for the Amyloidogenic behavior of the CAP superfamily proteins. Int J Mol Sci 21:6530
Shin Y, Berry J, Pannucci N, Haataja MP, Toettcher JE, Brangwynne CP (2017) Spatiotemporal control of intracellular phase transitions using light-activated optoDroplets. Cell 168:159–171. e114
Si K, Lindquist S, Kandel ER (2003) A neuronal isoform of the aplysia CPEB has prion-like properties. Cell 115:879–891
Si K, Choi YB, White-Grindley E, Majumdar A, Kandel ER (2010) Aplysia CPEB can form prion-like multimers in sensory neurons that contribute to long-term facilitation. Cell 140:421–435
Soria MA, Cervantes SA, Siemer AB (2022) Calmodulin binds the N-terminus of the functional amyloid Orb2A inhibiting fibril formation. PLoS One 17:e0259872
Sun D, Wu R, Li P, Yu L (2020) Phase separation in regulation of aggrephagy. J Mol Biol 432:160–169
Tauber D, Tauber G, Khong A, Van Treeck B, Pelletier J, Parker R (2020) Modulation of RNA condensation by the DEAD-box protein eIF4A. Cell 180:411–426.e416
Taylor JP, Brown RH Jr, Cleveland DW (2016) Decoding ALS: from genes to mechanism. Nature 539:197–206
Tian Z, Qian F (2021) Adenosine triphosphate-induced rapid liquid-liquid phase separation of a model IgG1 mAb. Mol Pharm 18:267–274
Tsang B, Pritišanac I, Scherer SW, Moses AM, Forman-Kay JD (2020) Phase separation as a missing mechanism for interpretation of disease mutations. Cell 183:1742–1756
Ulamec SM, Radford SE (2020) Spot the difference: function versus toxicity in amyloid fibrils. Trends Biochem Sci 45:635–636

Uversky VN (2017) Intrinsically disordered proteins in overcrowded milieu: membrane-less organelles, phase separation, and intrinsic disorder. Curr Opin Struct Biol 44:18–30

Van Langenhove T, van der Zee J, Sleegers K, Engelborghs S, Vandenberghe R, Gijselinck I, Van den Broeck M, Mattheijssens M, Peeters K, De Deyn PP et al (2010) Genetic contribution of FUS to frontotemporal lobar degeneration. Neurology 74:366–371

Vance C, Rogelj B, Hortobagyi T, De Vos KJ, Nishimura AL, Sreedharan J, Hu X, Smith B, Ruddy D, Wright P et al (2009) Mutations in FUS, an RNA processing protein, cause familial amyotrophic lateral sclerosis type 6. Science 323:1208–1211

Wang X, Arai S, Song X, Reichart D, Du K, Pascual G, Tempst P, Rosenfeld MG, Glass CK, Kurokawa R (2008) Induced ncRNAs allosterically modify RNA-binding proteins in cis to inhibit transcription. Nature 454:126–130

Ward JJ, Sodhi JS, McGuffin LJ, Buxton BF, Jones DT (2004) Prediction and functional analysis of native disorder in proteins from the three kingdoms of life. J Mol Biol 337:635–645

Wegmann S, Eftekharzadeh B, Tepper K, Zoltowska KM, Bennett RE, Dujardin S, Laskowski PR, MacKenzie D, Kamath T, Commins C et al (2018) Tau protein liquid-liquid phase separation can initiate tau aggregation. EMBO J 37:e98049

Wright PE, Dyson HJ (1999) Intrinsically unstructured proteins: re-assessing the protein structure-function paradigm. J Mol Biol 293:321–331

Yamashita M, Nonaka T, Hirai S, Miwa A, Okado H, Arai T, Hosokawa M, Akiyama H, Hasegawa M (2014) Distinct pathways leading to TDP-43-induced cellular dysfunctions. Hum Mol Genet 23:4345–4356

Yang Z, Lin F, Robertson CS, Wang KK (2014) Dual vulnerability of TDP-43 to calpain and caspase-3 proteolysis after neurotoxic conditions and traumatic brain injury. J Cereb Blood Flow Metab 34:1444–1452

Yang P, Mathieu C, Kolaitis R-M, Zhang P, Messing J, Yurtsever U, Yang Z, Wu J, Li Y, Pan Q et al (2020) G3BP1 is a tunable switch that triggers phase separation to assemble stress granules. Cell 181:325–345.e328

Yang Y, Arseni D, Zhang W, Huang M, Lovestam S, Schweighauser M, Kotecha A, Murzin AG, Peak-Chew SY, Macdonald J et al (2022) Cryo-EM structures of amyloidA-β42 filaments from human brains. Science 375:167–172

Yoo H, Bard JAM, Pilipenko EV, Drummond DA (2022) Chaperones directly and efficiently disperse stress-triggered biomolecular condensates. Mol Cell 82:741–755.e711

Yoshizawa T, Nozawa R-S, Jia TZ, Saio T, Mori E (2020) Biological phase separation: cell biology meets biophysics. Biophys Rev 12:519–539

Yu JF, Cao Z, Yang Y, Wang CL, Su ZD, Zhao YW, Wang JH, Zhou Y (2016) Natural protein sequences are more intrinsically disordered than random sequences. Cell Mol Life Sci 73:2949–2957

Zbinden A, Pérez-Berlanga M, De Rossi P, Polymenidou M (2020) Phase separation and neurodegenerative diseases: a disturbance in the force. Dev Cell 55:45–68

Zhang Z, Klionsky DJ (2022) CCT2, a newly identified aggrephagy receptor in mammals, specifically mediates the autophagic clearance of solid protein aggregates. Autophagy 18:1483–1485

Zhang Y, Gu J, Sun Q (2021) Aberrant stress granule dynamics and aggrephagy in ALS pathogenesis. Cells 10:2247

Chapter 6
Molecular Mechanisms Defining the Structural Basis for Self-Association of the FUS Low-Complexity Domain

Masato Kato

Abstract The disordered low-complexity domain of the fused-in-sarcoma (FUS) RNA-binding protein has been studied extensively over the past decade. Due to its pronounced capacity to self-associate, the low-complexity domain (LCD) of FUS is capable of forming dynamic droplets by liquid–liquid phase separation (LLPS) and solid hydrogels by phase transition (PT). Hydrogels formed from the FUS LCD are composed of labile cross-β polymers. Liquid-like droplets formed by the FUS LCD also utilize cross-β interactions to facilitate PS and mimic the behavior of subcellular non-membrane structures such as RNA granules. FUS hydrogel droplets have been found to selectively bind mRNAs and LCDs of other RNA-binding proteins found in RNA granules. Based on these observations, it is now understood that LLPS and PT qualify as useful representations of the mechanisms controlling RNA granule formation. As such the time has come for high resolution structural studies that will define the mechanisms of LCD self-association causal of LLPS and PT. Here, I summarize our findings relevant to how the FUS LCD that self-associates to form liquid-like droplets and hydrogels. Finally, I will discuss mutations within the FUS LCD have been reported for patients suffering from amyotrophic lateral sclerosis. Unlike most disease-causing mutations that enhance the localized strength of a cross-β core, the G156E variant of FUS significantly weakens the cross-β core in which it resides. Our studies of this disease-causing variant shed light on an unanticipated mechanism wherein the balanced strength of two distinct cross-β elements within the FUS LCD is critical for the protein to avoid run-away polymerization and consequent involvement in neurodegenerative disease.

M. Kato (✉)
Department of Biochemistry, University of Texas Southwestern Medical Center, Dallas, TX, USA

Institute for Quantum Life Science, National Institutes for Quantum Science and Technology (QST), Chiba, Japan
e-mail: masato.kato@utsouthwestern.edu

R. Kurokawa (ed.), *Phase Separation in Living Cells*,
https://doi.org/10.1007/978-981-99-4886-4_6

Keywords Low-complexity sequences · Phase separation · Liquid-like droplets · Phase transition · Hydrogels · Amyloid-like cross-β polymer · Neurodegenerative disease

6.1 Introduction

Most proteins use all 20 amino acids in their sequences to fold into a functional structure. However, the proteomes of eukaryotic cells also contain thousands of unusual protein sequences called low-complexity domains (LCDs), typified by having limited diversity in amino acid composition. Because of this sequence nature, LCDs are unable to fold into stable structures and have been recognized as intrinsically disordered regions (IDRs) of the proteome. Upwards of 20% of constituents of the human proteome contain LCDs longer than 50 amino acids (Rado-Trilla and Alba 2012; Toll-Riera et al. 2012). LC sequences are particularly enriched in regulatory proteins, such as RNA- and DNA-binding proteins. Indeed, the founding examples of LCDs were the activation domains of transcription factors (Ma and Ptashne 1987; Triezenberg et al. 1988). While LCDs were first discovered in studies of transcription factors, it is now clear that they are found in proteins involved in virtually all aspects of cell biology.

Many RNA-binding proteins, including fused-in-sarcoma (FUS), TAR DNA-binding protein 43 (TDP-43), and heterogeneous nuclear ribonucleoproteins (hnRNPs), also have LCDs. These proteins play many roles in RNA biogenesis, including pre-mRNA splicing, nuclear export, and localized translation. RNA-binding proteins are critical to the formation of subcellular non-membrane structures such as RNA granules. Finally, RNA-binding proteins have been extensively implicated in neurodegenerative diseases such as amyotrophic lateral sclerosis (ALS), frontotemporal dementia (FTD), and Alzheimer's disease (Mathieu et al. 2020).

The LCDs of RNA-binding proteins have been shown to self-associate in a manner leading to phase separation (Kato et al. 2012). Among many LCDs from RNA-binding proteins studied over the past decade, the FUS LCD has received the most attention. When studied in its purified form in vitro, the FUS LCD initially forms meta-stable liquid-like droplets that, with time, solidify into hydrogels (Kato et al. 2012; Patel et al. 2015). Because liquid-like droplets mimic the behavior of subcellular non-membrane structures such as RNA granules, LLPS has been proposed as the basis for molecular mechanisms of non-membrane subcellular structures formation. We have demonstrated that FUS LCD forms cross-β polymers and proposed that the molecular structure defining the protomer interface of these labile polymers represents the biological driving force of both PS and PT (Kato et al. 2012; Murray et al. 2017). In this review, we summarize what we have learned about how the LCD of FUS self-associates.

6.2 RNA-Binding Protein FUS

The FUS RNA-binding protein consists of 526 amino acid residues and is composed of domains including an N-terminal LCD, an RRM domain, a Zn-finger domain, and 3 arginine-glycine-glycine (RGG) motifs (Fig. 6.1a). FUS has been shown to bind RNA with diverse sequence specificity using all domains other than LC domain (Schwartz et al. 2015; Wang et al. 2015). FUS is involved in many cellular functions including RNA metabolism, pre-mRNA splicing, nuclear export of mRNAs, DNA repair, and transcriptional regulation (Polymenidou et al. 2012; Schwartz et al. 2012; Wang et al. 2013; Deng et al. 2014; Yang et al. 2014; Rhoads et al. 2018). FUS is primarily a nuclear protein, yet shuttles between nucleus and cytoplasm to assist in the export of mature mRNAs (Ederle and Dormann 2017).

FUS was originally discovered as part of a reciprocal translocation product in patients suffering a rare form of sarcoma cancers. The translocation forms an oncogene that encodes a fusion protein between the LCD of FUS and the DNA-binding domain of transcription factors including CHOP and ERG (Crozat et al. 1993; Panagopoulos et al. 1994). Similarly, the LCDs of the paralogous EWS and TAF15 proteins can also be fused to DNA-binding domains of FLI and CHOP to form disease-causing oncogenes (Guipaud et al. 2006). These chimeric transcription factors have been shown to be far more potent as transcriptional activators than the parental CHOP, ERG or FLI transcription factors (Bailly et al. 1994). These observations provide evidence that the LCDs of the FUS, EWS, and TAF15 proteins can function as activation domains despite being derived from RNA-binding proteins.

6.3 Anatomy of the FUS RNA-Binding Proteins

The sequence of the FUS LCD (residue 1–214) is composed primarily of only glycine, serine, glutamine, and tyrosine. These four amino acids account for 83% of the composition of the FUS LCD (Fig. 6.1b). In contrast, ten amino acids including most hydrophobic amino acids are completely absent from the FUS LCD. The sequence is quasi-repetitive in showing 27 tyrosine residues invariably flanked by glycine or serine. Either this [GS]Y[GS] motif or a variant motif using phenylalanine instead of tyrosine ([GS]F[GS]) are found in LCDs of many RNA-binding proteins including EWS, TAF15, and numerous hnRNPs. These motifs are reminiscent of the FG repeat motifs found in the proteins that form the filtration barrier of nuclear pores (Frey et al. 2006; Strambio-De-Castillia et al. 2010). Systematic mutational studies have shown that these motifs are essential for both self-association and biological function of these proteins (Frey et al. 2006; Kato et al. 2012). It is thought that the aromatic side chains of tyrosine and phenylalanine may interact by π–π stacking or by cation–π interactions between aromatic residues and positively charged side chains of arginines or lysines. These weak and non-specific interactions may help facilitate intermolecular collision between LCDs as a first step toward productive self-association.

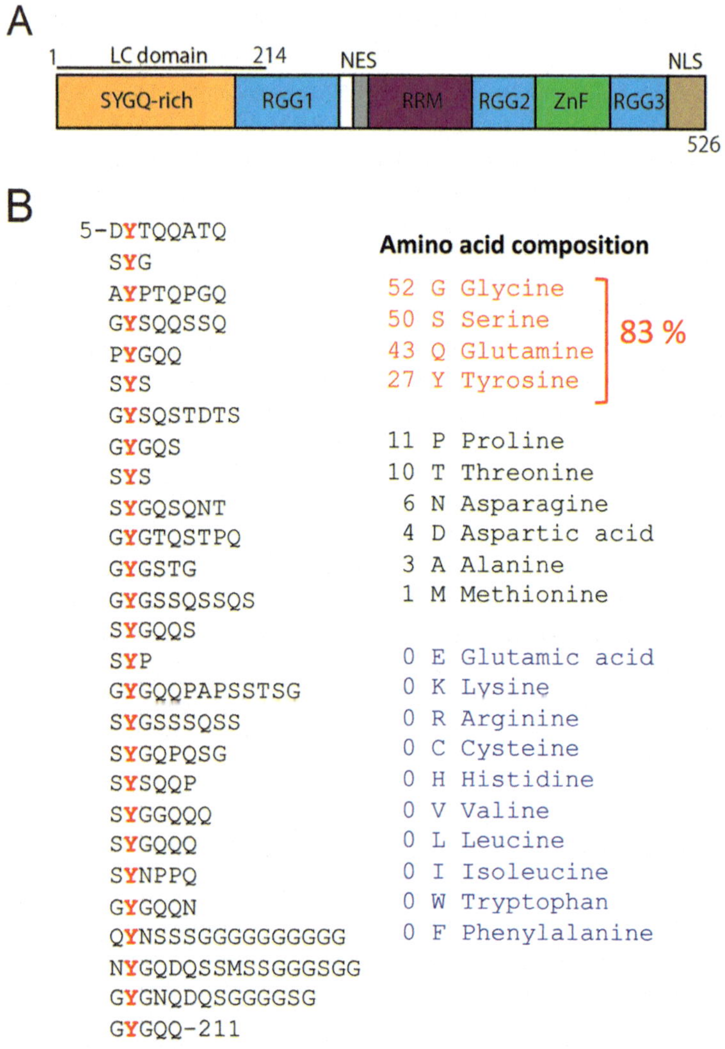

Fig. 6.1 The domain structure and the amino acid sequence of FUS. (**a**)The domain structure of FUS is shown with abbreviations as follows, *SYGQ-rich* serine, tyrosine, glycine, and glutamine-rich domain, *RGG* arginine-glycine-glycine domain, *NES* nuclear export signal, *RRM* RNA recognition motif, *ZnF* zinc-finger motif, *NLS* nuclear localization signal. (**b**)The amino acid sequence of the FUS LCD is shown with its amino acid composition. The sequence is shown with tyrosine residues aligned vertically. Four amino acids, glycine, serine, glutamine, and tyrosine, account for 83% of the total sequence. Ten amino acids (blue) are completely excluded from the sequence

FUS contains a C-terminal RGG domain that is enriched with arginine residues (Fig. 6.1a). The C-terminal RGG domain has been shown to interact with the N-terminal LCD, probably through the cation–π interactions. When this interaction takes place intramolecularly, FUS forms a closed ring (Hamad et al. 2020). If the N-terminal LCD of FUS is occupied intramolecularly via interaction with the C-terminal RGG motif, intermolecular self-association and phase separation are impeded.

6.4 Hydrogel Formation by the FUS LCD

In 2008–2009, Xinlin Du, a postdoctoral fellow in the McKnight lab, purified a recombinant protein fragment containing the FUS LCD (1–214 residues) linked to GST. He stored it as a concentrated solution at −80 °C. Another postdoctoral fellow in the McKnight lab, Shanhai Xie, observed that, upon thawing, concentrated preparation of the GST-FUS protein exhibited a gelatinous state. Upon hearing about this phenomenon, I tried to reproduce it with mCherry covalently linked to the FUS LCD. After purification and concentration (10–20 mg/mL), the mCherry-FUS LCD chimeric protein was dialyzed and incubated in a physiological buffer (pH 7.5 and 200 mM salt) at 4 °C. Several days later, I noticed that the protein solution solidified into a gel-like state (Fig. 6.2a).

Inspection of the hydrogel contents by transmission electron microscopy revealed uniform polymers (Kato et al. 2012) (Fig. 6.2b). X-ray diffraction studies of these hydrogels yielded a prominent diffraction ring at 4.7 Å. The combination of EM and X-ray analysis told us that the FUS LCD can form amyloid-like polymers. Although these polymers appeared indistinguishable from pathogenic amyloid fibers, the LCD polymers revealed one distinct difference. Pathogenic fibers require boiling temperature and a chemical denaturant to be disassembled. By contrast, the cross-β polymers formed from the FUS LCD are labile to disassembly. FUS LCD polymers can be melted by a low concentration of SDS or temperatures only mildly higher than 37 °C (Kato et al. 2012; Kato and McKnight 2021) (Fig. 6.2c).

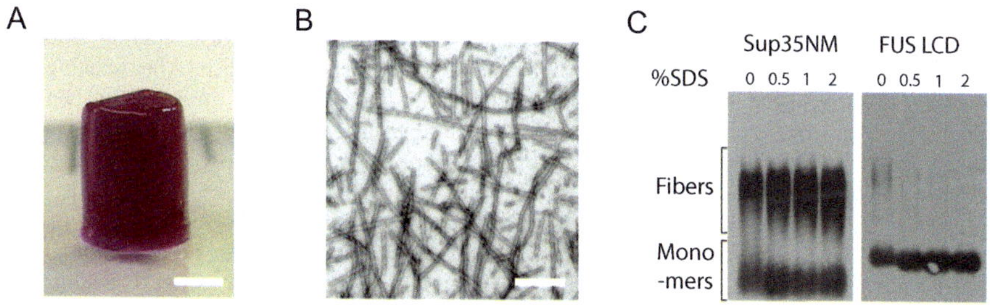

Fig. 6.2 A hydrogel and labile cross-β polymers of the FUS LCD. (**a**) A hydrogel of mCherry-linked to FUS LCD. A scale bar indicates 1 mm. (**b**) A transmission electron microscopy image of cross-β polymers of the FUS LCD. A scale bar indicates 200 nm. (**c**) Semi-desaturating detergent agarose gel electrophoresis (SDS-AGE) of cross-β polymers of the yeast Sup35 NM domain and the FUS LCD. The cross-β polymers were treated with different concentrations of SDS at 37 °C and then separated from dissociated monomers by an agarose gel electrophoresis. In contrast to the irreversible Sup35NM polymers, the cross-β polymers of the FUS LCD were completely degraded with 0.5% SDS

6.5 Hydrogel Binding Assays

Never having had a biochemical assay to study protein domains of low sequence complexity, we wondered whether we might take advantage of solid hydrogels formed from the FUS LCD (Kato et al. 2012) (Fig. 6.3a). As an initial test of this possibility, a preformed hydrogel droplet of mCherry-linked FUS LCD was immersed in solutions of either free GFP or GFP fusion proteins. No binding to the hydrogels was observed using free GFP. Strong hydrogel binding was observed when we tested GFP linked to the LCD of FUS itself (homotypic binding). Subsequent analysis using GFP-fused to the LCDs of other RNA-binding proteins revealed different levels of accumulation on mCherry-FUS LCD hydrogels. This

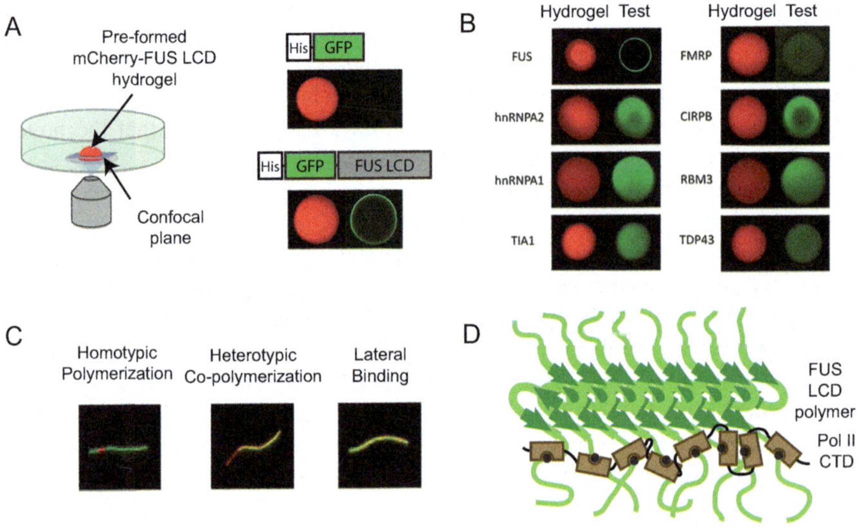

Fig. 6.3 Hydrogel binding assays. (**a**) Schematic drawing of the hydrogel binding assay. A preformed hydrogel of mCherry-FUS LCD is placed in a glass-bottom dish. After soaking in a solution of GFP-linked test protein, GFP signals in the hydrogels were measured by confocal fluorescence microscopy. GFP alone does not bind to the hydrogel, whereas GFP-FUS LCD binds to the periphery of the hydrogel. (**b**) LCDs from RNA granule proteins were tested on hydrogels of mCherry-FUS LCD. GFP-FUS LCD binds only to the peripheral part of the hydrogel because most of the GFP-FUS LCD molecules are quickly incorporated into the mCherry-FUS LCD polymers in the hydrogel. Therefore, GFP-FUS LCD cannot reach the center of the hydrogel droplet. LCDs of other RNA granule proteins are also captured by the mCherry-FUS LCD hydrogels, but the GFP signals are different from case to case. (**c**) Mechanisms of hydrogel binding assays. (Left) Homotypic binding represents a seeding effect. Polymer seeds of mCherry-FUS LCD are extended by GFP-FUS LCD. (Middle) Heterotypic binding between mCherry-FUS LCD and other LCDs that are capable of forming cross-b polymer by themselves. (Right) LCDs that are unable to form cross-b polymers bind uniformly to the lateral surface of cross-b polymers of mCherry-FUS LCD. (**d**) Repetitive binding between the repetitive surface of FUS LCD polymers and the heptapeptide repeats of the CTD of RNA pol II. An affinity of this repeated binding is higher than that of a single binding

heterotypic hydrogel binding was invariably of a lower signal level than the homotypic binding observed for GFP-FUS LCD (Fig. 6.3b).

One possible mechanism to account for homotypic binding would correspond to the incorporation of GFP-FUS LCD monomer molecules polymerizing onto the ends of pre-existing mCherry-FUS LCD polymers, of which the hydrogels themselves are composed (Fig. 6.3c). Evidence of polymer incorporation has been confirmed by fluorescence microscopic analysis of individual polymers.

Despite having observed this expected form of GFP-FUS LCd binding to mCherry-FUS LCD hydrogels, an unbiased approach to the problem has yielded unexpected results. We already knew that tyrosine residues are important for both hydrogel binding and in vivo recruitment of FUS to stress granules (Kato et al. 2012). To investigate GFP-FUS LCD homotypic binding to mCherry-FUS LCD hydrogels, we individually mutated each of 27 tyrosine residues to serine. Analysis of hydrogel binding by these 27 variants yielded the surprising observation that a small region localized between tyrosine residues 161 and 177 was of unusual importance for hydrogel binding. In order to make any sense of this observation, we had to perform structural studies on the hydrogel polymers themselves.

In the case of heterotypic binding, we envisioned two possible mechanisms. One first idea was some form of co-polymerization. If a test protein is capable of forming cross-β polymers on its own (such as hnRNPA1, hnRNPA2, or TDP-43), it might polymerize as its own fiber on the surface of mCherry-FUS LCD polymers. We indeed observed this form of co-polymerization. As shown in Fig. 6.3c, two separate polymers can be seen growing side by side (Fig. 6.3c). If a test LCD is unable to form cross-β polymers on its own, it might bind uniformly to the surface of the mCherry-FUS LCD polymers. We have observed this polymer surface binding of the C-terminal domain (CTD) of RNA polymerase II and the SR domains of RNA splicing factors (Kwon et al. 2013, 2014). It is interesting to note that the CTD has 52 repeats of a heptapeptide motif (YSPTSPS) and the SR domains have many repeats of an arginine-serine dipeptide. In the cross-β polymers, the FUS LCD molecules are stacked one on top of another every 4.7 Å along the polymer axis and exhibit a repeating surface. Therefore, we speculate that repetitive interactions between the surface of the cross-β polymers and the repeating CTD or SR motifs allow for hydrogel binding (Lin et al. 2016) (Fig. 6.3d). It will be an exciting challenge to resolve structures composed of LCD polymers bound by repetitive LCDs that cannot polymerize on their own such as the CTD of RNA polymerase II.

6.6 Structures of Cross-β Polymers Formed by the FUS LCD

The structures of cross-β polymers of FUS LCD (residues 2–214) were determined by solid-state NMR (ssNMR) and later by cryo-electron microscopy (cryo-EM) (Murray et al. 2017; Sun et al. 2022) (Fig. 6.4a). The two structures are generally similar. However, the cryo-EM structure has a more extended rigid portion in the C-terminal region of the LCD. The polymer core of the ssNMR structure spans from residues 37–95, while that of the cryo-EM structure spans from residues 34 to 124. The remaining N- and C-terminal regions of both structures are largely disordered.

As described above, the cross-β polymers of FUS LCD are labile to dissociation with a low concentration of SDS or at a mild temperature (40–50 °C). Analysis of our ssNMR structure of FUS LCD polymers reveals almost no hydrophobic amino acids at the protomer interface (Murray et al. 2017) (Fig. 6.4b). By contrast, pathogenic fibers, such as Aβ and α-synuclein fibrils, have between 20 and 30 hydrophobic residues at the protomer interface. These hydrophobic residues interact across the protomer interface to stabilize the polymer structure. Interestingly, the cryo-EM structure of the FUS LCD was shown to be more thermostable (Sun et al. 2022). The additional rigid part observed in the cryo-EM structure was formed by the C-terminal

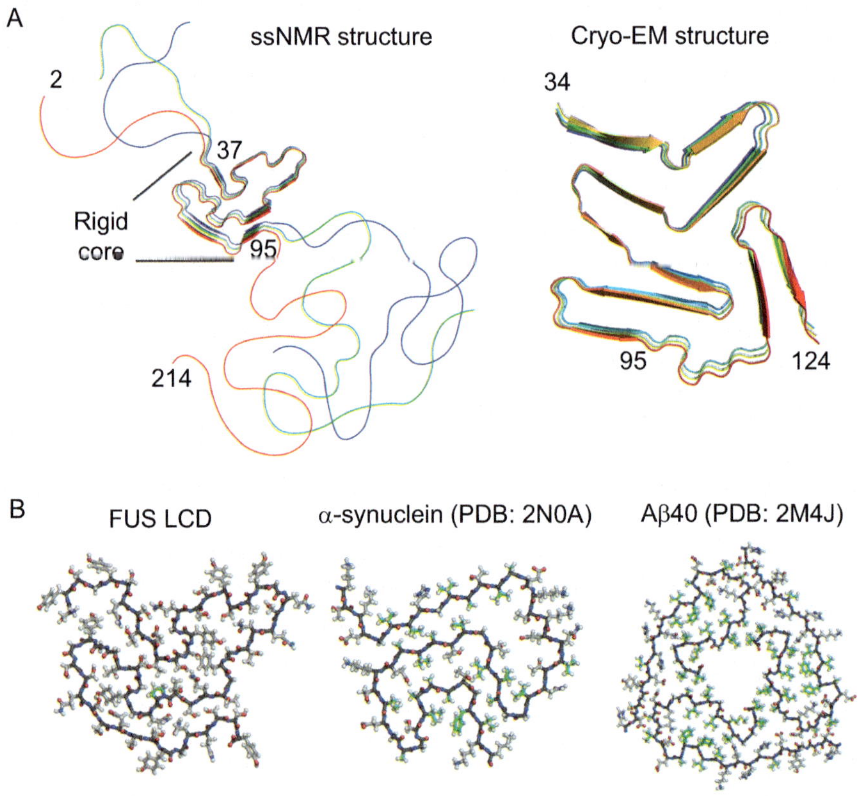

Fig. 6.4 Structures of the cross-β polymers formed by the FUS LCD. (**a**) Structures of the cross-β polymers formed by the FUS LCD determined by solid-state NMR (left: PDB code 5W3N) and cryo-EM (right: PDB code 7VQQ). For the ssNMR structure, disordered N- and C-terminal regions are shown as a cartoon. (**b**) Comparison of the cross-β core structures of the FUS LCD polymers and pathogenic fibers. For each of the structures, a single subunit within the polymer is shown. Hydrophobic residues are colored in green

~30 residues (residues 96–124). We simplistically believe that this additional rigidity contributes to the increased stability of the cryo-EM structure. This additional rigidity may reflect a transition from normal biological function to a pathogenic state of aggregation.

6.7 Liquid–Liquid Phase Separation by the FUS LCD

FUS LCD can form liquid-like droplets through LLPS in vitro (Fig. 6.5a) (Patel et al. 2015). These droplets fuse with each other to form larger droplets. LC domain molecules moved quickly back and forth between the inside and outside of the droplets, and different LCDs and RNAs were able to be captured within the droplets. These behaviors are similar to those of dynamic RNA granules inside the cells. Therefore, it has been proposed that RNA granules represent liquid-like structures resulting from LLPS.

Previous studies have shown that the mechanism behind the phase separation of LCDs involves the formation of a "3-dimentional mesh network structure" through "multivalent interactions" between LCD molecules (Fig. 6.5b). "Multivalent interactions" refers, at least in part, to the simultaneous interactions between one LCD molecule and multiple other molecules. To form a mesh network structure, a

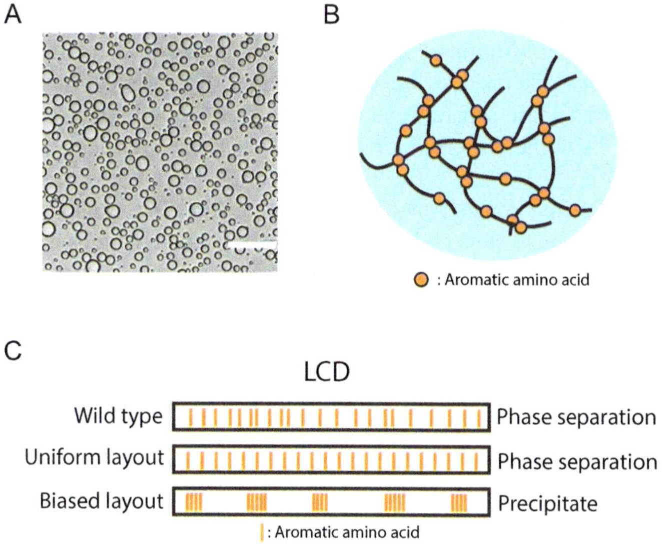

Fig. 6.5 Phase-separated liquid-like droplets formed by the FUS LCD. (**a**)A picture of liquid-like droplets of the FUS LCD. A scale bar indicates 25 μm. (**b**) A schematic drawing of the 3D mesh network formed by multivalent interaction of the FUS LCD molecules in a phase-separated liquid-like droplet. (**c**) Aromatic amino acids within the LCD sequence must be uniformly distributed to achieve phase separation

molecule should have at least three (trivalent) binding points (Martin and Holehouse 2020) (Fig. 6.5b). In other words, if a molecule binds to three or more different molecules simultaneously and these multivalent interactions occur in many molecules at the same time, a mesh network structure will be formed.

The FUS LCD contains 27 tyrosine residues that are periodically repeated in the sequence (Fig. 6.1b). The aromatic ring of the tyrosine side chain can interact with another aromatic ring by π–π interaction, which allows for 27-valencies open for π–π mediated interactions among FUS LCD molecules (Fig. 6.5b). The majority of tyrosines (Y) are flanked on both sides by either serine (S) or glycine (G), forming an [S/G]Y[S/G] motif (Fig. 6.1b). This motif is believed to expose the aromatic side chains, making them more likely to interact with other tyrosine side chains. While the interaction of a single aromatic ring is weak and quickly dissociates, the large number of tyrosine residues in FUS LCDs has been interpreted to give evidence that several of them are likely to interact with other molecules and trap them within the mesh network. Additionally, other reported binding points include the interaction of tyrosine and arginine residues by their side chains (cation–π interaction) and the interaction of acidic (negatively charged) and basic (positively charged) residues to form attractive charge–charge interaction.

It has been observed that not only the number (valency) of aromatic residues as binding points, but also their arrangement must be of a suitable pattern for phase separation. The 27 tyrosines in the FUS LCD are relatively evenly distributed in the sequence (Fig. 6.1b). This has been shown to be important for phase separation. An LCD variant, in which the aromatic residues were positioned at more evenly spaced intervals than in the wild-type sequence, formed phase-separated droplets as well as did the native protein (Martin et al. 2020) (Fig. 6.5d). A different variant, in which the number of aromatic residues was left unchanged, but were densely and unevenly arranged, failed to phase-separate and instead formed amorphous precipitates. These results indicate that the dispersed arrangement of tyrosine residues is important for phase separation.

This structure-free concept of LCD self-association has captured considerable attention. It represents a new and exciting idea about the dynamics of cell morphology. It raises several questions, however, that will require attention in the years to come. First, this structure-free concept often has no opportunity for specificity to be imparted upon LCD self-association. Second, it is hard to understand how human genetic mutations that change a single amino acid within an LCD might manifest their effect in the absence of any form of protein structure.

6.8 Maturation of Phase-Separated Liquid-like Droplets into Hydrogels

The phase-separated droplets of FUS LCD are initially dynamic, but over time they solidify and transform into a gel-like state. This process is referred to as "aging" or "maturation" and is thought to be at the heart of neurodegenerative diseases including ALS. It has been shown that this process involves the formation of cross-β polymers of FUS LCD within the droplets (Patel et al. 2015). The concentration of FUS LCD in the droplets is very high (25–30 mM) (Murthy et al. 2019), which favors polymer nucleation because nucleation is concentration dependent. Based on these observations, it has been proposed that the phase separation and phase transition of LCDs are related to each other as a series of state-changing reactions (Fig. 6.6). In summary, LCDs in the monomer state begin to self-associate, leading to the formation of phase-separated liquid-like droplets. During this process, a small amount of labile cross-β interactions already form and assist in the phase separation (Xiang et al. 2015; Lin et al. 2020). These cross-β interactions may provide more branching points to build the mesh network structure of LCDs in the droplets. As described earlier, the number of protomers within cross-β polymers increases over time in the droplets due to high protein concentration. Polymer growth, in turn, causes enhanced droplet viscosity and eventual transformation into a gel-like state. Normally, the formation of cross-β assemblies of LCDs in cells is regulated by phosphorylation, methylation, and other regulatory mechanisms that prevent them from growing too long or becoming too stable (Monahan et al. 2017; Murray et al. 2017; Hofweber et al. 2018; Qamar et al. 2018; Wang et al. 2018). When this regulatory mechanism fails, the cross-β polymers can grow to abnormal lengths. Disease-causing mutations within LCDs can cause the polymers to become abnormally stable as described for the asparagine-to-valine mutations in three different hnRNP proteins (Lin et al. 2016; Murray et al. 2018). Abnormal polymerization leads to the accumulation of pathogenic fibers in cells, which is now thought to be a central mechanism causing neurodegenerative diseases.

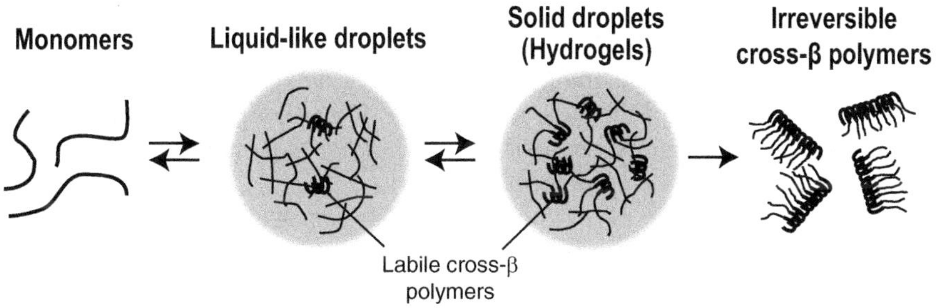

Fig. 6.6 A cascade of phase separation and phase transition of the FUS LCD. Phase separation is now considered to be a precursor state to a hydrogel state formed by phase transition. Formation of labile cross-β polymers of FUS LCD is one of the major driving forces for phase separation. Abnormal regulation of LCD self-association or pathogenic mutations in LCD would lead to the formation of irreversible polymers of LCDs

6.9 A Two-Core System Controls Formation of Labile Cross-β Interactions of the FUS LCD

Quite unexpectedly, we stumbled over a discovery showing that FUS LCD has its own internal regulatory mechanism to control the formation of labile cross-β structures. As described above, the C-terminal half of FUS LCD is largely flexible and is not visible in either the ssNMR or cryo-EM structures of FUS LCD cross-β polymers (Murray et al. 2017; Sun et al. 2022) (Fig. 6.4a). However, when separated from its N-terminal half, the C-terminal half of the FUS LCD can also self-associate to form cross-β polymers on its own (Lee et al. 2020; Kato and McKnight 2021). Furthermore, we have shown that the C-terminal region of FUS LCD polymers is essential for its incorporation as newly arriving monomers of the FUS LCD into the ends of preformed cross-β polymers (Kato and McKnight 2021). Based on these results, we hypothesized that the disordered C-terminal region of FUS LCD in the polymeric state would self-interact with the C-terminal region of an incoming FUS LCD monomer as if they were attempting to form a cross-β polymer (Fig. 6.7a). This interaction is hypothesized to momentarily trap the incoming molecule at the disordered C-terminal ends along the lateral sides of polymers. This transient interaction is then thought to allow the N-terminal regions to be incorporated into existing polymers by end localized polymer growth (Fig. 6.7a).

As observed in the two structures of FUS LCD polymers in which only N terminal core polymers are visible (Murray et al. 2017; Sun et al. 2022) (Fig. 6.4a), the C-terminal core polymer cannot coexist with the N-terminal core polymer. In other words, the N- and C-terminal core polymers are mutually exclusive. When two monomers of FUS LCD initially collide, we hypothesized that they self-associate via either the N-terminal core or the C-terminal core (Fig. 6.7b). Being mutually exclusive, the two modes of self-association are anticipated to compete. If the N-terminal cross-β core forms first, it leaves the region specifying the C-terminal cross-β core open and unstructured. When a third monomer is brought to the complex, we hypothesize that it will preferentially interact with the exposed, disordered C-terminal region of the existing dimer. If it succeeds in forming a C-terminal cross-β interaction with either molecule of the dimer, the forces of mutual exclusion will force disassembly of the N-terminal cross-β interaction. This concept ensures that wrong polymers are never built from either the N-terminal cross-β-forming region or the C-terminal cross-β-forming region.

What happens if either of these cross-β forming regions is inactivated by mutation? This is precisely the outcome of the ALS-causing G156E mutation within the C-terminal cross-β-forming region of the FUS LCD (see below).

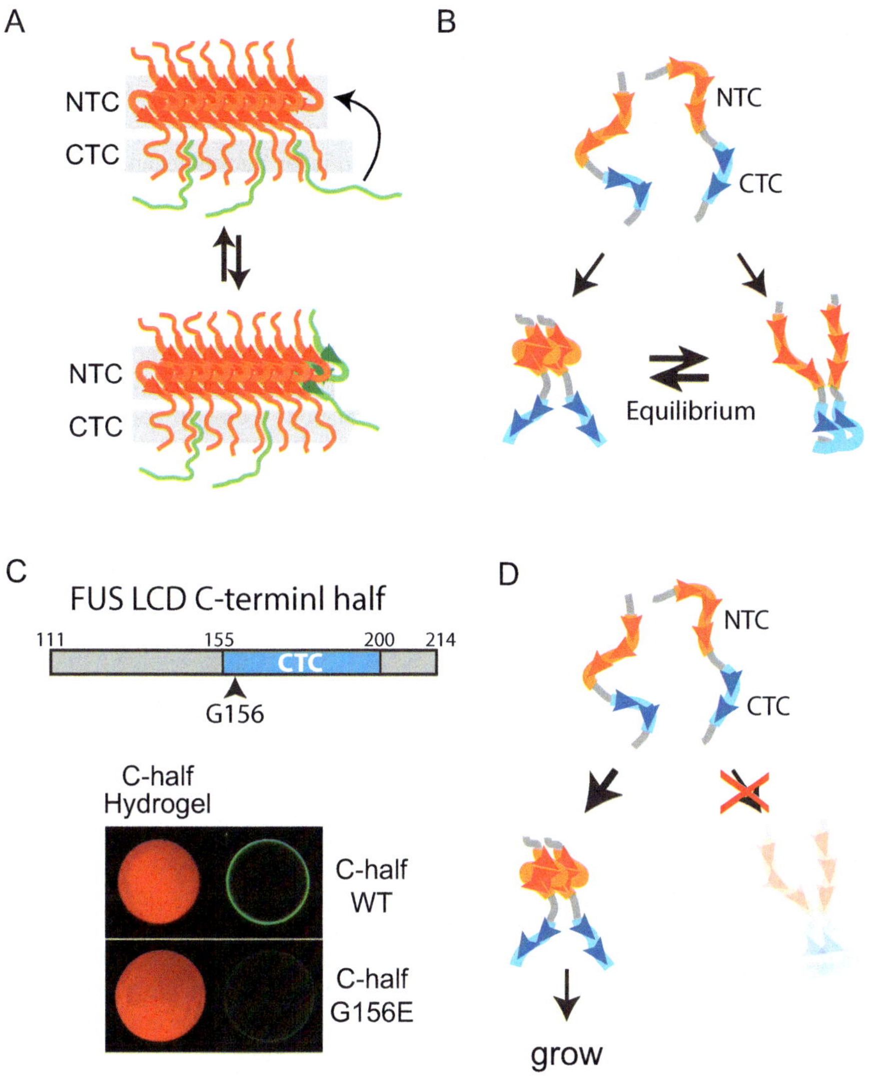

Fig. 6.7 An internal regulation mechanism for polymer formation of the FUS LCD. (**a**)The C-terminal core (CTC) region of cross-β polymers of FUS LCD can transiently bind to the CTC region of an incoming FUS LCD molecule (top). If this occurs at the end of the polymer, the N-terminal core (NTC) region of the incoming molecule would be readily incorporated into the NTC polymer end (bottom). (**b**) A schematic representation of a hypothetical equilibrium between a dimer formed by NTC interaction and a dimer formed by CTC interaction. (**c**) The ALS mutation, G156E, in the C-terminal half of the FUS LCD inhibits binding to hydrogel droplets, indicating that this mutant loses the ability to form cross-β polymers. (**d**) A hypothetical mechanism for formation of pathogenic fibers of the FUS LCD carrying the G156E mutation. When the CTC interaction is inhibited by this mutation, the equilibrium is lost and only the NTC interaction remains for self-association, resulting in rapid growth of the NTC polymers

6.10 Effects of Disease-Causing Mutations on Self-Association of FUS LC Domain

There are many mutations on FUS identified in patients with amyotrophic lateral sclerosis (ALS). Most such mutations are localized in the nuclear localization signal (NLS) located at the C-terminal end of FUS. A small number of disease-causing mutations map within the N-terminal LCD of FUS (Bentmann et al. 2013; Kapeli et al. 2017). The G156E mutation, one of the ALS mutations, is located within the C-terminal half of FUS LCD and has been shown to facilitate the "aging" process of the phase-separated liquid-like droplets (Murakami et al. 2015; Patel et al. 2015). G156 is far from the N-terminal core, which is the ultimate winner of the two-core system mentioned above. How does G156E affect the N-terminal core? We have shown that the G156E mutation inhibits the formation of cross-β polymers of the C-terminal half (Kato and McKnight 2021) (Fig. 6.7c). This results in a loss of competition, leading to the rapid formation of pathogenic N-terminal cross-β polymers (Fig. 6.7d). Although the net result of this mutation is increased formation of pathogenic N-terminal polymers, the underlying mechanism is inhibition of formation of the C-terminal cross-β polymers. This is the first example of a disease mechanism in which an ALS mutation inhibits formation of cross-β polymers.

6.11 Conclusion

Self-association of the FUS LCD is a bona fide function of the protein. The in vitro formation of phase-separated liquid-like droplets or reversible cross-β polymers represents the first description of how LCDs actively work. From a reductionist point of view, we believe that the mechanisms underlying these in vitro assays represent a useful reference point for considering the in vivo functions of LCDs. In particular, to understand the disease mechanisms caused by ALS mutations in LCDs, it is necessary to know how the mutations alter LCD function. In the past, our understanding of disease mechanism has been limited to mutations that strength the self-association of cross-β cores within LCDs because it made sense to do so with the facts that the mutant proteins accumulate in patient cells as irreversible polymer aggregates. We now show that the G156E mutation within the FUS LCD works by inactivating one of its two cross-β cores. This, in turn, leads to an imbalance between the strength of the two cross-β cores. This imbalance leads to hyper polymerization of the N-terminal cross-β core and consequent disease.

References

Bailly RA, Bosselut R, Zucman J, Cormier F, Delattre O, Roussel M, Thomas G, Ghysdael J (1994) DNA-binding and transcriptional activation properties of the EWS-FLI-1 fusion protein resulting from the t(11,22) translocation in Ewing sarcoma. Mol Cell Biol 14:3230–3241

Bentmann E, Haass C, Dormann D (2013) Stress granules in neurodegeneration—lessons learnt from TAR DNA binding protein of 43 kDa and fused in sarcoma. FEBS J 280:4348–4370

Crozat A, Aman P, Mandahl N, Ron D (1993) Fusion of CHOP to a novel RNA-binding protein in human myxoid liposarcoma. Nature 363:640–644

Deng Q, Holler CJ, Taylor G, Hudson KF, Watkins W, Gearing M, Ito D, Murray ME, Dickson DW, Seyfried NT et al (2014) FUS is phosphorylated by DNA-PK and accumulates in the cytoplasm after DNA damage. J Neurosci Off J Soc Neurosci 34:7802–7813

Ederle H, Dormann D (2017) TDP-43 and FUS en route from the nucleus to the cytoplasm. FEBS Lett 591:1489–1507

Frey S, Richter RP, Gorlich D (2006) FG-rich repeats of nuclear pore proteins form a three-dimensional meshwork with hydrogel-like properties. Science 314:815–817

Guipaud O, Guillonneau F, Labas V, Praseuth D, Rossier J, Lopez B, Bertrand P (2006) An in vitro enzymatic assay coupled to proteomics analysis reveals a new DNA processing activity for Ewing sarcoma and TAF(II)68 proteins. Proteomics 6:5962–5972

Hamad N, Watanabe H, Uchihashi T, Kurokawa R, Nagata T, Katahira M (2020) Direct visualization of the conformational change of FUS/TLS upon binding to promoter-associated non--coding RNA. Chem Commun (Camb) 56:9134–9137

Hofweber M, Hutten S, Bourgeois B, Spreitzer E, Niedner-Boblenz A, Schifferer M, Ruepp MD, Simons M, Niessing D, Madl T et al (2018) Phase separation of FUS is suppressed by its nuclear import receptor and arginine methylation. Cell 173(706–719):e713

Kapeli K, Martinez FJ, Yeo GW (2017) Genetic mutations in RNA-binding proteins and their roles in ALS. Hum Genet 136:1193–1214

Kato M, McKnight SL (2021) The low-complexity domain of the FUS RNA binding protein self-assembles via the mutually exclusive use of two distinct cross-beta cores. Proc Natl Acad Sci U S A 118:e2114412118

Kato M, Han TW, Xie S, Shi K, Du X, Wu LC, Mirzaei H, Goldsmith EJ, Longgood J, Pei J et al (2012) Cell-free formation of RNA granules: low complexity sequence domains form dynamic fibers within hydrogels. Cell 149:753–767

Kwon I, Kato M, Xiang S, Wu L, Theodoropoulos P, Mirzaei H, Han T, Xie S, Corden JL, McKnight SL (2013) Phosphorylation-regulated binding of RNA polymerase II to fibrous polymers of low-complexity domains. Cell 155:1049–1060

Kwon I, Xiang S, Kato M, Wu L, Theodoropoulos P, Wang T, Kim J, Yun J, Xie Y, McKnight SL (2014) Poly-dipeptides encoded by the C9orf72 repeats bind nucleoli, impede RNA biogenesis, and kill cells. Science 345:1139–1145

Lee M, Ghosh U, Thurber KR, Kato M, Tycko R (2020) Molecular structure and interactions within amyloid-like fibrils formed by a low-complexity protein sequence from FUS. Nat Commun 11: 5735

Lin Y, Mori E, Kato M, Xiang S, Wu L, Kwon I, McKnight SL (2016) Toxic PR poly-dipeptides encoded by the C9orf72 repeat expansion target LC domain polymers. Cell 167(789–802):e712

Lin Y, Zhou X, Kato M, Liu D, Ghaemmaghami S, Tu BP, McKnight SL (2020) Redox-mediated regulation of an evolutionarily conserved cross-beta structure formed by the TDP43 low complexity domain. Proc Natl Acad Sci U S A 117:28727–28734

Ma J, Ptashne M (1987) A new class of yeast transcriptional activators. Cell 51:113–119

Martin EW, Holehouse AS (2020) Intrinsically disordered protein regions and phase separation: sequence determinants of assembly or lack thereof. Emerg Top Life Sci 4:307–329

Martin EW, Holehouse AS, Peran I, Farag M, Incicco JJ, Bremer A, Grace CR, Soranno A, Pappu RV, Mittag T (2020) Valence and patterning of aromatic residues determine the phase behavior of prion-like domains. Science 367:694–699

Mathieu C, Pappu RV, Taylor JP (2020) Beyond aggregation: pathological phase transitions in neurodegenerative disease. Science 370:56–60

Monahan Z, Ryan VH, Janke AM, Burke KA, Rhoads SN, Zerze GH, O'Meally R, Dignon GL, Conicella AE, Zheng W et al (2017) Phosphorylation of the FUS low-complexity domain disrupts phase separation, aggregation, and toxicity. EMBO J 36:2951–2967

Murakami T, Qamar S, Lin JQ, Schierle GSK, Rees E, Miyashita A, Costa AR, Dodd RB, Chan FTS, Michel CH et al (2015) ALS/FTD mutation-induced phase transition of FUS liquid droplets and reversible hydrogels into irreversible hydrogels impairs RNP granule function. Neuron 88:678–690

Murray DT, Kato M, Lin Y, Thurber KR, Hung I, McKnight SL, Tycko R (2017) Structure of FUS protein fibrils and its relevance to self-assembly and phase separation of low-complexity domains. Cell 171(615–627):e616

Murray DT, Zhou X, Kato M, Xiang S, Tycko R, McKnight SL (2018) Structural characterization of the D290V mutation site in hnRNPA2 low-complexity-domain polymers. Proc Natl Acad Sci U S A 115:E9782–E9791

Murthy AC, Dignon GL, Kan Y, Zerze GH, Parekh SH, Mittal J, Fawzi NL (2019) Molecular interactions underlying liquid-liquid phase separation of the FUS low-complexity domain. Nat Struct Mol Biol 26:637–648

Panagopoulos I, Aman P, Fioretos T, Hoglund M, Johansson B, Mandahl N, Heim S, Behrendtz M, Mitelman F (1994) Fusion of the FUS gene with ERG in acute myeloid leukemia with t(16;21) (p11;q22). Genes Chromosomes Cancer 11:256–262

Patel A, Lee HO, Jawerth L, Maharana S, Jahnel M, Hein MY, Stoynov S, Mahamid J, Saha S, Franzmann TM et al (2015) A liquid-to-solid phase transition of the ALS protein FUS accelerated by disease mutation. Cell 162:1066–1077

Polymenidou M, Lagier-Tourenne C, Hutt KR, Bennett CF, Cleveland DW, Yeo GW (2012) Misregulated RNA processing in amyotrophic lateral sclerosis. Brain Res 1462:3–15

Qamar S, Wang G, Randle SJ, Ruggeri FS, Varela JA, Lin JQ, Phillips EC, Miyashita A, Williams D, Strohl F et al (2018) FUS phase separation is modulated by a molecular chaperone and methylation of arginine cation-pi interactions. Cell 173(720–734):e715

Rado-Trilla N, Alba M (2012) Dissecting the role of low-complexity regions in the evolution of vertebrate proteins. BMC Evol Biol 12:155

Rhoads SN, Monahan ZT, Yee DS, Leung AY, Newcombe CG, O'Meally RN, Cole RN, Shewmaker FP (2018) The prionlike domain of FUS is multiphosphorylated following DNA damage without altering nuclear localization. Mol Biol Cell 29:1786–1797

Schwartz JC, Ebmeier CC, Podell ER, Heimiller J, Taatjes DJ, Cech TR (2012) FUS binds the CTD of RNA polymerase II and regulates its phosphorylation at Ser2. Genes Dev 26:2690–2695

Schwartz JC, Cech TR, Parker RR (2015) Biochemical properties and biological functions of FET proteins. Annu Rev Biochem 84:355–379

Strambio-De-Castillia C, Niepel M, Rout MP (2010) The nuclear pore complex: bridging nuclear transport and gene regulation. Nat Rev Mol Cell Biol 11:490–501

Sun Y, Zhang S, Hu J, Tao Y, Xia W, Gu J, Li Y, Cao Q, Li D, Liu C (2022) Molecular structure of an amyloid fibril formed by FUS low-complexity domain. iScience 25:103701

Toll-Riera M, Rado-Trilla N, Martys F, Alba MM (2012) Role of low-complexity sequences in the formation of novel protein coding sequences. Mol Biol Evol 29:883–886

Triezenberg SJ, Kingsbury RC, McKnight SL (1988) Functional dissection of VP16, the trans-activator of herpes simplex virus immediate early gene expression. Genes Dev 2:718–729

Wang WY, Pan L, Su SC, Quinn EJ, Sasaki M, Jimenez JC, Mackenzie IR, Huang EJ, Tsai LH (2013) Interaction of FUS and HDAC1 regulates DNA damage response and repair in neurons. Nat Neurosci 16:1383–1391

Wang X, Schwartz JC, Cech TR (2015) Nucleic acid-binding specificity of human FUS protein. Nucleic Acids Res 43:7535–7543

Wang J, Choi JM, Holehouse AS, Lee HO, Zhang X, Jahnel M, Maharana S, Lemaitre R, Pozniakovsky A, Drechsel D et al (2018) A molecular grammar governing the driving forces for phase separation of prion-like RNA binding proteins. Cell 174(688–699):e616

Xiang S, Kato M, Wu LC, Lin Y, Ding M, Zhang Y, Yu Y, McKnight SL (2015) The LC domain of hnRNPA2 adopts similar conformations in hydrogel polymers, liquid-like droplets, and nuclei. Cell 163:829–839

Yang L, Gal J, Chen J, Zhu H (2014) Self-assembled FUS binds active chromatin and regulates gene transcription. Proc Natl Acad Sci U S A 111:17809–17814

Chapter 7
Winding and Tangling. An Initial Phase of Membrane-Less Organelle Formation

Hiroshi Maita and Shinichi Nakagawa

Abstract The formation of phase-separated condensates by many biomolecules is now widely accepted as the basic principle of membrane-less organelle (MLO) formation in the cell. However, various subcompartments form independently yet in parallel, which raises the question of how similar molecules can remain separated in different compartments but also join the same compartment when required. In this chapter, the assembly mechanism of the Cajal body, a classical but mysterious nuclear substructure, is discussed as an example. A characteristic pattern of charged patches in the intrinsically disordered regions (IDRs) of Cajal body components is highlighted as a key element explaining the behavior of biomolecule complexes involved in Cajal body formation.

Keywords Arginine dimethylation · Cajal body · Coilin · Intrinsically disordered region · Nopp140 · Perinucleolar cap · scaRNP · SMN · snRNP · Tudor domain

7.1 Introduction to Cajal Body

This chapter starts with a brief history of the Cajal body and introduces the functional versatility of the Cajal body, which involves the recruitment of multiple RNA-protein complexes.

H. Maita (✉) · S. Nakagawa
Graduate School of Life Science, Hokkaido University, Sapporo, Japan

Faculty of Pharmaceutical Sciences, Hokkaido University, Sapporo, Japan
e-mail: maimai@pharm.hokudai.ac.jp

R. Kurokawa (ed.), *Phase Separation in Living Cells*,
https://doi.org/10.1007/978-981-99-4886-4_7

7.1.1 *Nucleolar Accessory Body, Coiled Body, and Cajal Body*

Cajal bodies are subnuclear structures reported by Ramon y Cajal (Cajal 1903) and initially described as "accessory bodies," that appear as small, dot-like structures, often located close to the nucleoli, upon silver staining. In the 1960s, electron microscopy studies of cellular structures rediscovered these structures and named them "coiled bodies" because of their notable structural feature of appearing to be composed of many coiled threads (Lafontaine 1965). A quarter century after these coiled bodies were observed by electron microscopy, studies investigating the reactivity of antibodies in sera from patients with autoimmune disease found that several sera stained small, dot-like structures in the nucleus (Andrade et al. 1991). An antibody raised against the recombinant protein synthesized from the coding sequence of the common antigenic 80 kDa protein also stained these nuclear dot-like structures. Furthermore, immunoelectron microscopy using the antibody showed that this subnuclear structure is full of coiled threads, which indicated that the antigenic protein was localized in the coiled body in mammalian cells (Raška et al. 1991). Thus, the antigenic protein was named "p80 coilin" and is now referred to simply as "coilin." The diagnostic value of the presence of autoantibodies against coilin in autoimmune diseases is unclear because this feature was not correlated with any single disease. However, coilin is now recognized as the molecular marker of coiled bodies. In 1999 coiled bodies were renamed "Cajal body" in memory of the first discovery of the structure by Ramon y Cajal (Gall 2000), so this structure will be referred to as the Cajal body hereafter in this review.

7.1.2 *Functions of Cajal Body*

7.1.2.1 snRNP Biogenesis

Studies focusing on other sera from patients with autoimmune disease, specifically lupus erythematosus (LE), provided the first glimpses of the potential function of Cajal bodies (Lerner and Steitz 1979). These sera reacted with the Sm protein, which binds to a specific motif in the 3′ end of several small nuclear RNAs (snRNAs) to form a snRNA-protein complex called snRNP. At the time, snRNPs had been recognized as comprising the spliceosome, a macromolecular machine that catalyzes RNA splicing (Maniatis and Reed 1987; Steitz et al. 1988). Antibodies against snRNP and probes for snRNAs label nuclear speckles as well as Cajal bodies, suggesting that Cajal body may have a function related to snRNP storage or production (Carmo-Fonseca et al. 1991). Studies of spherical organelles that assemble in *Xenopus* oocyte extracts support this hypothesis (Bauer et al. 1994). All snRNAs except for U6, which is transcribed by RNA polymerase III, have to be exported from the nucleus to the cytoplasm to be hypermethylated on the 5′ cap and

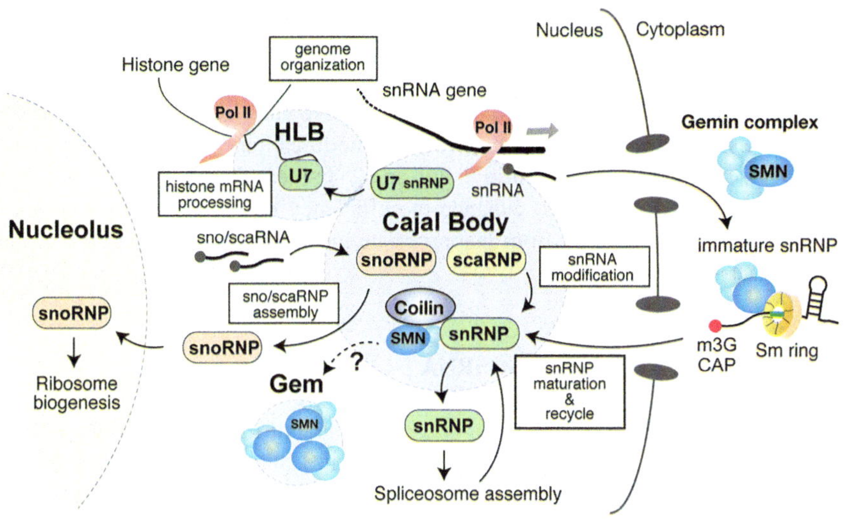

Fig. 7.1 Multiple roles of the Cajal body. The Cajal body brings together nascent short RNAs and their binding proteins. After assembly of the RNA-protein complexes (RNPs), scaRNPs and the U7 snRNP stay in the Cajal body to carry out their functions, while the snRNPs and snoRNPs relocate to other subcompartments, such as nuclear speckles and the nucleolus. All snRNPs, except for U6, mature via a unique pathway involving a cytoplasmic phase, as shown on the right side of the figure. The Gemin complex comprising SMN and other subunits functions as a chaperone for snRNPs

bind Sm protein complex on the 3′ end. After these maturation steps, the snRNA returns to the nucleus and is incorporated into the Cajal body (Fig. 7.1). Therefore, the Cajal body has been suggested to function as a "melting pot" that mediates the final step of snRNP assembly (Stanek and Neugebauer 2006; Staněk 2017). This hypothesis is supported by experimental evidence showing significant accumulation of snRNP intermediates within Cajal body in cells lacking snRNP maturation factors.

7.1.2.2 Biogenesis of Other RNPs; snoRNP, scaRNP, and Telomerase RNP

In addition to snRNPs, small non-coding guide RNAs associated with RNA modification enzymes have also been found in Cajal bodies. Large number of small nucleolar RNA (snoRNA) have been identified (Kiss 2002), and many of them are thought to pass through the Cajal body as they mature (Machyna et al. 2014). snoRNAs are divided in two classes—those that contain the H/ACA box motif and those that contain the C/D box motif—and members of both classes have been identified as localizing to the Cajal body; these snoRNAs are therefore referred to as small Cajal body-specific RNAs (scaRNAs) (Darzacq et al. 2002). hTR, the telomerase RNA, is also classified as H/ACA box-containing scaRNA because it localizes to the Cajal body. scaRNA Cajal body localization requires binding of TCAB1/WRAP53 to specific sequences in scaRNA (Richard et al. 2003). The presence of scaRNP, which catalyze RNA modifications, in the Cajal body correlates well with the fact that spliceosomal snRNAs require various RNA modifications for precise assembly and function. Over 40 RNA modifications are known to be made to snRNAs, even only for pseudouridylation and 2′-O-methylation (Meier 2017). Thus, the Cajal body is thought to enhance the efficiency of RNA modification and to protect other nuclear RNA from inappropriate modifications by spatial partitioning (Jády et al. 2003).

7.1.2.3 Processing of Histone mRNA

Another role of the Cajal body is atypical processing of intron-less histone mRNAs. Earlier studies showed that Cajal bodies often colocalize with histone genes (Frey and Matera 1995), and later the histone locus body (HLB), a nuclear body that forms on histone genes, was found to transiently associate with Cajal body (Imada et al. 2021). NPAT and FLASH, identified as Cajal body components, have been thought to mediate the interaction between Cajal bodies and HLBs. Recent reports indicate that this transient association between two independent subnuclear structures is required for transcriptional pausing and 3′ end processing of histone mRNAs. The role of the Cajal body in this process is to provide U7 snRNP, which is stored in Cajal bodies (Suzuki et al. 2022). U7 snRNP is a unique snRNP variant that mediates 3′ end processing of histone mRNA, but accumulates in Cajal bodies like other snRNPs. U2 snRNP also mediates transcriptional elongation of histone mRNA. These findings show that the Cajal body associates with the HLB to supply essential factors for producing correct histone mRNA.

7.1.2.4 Cajal Body as a Hub for Genome Organization

Cajal Body Is Assembled on an Actively Transcribed snRNA Locus

Cajal bodies associate with snRNA genes, as well as histone genes, and this association requires nascent transcripts from the locus (Frey and Matera 1995; Frey et al. 1999). In agreement with this, Cajal body efficiently absorbs exogenous small non-coding RNA. Moreover, artificial expressing snRNA from a genomic locus resulted in de novo formation of a Cajal body (Shevtsov and Dundr 2011), which is consistent with the role of nascent snRNA in establishing the spatial positioning of Cajal bodies. Artificial tethering of Cajal body components to the genome also leads to de novo Cajal body formation (Kaiser et al. 2008); however, it will commonly occur for other MLOs such as PML body (Wang et al. 2018a). Cajal bodies are unique in that they require small non-coding RNAs in order to form.

Cajal Body Assembly Organizes a Higher Order Structure of the Genome

snRNA-targeted Cajal body deposition on genomic loci results in higher order organization of the genome that contributes to transcriptional regulation. Indeed, disruption of the Cajal bodies upon coilin knockdown causes decreased expression of sn/snoRNA and histone genes that are normally clustered in the Cajal body, which results in an increase in noisy splicing (Wang et al. 2016). On the other hand, de novo Cajal body production by the CRISPR-Go system caused transcriptional repression of associated genes (Wang et al. 2018a). The genomic organization would likely be evolutionarily optimized. Allopatric de novo formation of the Cajal body might be harmful to gene expression, especially in case of mRNA-coding genes.

7.1.3 Physiological Relevance of Cajal Body

As described above, research studies on molecular functions of the Cajal body made significant advances. However, studies about the physiological roles of Cajal body and Coilin were relatively few. Although both mice and flies lacking functional coilin are viable (Tucker et al. 2001; Liu et al. 2009; Walker et al. 2009), coilin knockdown in early zebrafish embryos resulted in defective snRNP biogenesis, which aborted further development (Strzelecka et al. 2010). This lethal outcome was rescued by supplementing with snRNPs, which supports a role for the Cajal body in snRNP biogenesis.

However, a coilin knockout study showed that post-transcriptional snRNA modification was unchanged in coilin-deficient flies (Deryusheva and Gall 2009). It is thought that a sufficient level of RNA modification occurs in the nucleoplasm in the absence of Cajal bodies, and this is in consistent with the observation that there was no alteration in RNA modification in mouse cells lacking full-length coilin (Jády et al. 2003). Nonetheless, Coilin knockout mice showed slight abnormalities in that the number of offspring decreased by about a half compared with wild-type mice between E13.5 and birth (Walker et al. 2009). The reason for this incomplete penetrance of lethality is unclear based on current knowledge. Thus, the physiological roles of the Cajal body and why coilin is evolutionary conserved among metazoans remain largely unknown.

7.2 Multiple Faces of Coilin and the Cajal Body

Corresponding to the diverse roles of Cajal bodies, coilin localizes to cellular compartments other than the Cajal body, and this varied localization is regulated by post-translational modifications.

7.2.1 The Close Relationship of Coilin with the Nucleolus

Important aspect of the Cajal body is that it has a close relationship with the nucleolus. Cajal body integration into the nucleolus has been shown to occur in breast cancer cell lines, and dormouse hepatocytes and brown adipose tissue similarly exhibit intranucleolar Cajal bodies under physiological conditions, but only during hibernation (Malatesta et al. 1994). Because the disturbance to normal nucleolar structure caused by the formation of this type of aberrant nucleolar compartment did not interrupt transcription by RNA polymerase I, it appears that the nucleolus and the Cajal body are open structures that can accommodate a range of components. On the other hand, these two structures sometimes associate each other, without mixing components, in rat neurons.

In addition, treating cells with small-molecule RNA polymerase inhibitors such as actinomycin D and DRB causes rapid (<1 h) recruitment of coilin to the perinucleolar compartment (PNC), which suggests that the blockage of ribosomal RNA or snRNA production induces coilin relocation to the PNC. However, assembly of PNC is not always associated with defects in RNA-related processes, as a similar structure was also observed in NGF-treated PC12 cells.

7.2.2 Cajal Body and Gem

SMN is a component of the Gemin complex that promotes snRNP maturation in the cytoplasm and escorts premature snRNPs to the nucleus, as shown in Fig. 7.1. SMN is thought to be a key factor in Cajal body assembly because it links snRNP to coilin and other Cajal body components. The ability of SMN to form oligomers is essential to its function. Furthermore, the tudor domain of SMN recognizes dimethylated arginine-glycine (RG) repeats found in coilin and some Sm and Lsm proteins (snRNP components). Recently, this interaction between the tudor domain and dimethylated RG repeat has been shown to trigger the assembly of phase-separated condensates(Courchaine et al. 2021), which would indicate a role for SMN in initiation of Cajal body assembly. It is hypothesized that the nuclear imported SMN complex, including Gemin and snRNP, multimerize to form a Cajal body seed that may recruit coilin and other Cajal body components (Fig. 7.2b). However, depending on culturing conditions, coilin and SMN sometimes localize to separated structures; the nuclear body involving SMN but not coilin is called Gems (Gemini of Cajal bodies). Although the cellular function of Gems is still unclear, it has been proposed that Gems are a place to keep the Gemin complex for recycling because snRNP is absent in Gems (Morris 2008).

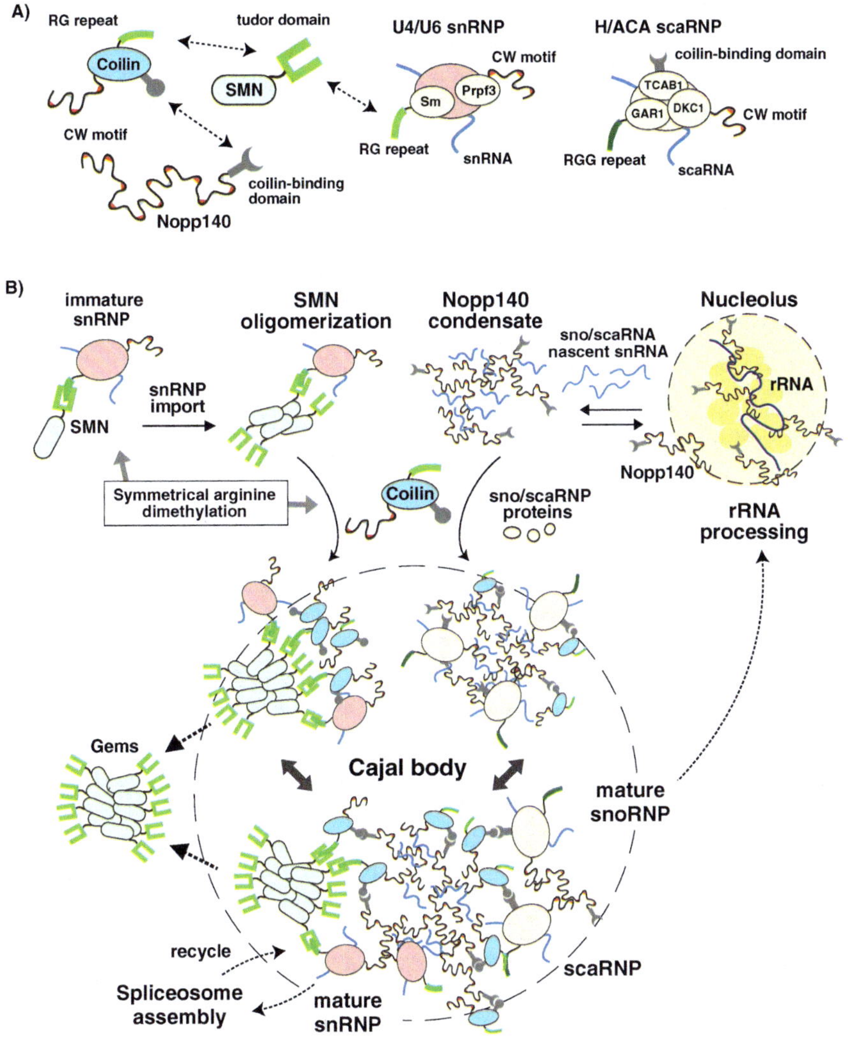

Fig. 7.2 (**a**) Schematic illustration of Cajal body components. (**b**) A dual-pathway model of Cajal body assembly (see Sect. 7.3.6)

7.2.3 Residual Bodies in the Absence of Coilin

In Coilin knockout mice or depleted cells, residual Cajal bodies have been observed. Although coilin is considered to be a necessary component of the Cajal body, molecules such as fibrillarin, Nopp140, and snoRNAs that normally localize to both the Cajal body and the nucleolus formed nuclear bodies in the absence of coilin; and SMN and other snRNP components also form residual Cajal bodies in the absence of coilin (Tucker et al. 2001). However, those two types of residual bodies never overlap, even though both sets of proteins frequently colocalize within the same Cajal body in the presence of coilin. Similar results were obtained when coilin was depleted by RNAi. In contrast, knockdown of SMN resulted in the disappearance of snRNP-containing Cajal bodies, although coilin still colocalized with fibrillarin and scaRNA at numerous small nuclear foci (Girard et al. 2006). These observations suggest that SMN-mediated nuclear snRNP import is required for the formation of snRNP-containing Cajal bodies, but not for sno/scaRNP-containing Cajal bodies. Again, these findings imply that Cajal body composition and assembly are not uniform, which is in agreement with a study showing de novo assembly of Cajal bodies (Kaiser et al. 2008).

7.2.4 Phosphorylation Regulates Coilin Localization

The first notice of the role of phosphorylation on coilin function was that accumulation of coilin and snRNP within nucleolus as an electron-dense novel structure upon inhibition of dephosphorylation by okadaic acid, and phospho-mimic coilin S202D mutant protein also formed similar intranucleolar structures (Lyon et al. 1997). In addition, the CDK2-Cyclin E complex localizes to the Cajal body in a cell cycle-dependent manner and phosphorylates coilin at Ser184. Localization of coilin S184A to the nucleolus may reflect a cell cycle-dependent Cajal body regulatory mechanism (Hebert and Matera 2000; Liu et al. 2000). VRK1, a kinase that phosphorylates many nuclear factors related to chromatin organization (Campillo-Marcos et al. 2021), also phosphorylates coilin at Ser184 (Sanz-García et al. 2011). Phosphorylation of Ser184 by VRK1 prevents degradation by the ubiquitin-proteasome system, which results in the stabilization of coilin. Importantly, an S184A phospho-defective mutant exhibited atypical localization, while an S184D phospho-mimic mutant showed no obvious change in localization (Hebert and Matera 2000), suggesting that cellular S184 residue is mostly phosphorylated. In addition, although coilin directly binds Sm proteins and SMN, these interactions are modulated by phosphorylation. When JNK, a stress-responsive kinase, was activated by the translation elongation inhibitor cycloheximide, coilin relocated from Cajal bodies to numerous smaller foci within a few hours. Although it is unclear if the rapid destruction of Cajal body provides any advantage that allows cells to adapt to these stressors, it is evident that the Cajal body is strictly regulated by cell signaling pathways.

7.2.5 Roles of Arginine (di)Methylation

SMN directly binds to the RG repeat in coilin via its tudor domain, and the symmetrically dimethylated arginine (sDMA) levels in these RG repeats are considered to be another post-translational modification that may regulate Cajal body formation in addition to phosphorylation (Hebert et al. 2002; Boisvert et al. 2002).

SMN also binds Sm proteins as a Gemin complex to assist deposition of the ring-shaped heptameric Sm protein complex on snRNA, and this interaction depends on recognition of the dimethylated RG repeat by SMN tudor domain (reviewed in (Paushkin et al. 2002; Meister et al. 2002).

When cells were treated with Methylthioadenosine, an inhibitor of dimethylation enzyme PRMT5, coilin dimethylation levels decreased, and coilin relocated from Cajal bodies to the dense fibrillar components (DFC) in the nucleolus. In agreement with this, knockdown of PRMT5 also induced the DFC localization of coilin. These results indicate that dimethylation of the coilin RG repeat is required for escape from the nucleolus via interaction with SMN.

Thus, it is hypothesized that sDMA modification has two roles in snRNP biogenesis: first, dimethylation of the RG repeats in Sm proteins initiates assembly of the Gemin complex by enhancing recognition of Sm proteins by the SMN tudor domain; and second, the dimethylated RG repeat in coilin is also bound by SMN, and this interaction could initiate phase separation, leading to the formation of the Cajal body (Courchaine et al. 2021). In addition, prolonged inhibition of PRMT5 caused the complete separation of coilin-labeled Cajal body and Gems. Because inhibition of dimethylation could lead to decreased import of newly assembled snRNPs, this result further supports that Gems are the storage site for the Gemin complex until recycling.

7.3 Possible Mechanisms of Cajal Body Formation

Observation of the dynamics of fluorescently labeled coilin revealed that coilin is recruited to the Cajal body but also released after an unexpectedly short period of time. Early studies from the Gall lab suggested that spherical organelles/Cajal bodies that form in *Xenopus* oocyte extract can segregate and fuse, which is a typical droplet-like behavior exhibited by phase-separated condensates. Accumulating evidence indicates that the physical basis for the assembly of these submicron-scale structures is liquid–liquid phase separation. Although there is still some uncertainty regarding whether Cajal body formation is driven purely by liquid–liquid phase separation, this theory would help explain how coilin transits among the Cajal body, nucleolus, and PNC.

7.3.1 Nopp140, an Essential Component for Canonical Cajal Body Assembly

Nopp140 is a protein that localizes to both the nucleolus and the Cajal body and is thought to assist with RNA polymerase I-mediated nucleolar transcription (Chen et al. 1999). Despite its role in nucleolar transcription, Nopp140 interacts with coilin and is essential for preserving the integrity of the Cajal body (Isaac et al. 1998;

Renvoisé et al. 2009; Bizarro et al. 2019, 2021; Courchaine et al. 2022). When Nopp140 was depleted from cells by CRISPR/Cas9, snoRNPs, and scaRNPs, but not snRNPs, were excluded from the Cajal body (Bizarro et al. 2019). Electron microscopy analysis of Cajal bodies lacking snoRNPs and scaRNPs in Nopp140-depleted cells showed smaller coiled threads, but the number of granules was not changed, which suggests that the coiled threads evident within the Cajal body are complexes containing scaRNPs and their target snRNPs (Bizarro et al. 2019). While snoRNP and scaRNP were depleted from the Cajal body in the absence of Nopp140, snRNPs were still present within the Cajal body; however, scaRNP depletion caused reduced snRNA modification in cells partially lacking Nopp140 (Bizarro et al. 2019).

It was recently shown that scaRNAs form scaRNA bodies in ES cells that lack obvious Cajal bodies, as identified by coilin labeling (Quinodoz et al. 2021), which suggests that numerous unidentified nuclear bodies can compensate for the absence of Cajal bodies.

7.3.2 Nopp140 Forms the Condensate

Notably, Nopp140 localizes to both the nucleolus and the Cajal body(Thomas Meier and Blobel 1992; Isaac et al. 1998); however, its function in these two closely related membrane-less organelles remains unclear. Nopp140 has a highly conserved motif in its C-terminus, and deletion of this motif caused a massive reduction in Nopp140 binding to coilin in Y2H liquid assay; furthermore, the mutant protein lacking the C-terminal domain only localized to the nucleolus in a fibrillar shape (Isaac et al. 1998). Thus, the conserved C-terminal domain is important for forming an isolated, subnuclear Cajal body that is independent of the nucleolus, and the role of this motif is likely conserved in yeast (Ikonomova et al. 1997). This may explain why a residual Cajal body containing Nopp140 formed in cells derived from coilin knock-out mice. Interestingly, it has been shown that Nopp140 forms condensates when dimer formation is forcibly induced using the opto-switch system, while coilin did not show such the activity (Courchaine et al. 2022); this is consistent with overexpressed Srp40p C-terminal domain forming punctate structures in yeast.

7.3.3 The Role of Interaction between Coilin and Nopp140 for Cajal Body Assembly

Mutant coilin proteins, in which conserved N-terminal residues predicted to be important for binding to Nopp140 are altered, show diffuse localization throughout the nucleus, which suggests that Cajal body formation and coilin localization to the Cajal body require coilin interaction with Nopp140. In addition, one study showed that the coilin N-terminal domain (Coil-NTD) formed large fibrils in the cytoplasm (Courchaine et al. 2022). The largest fibril formed by Coil-NTD was over 10 μm in length, and the fibrils did not stain with Congo red or thioflavin T, indicating that they are not amyloid fibers. The fibrils look like coalescent granules by STED imaging, suggesting that they may be derived from an RNA-bound multimerized coilin NTD complex. The coilin N-terminus binds RNA (Bellini and Gall 1998), and RNA-bound AtCoilin forms large multimer that is visible by atomic force microscopy (Makarov et al. 2013).

This fibrillar property was suppressed when Coil-NTD was restricted to the nucleus by adding an NLS (Courchaine et al. 2022). Moreover, when the mutations that caused reduced interaction with Nopp140 were introduced into the NLS-Coil-NTD, the mutated proteins formed fibrils in the nucleus. Thus, Nopp140 binding to the coilin NTD not only mediates coilin localization to the Cajal body, but also regulates coilin multimerization in a coordinated manner with RNA binding. Importantly, in the absence of Nopp140, full-length coilin localizes diffusely throughout the nucleus and never forms fibrils, suggesting that a region outside of the NTD suppresses fibril formation. Indeed, a deletion mutant of coilin composed of 1–291 aa residues with the NTD and most of the IDR formed a dense round structure sequestering Nopp140 (Bohmann et al. 1995).

7.3.3.1 Characteristic Features of the IDRs of some Cajal Body Components

Roles of IDR Charged Residues in Nuclear Body Assembly

As our understanding of membrane-less organelle assembly through phase separation increases, the question has arisen of how multiple types of MLOs can exist independently even though they are formed by similar components such as proteins and RNA. Currently, it is thought that these different MLOs are distinguished by differences in the primary amino acid sequences of their component proteins that mediate the weak multivalent interactions required for phase separation. An important feature of proteins that form phase-separated condensate is that they often contain IDRs, and in many cases the amino acid composition of these IDRs shows low complexity. Moreover, these regions often contain a characteristic pattern of residues with mixed charges. Electrostatic interactions via these charged domains contribute to assembly of phase-separated condensates by forming weak multivalent interactions in addition to other types of molecular interactions (Nott et al. 2015; Brangwynne et al. 2015; Pak et al. 2016; Wang et al. 2018b; Greig et al. 2020; Lyons et al. 2023). Notably, the roles of the charged residues differ among amino acid species. Arginine is often found in the IDRs of proteins that form nuclear speckles, and substitution of these arginine residues with lysine residues did not change the net charge but significantly altered the localization of the mutated protein (Greig et al. 2020). In addition, low complexity IDRs often show repeats of stretches of positively and negatively charged residues, and altering the length of the charged residue stretches also changes MLO localization in the nucleus, even if it does not change the net charge or the number of charged residues (Greig et al. 2020; Miyagi et al. 2022).

7.3.3.2 Characteristic Features of Cajal Body Component Proteins

Many Cajal body components contain low complexity IDRs, as seen in other nuclear body components, and these IDRs appear to be specifically enriched in lysine. Nopp140 contains the largest charged IDR among known Cajal body components. Over 90% of the Nopp140 is predicted to be disordered, and this region is composed of 13 repeats of stretches of positively and negatively charged residues (Fig. 7.3b). The coilin IDR also contains a lysine-rich positively charged stretch followed by a negatively charged stretch, which suggests that Cajal body component proteins contain a common pattern of charged residues. Plotting localized charges within the IDRs of Cajal body components using the R package "idpr" (McFadden and Yanowitz 2022) showed that, while the length of the IDR differed widely, many of these proteins exhibited a unique wave-like pattern of positively and negatively charged stretches within their IDRs (Figs. 7.3 and 7.4). Hereafter it will be referred to the charged wave (CW) motif. While the position of the CW motif differed in each protein, this motif was not present in any of over 60 other nuclear body components that were surveyed. Nopp140 contains 13 repeats of the wave-like pattern, and other proteins contained significantly fewer repeats of this pattern. The charge values at the peak position are around 0.5.

7.3.3.3 The Amplitude of the CW Motif Is Vital to Sort Proteins to the Specific Nuclear Body

A recent study investigated the effect of the length of the stretch of charged residues using model peptide sequences composed of aspartic acid (D) and arginine (R) (Miyagi et al. 2022). The study showed that a peptide comprising 50 repeats of DR was diffusedly localized in the nucleus and overlapped with SRSF; however, as the length of each repeat unit was extended to D4R4 (numbers indicate the length of the aspartic acid and arginine stretches), D8R8, and D16R16, the peptides localized to the nucleolus (schematically illustrated in Fig. 7.4). Surprisingly, D8R8 localized not only to the nucleolus but also to nuclear puncta that partially overlapped with SRSF1. This half-and-half staining of two different structures in the same nucleus closely resembles the localization of Nopp140 to both Cajal bodies and the nucleolus. Analysis of the DR peptide sequences using the idpr package showed a wave-like pattern of local charges very similar to that exhibited by Nopp140 and indicated that the average charge value for D8R8 is around 0.5 (Fig. 7.4a).

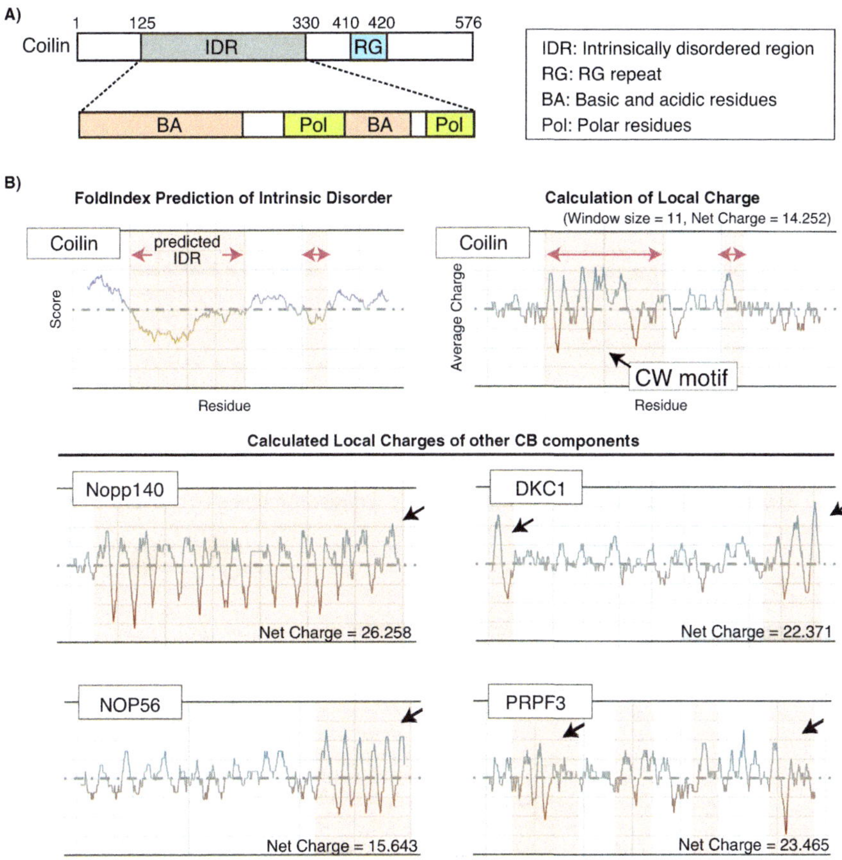

Fig. 7.3 Identification of common features among Cajal body proteins. (**a**) Coilin structure (residue numbers are denoted according to UniProt). (**b**) Average local charges and putative IDRs for coilin and four other Cajal body components, Nopp140, DKC1, NOP56 and PRPF3 were predicted using "idpr." These proteins all contain the charged wave (CW) motif (see Sect. 7.3.3.2)

7.3.4 Roles of RNA in Phase Separation of Proteins with Charged Patches

Electric repulsion is thought to disrupt the assembly of condensates mediated by charged IDRs, and it was recently suggested that neutralization of positive charges by interaction with RNA is important to avoid repulsion between charged IDRs (Miyagi et al. 2022; Toyama et al. 2022). Cajal body components often show a highly positive net charge, which suggests that RNA absorption is an essential function of these proteins. This is consistent with the observation that AtCoilin forms condensates only in the presence of RNA. However, given that longer stretches of positively charged residues favor nucleolar localization (Greig et al.

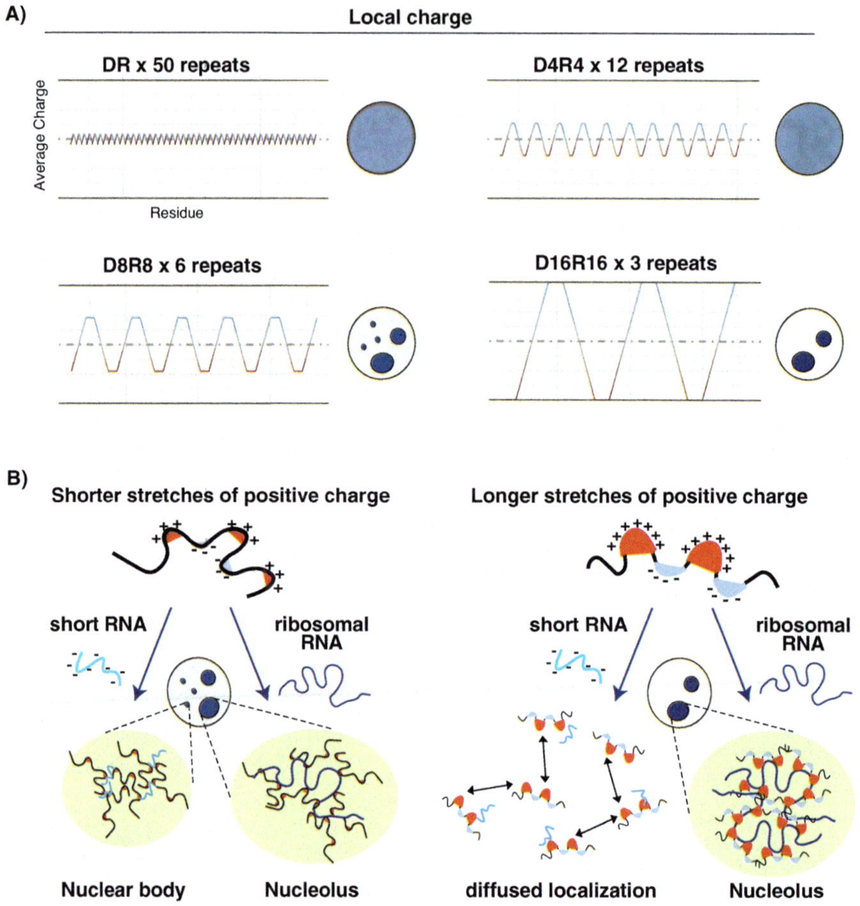

Fig. 7.4 (**a**) The pattern of average local charges and a localization schematic are presented for model peptides based on the report from Miyagi et al. (2022). (**b**) A possible explanation for how RNA cooperatively induces the formation of phase-separated condensates via neutralizing interactions with positive charges in the CW motif

2020; Miyagi et al. 2022), abundant ribosomal RNA could bind longer positively charged stretches. Conversely, shorter stretches of positively charged residues may accommodate long rRNAs but favor shorter RNAs to form condensates, which would explain why Cajal bodies can absorb various short nuclear RNAs (Fig. 7.4b). Abrogation of snRNA production by knockdown of INST4, a component of the integrator complex required for snRNA transcription, relocates coilin to the nucleolus, which suggests that ongoing production of short RNA indeed keeps coilin away from the nucleolus (Takata et al. 2012). Thus, the length of the stretch of positively and negatively charged residues may determine protein distribution, probably via arranging average local charge values, and this could explain how native proteins localize to the Cajal body (Fig. 7.4b).

7.3.5 Effects of Post-Translational Modifications on the Property of IDR in Coilin

The nonphosphorylatable S184A coilin mutant accumulates in the nucleolus. This change increases the local charge around the site, which flattens the wave-like pattern of charges in this region, thereby preventing the phosphorylation that allows coilin to localize to the nucleolus. Phosphorylation of S184 by VRK1 is important for coilin stability and could also prevent inappropriate nucleolar accumulation of coilin by reducing local charge.

Arginine methylation changes a protein's RNA/DNA-binding capacity and the efficiency of phase separation; however, the direction of change (that is, whether it increases or decreases these activities) varies among tested proteins and experimental systems (Chong et al. 2018); thus, it is difficult to integrate the effects of methylation into a simple model. However, because low methylation causes the accumulation of Cajal body components in the nucleolus, methylation would partly reduce interactions between RGG/RG repeats and ribosomal RNA, which is abundant in the nucleolus. In contrast, symmetrical dimethylation of RG repeat functions as a binding platform for the tudor domain of SMN (Côté and Richard 2005), which never localizes to the nucleolus. In the case of Cajal body assembly, arginine methylation could balance distribution of Cajal body components within and outside of the nucleolus. Sumoylation of coilin also regulates Cajal body formation (Sun et al. 2005; Navascues et al. 2008; Schulz et al. 2012), and reduced expression of the isopeptidase that catalyzes desumoylation leads to coilin relocalization to both small nuclear foci and the nucleolus(Schulz et al. 2012). Because sumoylation occurs on lysine residues, it is hypothesized that sumoylation modulates coilin localization by abrogating functionality of the charged stretch.

7.3.6 A Dual-Pathway Model of Cajal Body Assembly

Based on the evidence regarding the roles of IDRs in Cajal body assembly and observations about the general behavior of IDR-containing proteins, it seems likely that the CW motif participates in Cajal body assembly in the presence of short nuclear RNAs. Similar to the D8R8 peptide, Nopp140 could accommodate short nuclear RNAs and long nucleolar RNAs, although it would probably have better affinity for rRNAs due to the number of repeats in the CW motif. In contrast, short RNAs would be sufficient to neutralize the positive charge of coilin and other components with fewer repeats of the CW motif. Because Nopp140 can form phase-separated condensates by itself, a spontaneously-formed initial Nopp140 condensate could grow larger by responding to active transcription of short RNAs in the nucleus. Following the formation of a small Nopp140 condensate, scaRNP components and coilin would join the condensate because of abundant short RNAs in the compartment, leading to stable and robust interactions among them via specific interaction motifs. Formation of these robust interactions may contribute to keeping the condensate independent of the nucleolus.

In addition to the Nopp140-dependent pathway, SMN oligomerization is thought to be another pathway involved in Cajal body assembly. When SMN returns to the nucleus, escorting partially matured snRNPs, it multimerizes and recruits coilin. Because both coilin and SMN can bind the dimethylated RG repeats of Sm proteins, and SMN can also bind the RG repeat in coilin, coilin accumulation on the SMN oligomer would trigger not only formation of the phase-separated condensate but also the release of snRNPs from SMN. The released snRNPs could remain within the newly assembled coilin condensates via interactions with their CW motifs. This would lead to the formation of a condensate containing coilin and snRNPs, which could then join the Nopp140-containing Cajal body (Fig. 7.2).

7.4 Perspective

7.4.1 RNPs in the Cajal Body Are a Mosaic Complex of Proteins with Different Motifs

While many Cajal body proteins contain the wave-like pattern of local charges, there are many components without such a motif (Fig. 7.5). Some of these proteins contain different characteristic IDRs incorporating other motifs, such as RGG/RG repeat. For example, H/ACA snoRNPs include DKC1, which contains the CW motif, and GAR1, which contains an RGG motif. Similarly, box C/D snoRNPs include NOP56/58, which contains the CW motif, and fibrillarin, which contains the RGG motif. The RGG/RG repeats are expected to mediate interactions with different binding partners than the CW motif, because it binds G-quadruplexes that are abundant in rRNAs. This modular design of RNPs might function to ensure RNP maturation. snoRNPs are known to pass through the Cajal body to reach the nucleolus. Because the RNA binds subunits in a stepwise manner, each protein may mask or hinder the activity of the IDRs of other proteins to prevent multivalent interactions, which could lead to relocalization of the complex to other bodies that can accommodate the newly displayed IDR.

7.4.2 How scaRNPs Mediate Cajal Body Localization

DKC1 and NOP56/58 help incorporate RNPs into Cajal bodies through the CW motif carried by H/ACA RNPs and box C/D RNPs, respectively. However, snoRNPs, which have a similar protein composition to scaRNPs, never stay in the Cajal body. The difference between snoRNAs and scaRNAs is that scaRNAs contain a specific sequence that is bound by TCAB1/WRAP53, thereby keeping

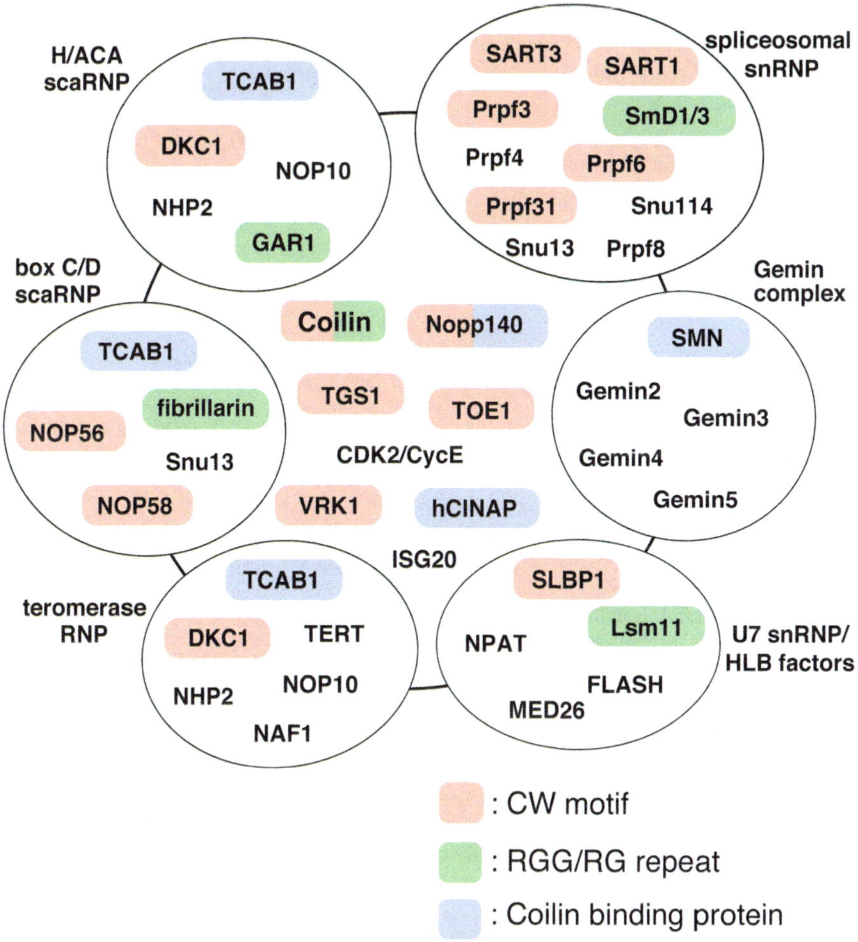

Fig. 7.5 Cajal body proteins can be divided in several groups based on the subcomplexes in which each protein is involved. Proteins are labeled with the localization-related motifs that they contain

them within the Cajal body. TCAB1 has no CW motif; however, it directly interacts with coilin and SMN (Richard et al. 2003; Tycowski et al. 2009; Marnef et al. 2014). This means that at least one structurally specific robust interaction—in this case between TCAB1 and coilin—helps control snoRNP movement. Similar to TCAB1, Nopp140 contains a conserved domain in its C-terminus that is necessary for interaction with coilin. This may explain why Treacle, a close homolog of Nopp140 that lacks the C-terminal coilin-interacting domain, has never been found within the Cajal body (Isaac et al. 2000). This modular RNP design would be evolutionarily advantageous for generating novel RNPs with regulated localization, and could also be useful for engineering proteins with targeted localization. This mechanism could also help explain how a particular nuclear body forms independently in the cell using similar constituents (such as proteins and RNA) as other, distinct nuclear bodies, to avoid crossing wires in the crowded environment of the cell without requiring additional energy consumption.

Acknowledgments We are grateful to the members of our laboratory, especially Ryota Yokoyama, Saki Ohazama, Miyu Ichii, and Hikaru Toya, for their helpful discussions and comments on the manuscript. This study was supported by JSPS KAKENHI Grant Numbers 20 K07028, 20H04687, 16H06279 (PAGS) to HM and KAKENHI Grant-in-Aid for Transformative Research Areas 21H05274 to SN.

References

Andrade LE, Chan EK, Raska I, Peebles CL, Roos G, Tan EM (1991) Human autoantibody to a novel protein of the nuclear coiled body: immunological characterization and cDNA cloning of p80-coilin. J Exp Med 173:1407–1419

Bauer DW, Murphy C, Wu Z, Wu CH, Gall JG (1994) In vitro assembly of coiled bodies in Xenopus egg extract. Mol Biol Cell 5:633–644

Bellini M, Gall JG (1998) Coilin can form a complex with the U7 small nuclear ribonucleoprotein. Mol Biol Cell 9:2987–3001

Bizarro J, Bhardwaj A, Smith S, Meier UT (2019) Nopp140-mediated concentration of telomerase in Cajal bodies regulates telomere length. Mol Biol Cell 30:3136–3150

Bizarro J, Deryusheva S, Wacheul L, Gupta V, Ernst FGM, Lafontaine DLJ, Gall JG, Thomas Meier U (2021) Nopp140-chaperoned 2′-O-methylation of small nuclear RNAs in Cajal bodies ensures splicing fidelity. Genes Dev 35:1123–1141

Bohmann K, Ferreira JA, Lamond AI (1995) Mutational analysis of p80 coilin indicates a functional interaction between coiled bodies and the nucleolus. J Cell Biol 131:817–831

Boisvert F-M, Cote J, Boulanger M-C, Cleroux P, Bachand F, Autexier C, Richard S (2002) Symmetrical dimethylarginine methylation is required for the localization of SMN in Cajal bodies and pre-mRNA splicing. J Cell Biol 159:957–969

Brangwynne CP, Tompa P, Pappu RV (2015) Polymer physics of intracellular phase transitions. Nat Phys 11:899–904

Cajal (1903) Un sencillo metodo de coloracion seletiva del reticulo protoplasmatico y sus efectos en los diversos organos nerviosos de vertebrados e invertebrados. Trab Lab Invest Biol (Madrid) 1: 63

Campillo-Marcos I, García-González R, Navarro-Carrasco E, Lazo PA (2021) The human VRK1 chromatin kinase in cancer biology. Cancer Lett 503:117–128

Carmo-Fonseca M, Tollervey D, Pepperkok R, Barabino SM, Merdes A, Brunner C, Zamore PD, Green MR, Hurt E, Lamond AI (1991) Mammalian nuclei contain foci which are highly enriched in components of the pre-mRNA splicing machinery. EMBO J 10:195–206

Chen HK, Pai CY, Huang JY, Yeh NH (1999) Human Nopp140, which interacts with RNA polymerase I: implications for rRNA gene transcription and nucleolar structural organization. Mol Cell Biol 19:8536–8546

Chong PA, Vernon RM, Forman-Kay JD (2018) RGG/RG motif regions in RNA binding and phase separation. J Mol Biol 430:4650–4665

Côté J, Richard S (2005) Tudor domains bind symmetrical dimethylated arginines. J Biol Chem 280:28476–28483

Courchaine EM, Barentine AES, Straube K, Lee D-R, Bewersdorf J, Neugebauer KM (2021) DMA-tudor interaction modules control the specificity of in vivo condensates. Cell 184:3612–3625.e17

Courchaine E, Gelles-Watnick S, Machyna M, Straube K, Sauyet S, Enright J, Neugebauer KM (2022) The coilin N-terminus mediates multivalent interactions between coilin and Nopp140 to form and maintain Cajal bodies. Nat Commun 13:6005

Darzacq X, Jády BE, Verheggen C, Kiss AM, Bertrand E, Kiss T (2002) Cajal body-specific small nuclear RNAs: a novel class of 2'-O-methylation and pseudouridylation guide RNAs. EMBO J 21:2746–2756

Deryusheva S, Gall JG (2009) Small Cajal body–specific RNAs of drosophila function in the absence of Cajal bodies. Mol Biol Cell 20:5250–5259

Frey MR, Matera AG (1995) Coiled bodies contain U7 small nuclear RNA and associate with specific DNA sequences in interphase human cells. Proc Natl Acad Sci U S A 92:5915–5919

Frey MR, Bailey AD, Weiner AM, Matera AG (1999) Association of snRNA genes with coiled bodies is mediated by nascent snRNA transcripts. Curr Biol 9:126–135

Gall JG (2000) Cajal bodies: the first 100 years. Annu Rev Cell Dev Biol 16:273–300

Girard C, Neel H, Bertrand E, Bordonné R (2006) Depletion of SMN by RNA interference in HeLa cells induces defects in Cajal body formation. Nucleic Acids Res 34:2925–2932

Greig JA, Nguyen TA, Lee M, Holehouse AS, Posey AE, Pappu RV, Jedd G (2020) Arginine-enriched mixed-charge domains provide cohesion for nuclear speckle condensation. Mol Cell 77:1237–1250.e4

Hebert MD, Matera AG (2000) Self-association of coilin reveals a common theme in nuclear body localization. Mol Biol Cell 11:4159

Hebert MD, Shpargel KB, Ospina JK, Tucker KE, Matera AG (2002) Coilin methylation regulates nuclear body formation. Dev Cell 3:329–337

Ikonomova R, Sommer T, Képès F (1997) The Srp40 protein plays a dose-sensitive role in preribosome assembly or transport and depends on its carboxy-terminal domain for proper localization to the yeast nucleoskeleton. DNA Cell Biol 16:1161–1173

Imada T, Shimi T, Kaiho A, Saeki Y, Kimura H (2021) RNA polymerase II condensate formation and association with Cajal and histone locus bodies in living human cells. Genes Cells 26:298–312

Isaac C, Yang Y, Meier UT (1998) Nopp140 functions as a molecular link between the nucleolus and the coiled bodies. J Cell Biol 142:319–329

Isaac C, Marsh KL, Paznekas WA, Dixon J, Dixon MJ, Jabs EW, Meier UT (2000) Characterization of the nucleolar gene product, treacle, in Treacher Collins syndrome. Mol Biol Cell 11:3061–3071

Jády BE, Darzacq X, Tucker KE, Matera AG, Bertrand E, Kiss T (2003) Modification of Sm small nuclear RNAs occurs in the nucleoplasmic Cajal body following import from the cytoplasm. EMBO J 22:1878–1888

Kaiser TE, Intine RV, Dundr M (2008) De novo formation of a subnuclear body. Science 322: 1713–1717

Kiss T (2002) Small nucleolar RNAs: an abundant group of noncoding RNAs with diverse cellular functions. Cell 109:145–148

Lafontaine JG (1965) A light and electron microscope study of small, spherical nuclear bodies in meristematic cells of Allium cepa, Vicia faba, and Raphanus sativus. J Cell Biol 26:1–17

Lerner MR, Steitz JA (1979) Antibodies to small nuclear RNAs complexed with proteins are produced by patients with systemic lupus erythematosus. Proc Natl Acad Sci U S A 76:5495–5499

Liu J, Hebert MD, Ye Y, Templeton DJ, Kung H, Matera AG (2000) Cell cycle-dependent localization of the CDK2-cyclin E complex in Cajal (coiled) bodies. J Cell Sci 113(Pt 9): 1543–1552

Liu J-L, Wu Z, Nizami Z, Deryusheva S, Rajendra TK, Beumer KJ, Gao H, Matera AG, Carroll D, Gall JG (2009) Coilin is essential for Cajal body organization in Drosophila melanogaster. Mol Biol Cell 20:1661–1670

Lyon CE, Bohmann K, Sleeman J, Lamond AI (1997) Inhibition of protein dephosphorylation results in the accumulation of splicing snRNPs and coiled bodies within the nucleolus. Exp Cell Res 230:84–93

Lyons H, Veettil RT, Pradhan P, Fornero C, De La Cruz N, Ito K, Eppert M, Roeder RG, Sabari BR (2023) Functional partitioning of transcriptional regulators by patterned charge blocks. Cell 186: 327–345.e28

Machyna M, Kehr S, Straube K, Kappei D, Buchholz F, Butter F, Ule J, Hertel J, Stadler PF, Neugebauer KM (2014) The coilin interactome identifies hundreds of small noncoding RNAs that traffic through Cajal bodies. Mol Cell 56:389–399

Makarov V, Rakitina D, Protopopova A, Yaminsky I, Arutiunian A, Love AJ, Taliansky M, Kalinina N (2013) Plant coilin: structural characteristics and RNA-binding properties. PLoS One 8:e53571

Malatesta M, Zancanaro C, Martin TE, Chan EK, Amalric F, Lührmann R, Vogel P, Fakan S (1994) Is the coiled body involved in nucleolar functions? Exp Cell Res 211:415–419

Maniatis T, Reed R (1987) The role of small nuclear ribonucleoprotein particles in pre-mRNA splicing. Nature 325:673–678

Marnef A, Richard P, Pinzón N, Kiss T (2014) Targeting vertebrate intron-encoded box C/D 2'-O-methylation guide RNAs into the Cajal body. Nucleic Acids Res 42:6616–6629

McFadden WM, Yanowitz JL (2022) Idpr: a package for profiling and analyzing intrinsically disordered proteins in R. PLoS One 17:e0266929

Meier UT (2017) RNA modification in Cajal bodies. RNA Biol 14:693–700

Meister G, Eggert C, Fischer U (2002) SMN-mediated assembly of RNPs: a complex story. Trends Cell Biol 12:472–478

Miyagi T, Yamazaki R, Ueda K, Narumi S, Hayamizu Y, Uji-I H, Kuroda M, Kanekura K (2022) The patterning and proportion of charged residues in the arginine-rich mixed-charge domain determine the membrane-less organelle targeted by the protein. Int J Mol Sci 23:7658. https://doi.org/10.3390/ijms23147658

Morris GE (2008) The Cajal body. Biochim Biophys Acta 1783:2108–2115

Navascues J, Bengoechea R, Tapia O, Casafont I, Berciano MT, Lafarga M (2008) SUMO-1 transiently localizes to Cajal bodies in mammalian neurons. J Struct Biol 163:137–146

Nott TJ, Petsalaki E, Farber P, Jervis D, Fussner E, Plochowietz A, Craggs TD, Bazett-Jones DP, Pawson T, Forman-Kay JD, Baldwin AJ (2015) Phase transition of a disordered nuage protein generates environmentally responsive membraneless organelles. Mol Cell 57:936–947

Pak CW, Kosno M, Holehouse AS, Padrick SB, Mittal A, Ali R, Yunus AA, Liu DR, Pappu RV, Rosen MK (2016) Sequence determinants of intracellular phase separation by complex Coacervation of a disordered protein. Mol Cell 63:72–85

Paushkin S, Gubitz AK, Massenet S, Dreyfuss G (2002) The SMN complex, an assemblyosome of ribonucleoproteins. Curr Opin Cell Biol 14:305–312

Quinodoz SA, Jachowicz JW, Bhat P, Ollikainen N, Banerjee AK, Goronzy IN, Blanco MR, Chovanec P, Chow A, Markaki Y, Thai J, Plath K, Guttman M (2021) RNA promotes the formation of spatial compartments in the nucleus. Cell 184:5775–5790.e30

Raška I, Andrade LEC, Ochs RL, Chan EKL, Chang C-M, Roos G, Tan EM (1991) Immunological and ultrastructural studies of the nuclear coiled body with autoimmune antibodies. Exp Cell Res 195:27–37

Renvoisé B, Colasse S, Burlet P, Viollet L, Meier UT, Lefebvre S (2009) The loss of the snoRNP chaperone Nopp140 from Cajal bodies of patient fibroblasts correlates with the severity of spinal muscular atrophy. Hum Mol Genet 18:1181–1189

Richard P, Darzacq X, Bertrand E, Jády BE, Verheggen C, Kiss T (2003) A common sequence motif determines the Cajal body-specific localization of box H/ACA scaRNAs. EMBO J 22: 4283–4293

Sanz-García M, Vázquez-Cedeira M, Kellerman E, Renbaum P, Levy-Lahad E, Lazo PA (2011) Substrate profiling of human vaccinia-related kinases identifies coilin, a Cajal body nuclear protein, as a phosphorylation target with neurological implications. J Proteome 75:548–560

Schulz S, Chachami G, Kozaczkiewicz L, Winter U, Stankovic-Valentin N, Haas P, Hofmann K, Urlaub H, Ovaa H, Wittbrodt J, Meulmeester E, Melchior F (2012) Ubiquitin-specific protease-like 1 (USPL1) is a SUMO isopeptidase with essential, non-catalytic functions. EMBO Rep 13: 930–938

Shevtsov SP, Dundr M (2011) Nucleation of nuclear bodies by RNA. Nat Cell Biol 13:167–173

Staněk D (2017) Cajal bodies and snRNPs - friends with benefits. RNA Biol 14:671–679

Stanek D, Neugebauer KM (2006) The Cajal body: a meeting place for spliceosomal snRNPs in the nuclear maze. Chromosoma 115:343–354

Steitz JA, Black DL, Gerke V, Parker KA, Krämer A, Frendewey D, Keller W (1988) Functions of the abundant U-snRNPs. In: Birnstiel ML (ed) Structure and function of major and minor small nuclear ribonucleoprotein particles. Springer, Berlin, pp 115–154

Strzelecka M, Trowitzsch S, Weber G, Lührmann R, Oates AC, Neugebauer KM (2010) Coilin-dependent snRNP assembly is essential for zebrafish embryogenesis. Nat Struct Mol Biol 17: 403–409

Sun J, Xu H, Subramony SH, Hebert MD (2005) Interactions between coilin and PIASy partially link Cajal bodies to PML bodies. J Cell Sci 118:4995–5003

Suzuki H, Abe R, Shimada M, Hirose T, Hirose H, Noguchi K, Ike Y, Yasui N, Furugori K, Yamaguchi Y, Toyoda A, Suzuki Y, Yamamoto T, Saitoh N, Sato S, Tomomori-Sato C, Conaway RC, Conaway JW, Takahashi H (2022) The 3' pol II pausing at replication-dependent histone genes is regulated by mediator through Cajal bodies' association with histone locus bodies. Nat Commun 13:2905

Takata H, Nishijima H, Maeshima K, Shibahara K-I (2012) The integrator complex is required for integrity of Cajal bodies. J Cell Sci 125:166–175

Thomas Meier U, Blobel G (1992) Nopp 140 shuttles on tracks between nucleolus and cytoplasm. Cell 70:127–138

Toyama Y, Rangadurai AK, Forman-Kay JD, Kay LE (2022) Mapping the per-residue surface electrostatic potential of CAPRIN1 along its phase-separation trajectory. Proc Natl Acad Sci U S A 119:e2210492119

Tucker KE, Berciano MT, Jacobs EY, LePage DF, Shpargel KB, Rossire JJ, Chan EK, Lafarga M, Conlon RA, Matera AG (2001) Residual Cajal bodies in coilin knockout mice fail to recruit Sm snRNPs and SMN, the spinal muscular atrophy gene product. J Cell Biol 154:293–307

Tycowski KT, Shu M-D, Kukoyi A, Steitz JA (2009) A conserved WD40 protein binds the Cajal body localization signal of scaRNP particles. Mol Cell 34:47–57

Walker MP, Tian L, Matera AG (2009) Reduced viability, fertility and fecundity in mice lacking the cajal body marker protein, coilin. PLoS One 4:e6171

Wang Q, Sawyer IA, Sung M-H, Sturgill D, Shevtsov SP, Pegoraro G, Hakim O, Baek S, Hager GL, Dundr M (2016) Cajal bodies are linked to genome conformation. Nat Commun 7:10966

Wang H, Xu X, Nguyen CM, Liu Y, Gao Y, Lin X, Daley T, Kipniss NH, La Russa M, Qi LS (2018a) CRISPR-mediated programmable 3D genome positioning and nuclear organization. Cell 175:1405–1417.e14

Wang J, Choi J-M, Holehouse AS, Lee HO, Zhang X, Jahnel M, Maharana S, Lemaitre R, Pozniakovsky A, Drechsel D, Poser I, Pappu RV, Alberti S, Hyman AA (2018b) A molecular grammar governing the driving forces for phase separation of prion-like RNA binding proteins. Cell 174:688–699.e16

Chapter 8
Formation and Function of Phase-Separated Nuclear Bodies Directed by Architectural Noncoding RNA

Hiro Takakuwa, Tomohiro Yamazaki, and Tetsuro Hirose

Abstract The cell nucleus contains numerous membraneless organelles (MLOs) and biomolecular condensates that lack surrounding lipid membranes. Growing evidence suggests that these MLOs are formed through phase separation and contribute to various biological processes and regulatory mechanisms. RNAs are a major constituent of MLOs and contribute to MLO formation by serving as scaffolds for molecular interactions of MLO proteins. A class of long noncoding RNAs called architectural RNAs (arcRNAs) function as the essential structural scaffold of phase-separated MLOs in multiple eukaryotic species. These arcRNAs are temporarily upregulated under specific stress conditions or physiological/pathological conditions, which can modulate gene expression patterns by sequestering or concentrating regulatory proteins and RNAs into MLOs. NEAT1_2 is a well-characterized arcRNA that functions as an essential scaffold of paraspeckle nuclear MLOs. This review describes the molecular mechanisms underlying the formation and functions of paraspeckles with a particular focus on the functions of NEAT1_2 arcRNAs.

Keywords Long noncoding RNA · Membraneless organelle · Biomolecular condensates · Phase separation · RNA-binding protein · RNA domain

---

H. Takakuwa
Graduate School of Frontier Biosciences, Osaka University, Suita, Japan

Institute for Genetic Medicine, Hokkaido University, Sapporo, Japan

T. Yamazaki
Graduate School of Frontier Biosciences, Osaka University, Suita, Japan

T. Hirose (✉)
Graduate School of Frontier Biosciences, Osaka University, Suita, Japan

Institute for Open and Transdisciplinary Research Initiatives (OTRI), Osaka University, Suita, Japan
e-mail: hirose.tetsuro.fbs@osaka-u.ac.jp

R. Kurokawa (ed.), *Phase Separation in Living Cells*,
https://doi.org/10.1007/978-981-99-4886-4_8

8.1 Introduction

Eukaryotic cells contain numerous, molecularly distinct intracellular organelles that play roles in fundamental biological processes. Membranous organelles surrounded by lipid bilayer membranes, such as the endoplasmic reticulum, mitochondria, and the Golgi apparatus, create discrete chemical environments within the cell. However, cells also contain organelles that lack surrounding membranes, which are called membraneless organelles (MLOs) or biomolecular condensates. Examples include nuclear bodies such as nucleoli, Cajal bodies, nuclear speckles, and many cytoplasmic structures including P granules, stress granules, and processing bodies (P-bodies). Such MLOs play essential roles in various biological processes through multiple modes of action, including acting as reaction crucibles, sequestration, and chromatin hubs (Banani et al. 2017; Hirose et al. 2022; Shin and Brangwynne 2017). Accumulating evidence suggests that various MLOs are formed through phase separation and behave as fluid droplets in cells (Courchaine et al. 2021). Most phase-separated MLOs contain proteins with intrinsically disordered regions (IDRs) and nucleic acids, mainly RNAs. These components dictate the physical properties of MLOs as a consequence of their intra- and intermolecular interactions and operate the MLO functions by controlling them. While the high concentration of RNA in the nucleus suppresses excessive phase separation of prion-like RNA-binding proteins (RBPs) (Maharana et al. 2018), RNAs contribute to the formation of MLOs (Roden and Gladfelter 2021). The sum of multivalent intermolecular interactions, including RNA–RNA, RNA–protein, and protein–protein interactions, dictates the formation of MLOs through phase separation (Van Treeck and Parker 2018). A specific class of long noncoding RNAs (lncRNAs) termed architectural RNAs (arcRNAs) can act as essential structural scaffolds of nuclear bodies in a variety of eukaryotes from yeast to humans (Chujo et al. 2016; Yamazaki et al. 2019). To date, several arcRNAs and many of their candidate RNAs have been identified. In this review, we first describe the definition and general features of arcRNAs, and then focus on NEAT1_2 as a model arcRNA and introduce the molecular mechanisms of MLO formation through NEAT1_2 arcRNA-mediated phase separation. Finally, we discuss the functional significance of the internal core-shell structure of MLOs formed by NEAT1_2 arcRNA.

8.2 General Molecular Features of arcRNAs

Over the past few decades, multiple lncRNAs have been found to have architectural roles in the assembly of specific MLOs as arcRNAs. In this section, we introduce representative arcRNAs and the MLOs that they construct, along with the general importance of using RNAs as architectural scaffolds of MLOs.

8.2.1 Definition and Molecular Species of arcRNAs

ArcRNAs are involved in the formation of various nuclear bodies as their essential scaffolds (Fig. 8.1). For a lncRNA to be classified as an arcRNA, the lncRNA must be localized or enriched in a specific MLO and its removal must lead to disassembly of the MLO, causing the dispersion or change in localization of marker proteins (Chujo et al. 2016). As known arcRNA species, NEAT1_2, the long isoform of nuclear paraspeckle assembly transcript 1 (NEAT1), is an essential scaffold of the paraspeckle nuclear body; primate-specific highly repetitive human satellite III (HSATIII) lncRNA is transcribed upon thermal stress and promotes the formation of the nuclear stress body (nSB); ribosomal intergenic spacer (rIGS) lncRNAs construct the amyloid-like solid structures called amyloid bodies (A-bodies); pyrimidine-rich noncoding transcript (PNCTR) is critical for the structural integrity of the perinucleolar compartment (PNC); Hsrω-n RNAs function as the scaffold of omega-speckle in *Drosophila melanogaster*; and the fission yeast Sme2 RNA forms the Mei2 dot nuclear body. The formation of the Sam68 nuclear body is also RNA-dependent, but their arcRNAs have not yet been identified. Conversely, tumor-associated NBL2 transcripts (TNBL) form perinucleolar aggregates adjacent

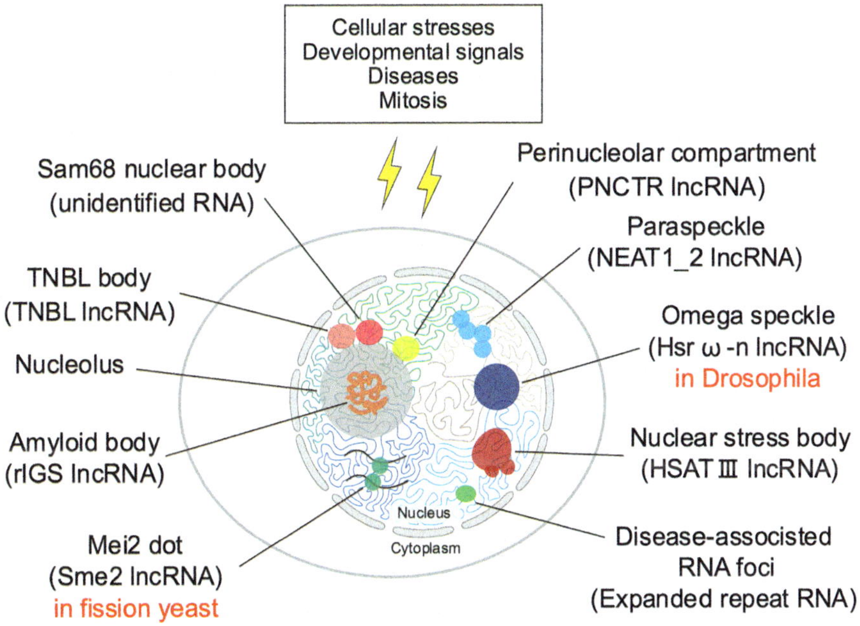

Fig. 8.1 Membraneless organelles (MLOs) scaffolded by architectural RNAs (arcRNAs) in the nucleus. A schematic illustration of MLOs that are constructed by arcRNAs in various organisms. All arcRNAs are transcribed in response to various stresses, developmental signals, and diseases. The scaffold RNAs are described in parentheses. Some arcRNAs are only observed in species other than humans (shown in red)

to Sam68 nuclear bodies, but the major protein components have not yet been identified. Pathologically expanded, repeat-containing RNAs can also be considered as a type of arcRNA. The number of newly identified arcRNAs has increased over the years, suggesting that MLO formation by arcRNA is one of the fundamental mechanisms through which RNAs function.

8.2.2 *Advantages of Using RNAs as Architectural Scaffolds of MLOs*

Most MLOs contain RNAs that widely act as MLO scaffolds in a variety of eukaryotes, suggesting that there are advantages to using RNAs as structural scaffolds. This section discusses why RNAs are suitable biomolecules for scaffolding MLOs.

First, RNAs can induce phase separation in a spatiotemporally controlled manner. In many cases, MLOs are formed at the transcription sites of arcRNAs, so the spatial distribution of the MLOs can be dictated by the spatial positions of the chromosomal loci of arcRNA genes within the nucleus. The transcription of known arcRNAs is induced or upregulated upon cellular stresses or in specific developmental and disease conditions (Fig. 8.2a) (Chujo et al. 2016; Yamazaki et al. 2019). For example, the transcription of NEAT1_2 arcRNA is upregulated under various cellular stress conditions and at specific developmental stages (Hirose et al. 2014; Imamura et al. 2014; Mohammad Lellahi et al. 2018; Nakagawa et al. 2014; Standaert et al. 2014). The expression of rIGS arcRNA is induced by thermal stress and acidosis (Audas et al. 2012). Thermal stress also induces the expression of HSATIII arcRNA (Jolly et al. 2004; Rizzi et al. 2004) and *Drosophila* Hsrω arcRNA (Ashburner 1970; Garbe and Pardue 1986). PNCTR arcRNA is markedly overexpressed in a wide range of cancer cells (Yap et al. 2018). TNBL arcRNA is expressed in a subset of colorectal cancer cells (Dumbović et al. 2018). In the nucleus, sufficiently high levels of arcRNA transcription markedly increase local concentrations of arcRNA and RBPs at the site of transcription, eventually inducing phase separation. Conversely, the arrest of arcRNA transcription induces rapid degradation of arcRNA and concomitantly leads to the disassembly of MLOs (Fig. 8.2a). Indeed, the inhibition of global RNA polymerase II transcription was shown to cause rapid disassembly of paraspeckles, suggesting that paraspeckle maintenance is coupled with ongoing NEAT1_2 transcription (Fox et al. 2005; Mao et al. 2011; Sasaki et al. 2009; Sunwoo et al. 2009). NEAT1_2 is a relatively unstable RNA and its half-life is approximately 60 min (Clark et al. 2012). When the degradation of NEAT1_2 by RNA exosome is inhibited, the half-life of NEAT1_2 significantly increased (Imamura et al. 2018; Machitani et al. 2020; Tanu et al. 2021). It was also reported that the expression level of Hsrω-n RNAs increases by eight-fold within 1 h of heat shock, and it rapidly decreases to normal conditions within 2 h after stress removal (Hogan et al. 1994). In contrast, HSATIII lncRNAs

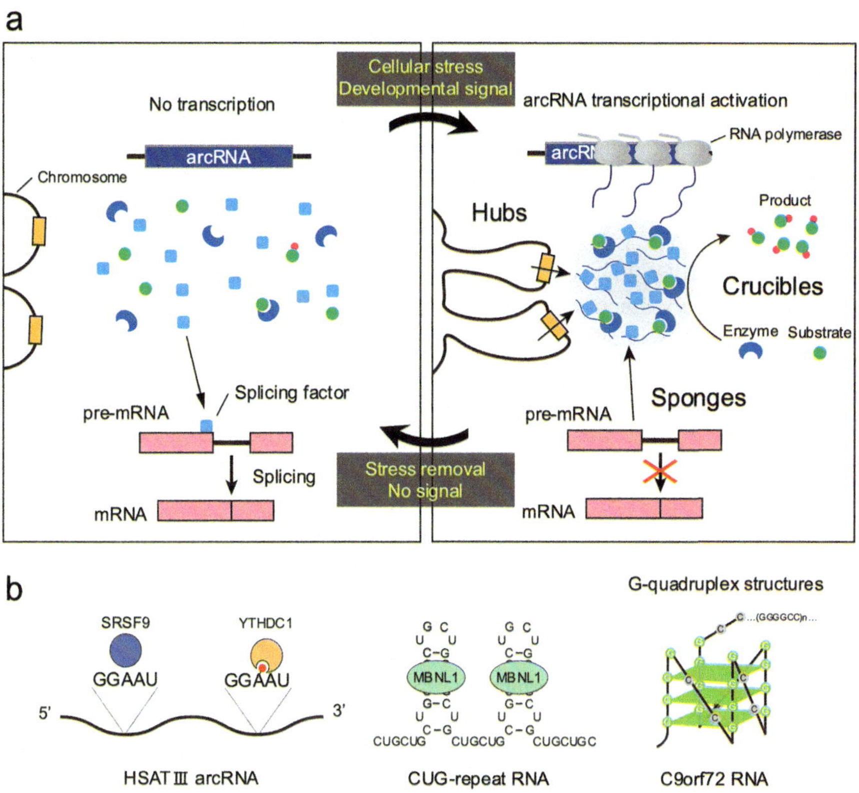

Fig. 8.2 Advantage of using RNA as an architectural scaffold of MLOs. (**a**) A model of the mode of action of MLOs scaffolded by arcRNAs. arcRNAs are temporarily upregulated in response to cellular stress and/or developmental signals. Transcription of arcRNAs induces phase-separated MLOs by increasing their own local concentration and that of RBPs. MLOs play diverse regulatory roles through various modes of action, including acting as hubs, crucibles, and sponges. When stresses and signals are no longer present, arcRNAs can be rapidly degraded, and trapped proteins disperse into the nucleoplasm. (**b**) Some modes of RNA-protein and RNA-RNA interactions. SRSF9 specifically binds to the GGAAU repeats of HSATIII. YTHDC1 recognizes m^6A-modified GGAAU repeat sequence of HSATIII. A methyl group is illustrated as a red circle (left). CUG-repeat RNAs fold into Watson-Crick hairpin structures. MBNL1 proteins seem to recognize both the sequence and the structure of CUG-repeat RNAs (middle). G-rich RNAs such as C9orf72 form G-quadruplex structures in vitro and in vivo (right)

are relatively stable and many nSBs are still detectable even 4 h after stress removal (Ninomiya et al. 2021). This feature may be linked to the function of nSB that promotes the intron retention of hundreds of target pre-mRNAs during stress recovery. These findings suggest that arcRNAs induce MLO formation in response to various stresses and their stability may affect the length of the period during which MLOs remain functional.

Second, RNA can capture RBPs more effectively than proteins. To induce phase separation and form nuclear bodies, architectural molecules need to trap a large number of proteins (often containing IDRs or PLDs) and increase their local concentration. One protein can typically be captured by 4–17 nucleotides or a secondary structure of RNA (Lunde et al. 2007; Prikryl et al. 2011). Therefore, a single RNA molecule, ranging from hundreds of bases to tens of kilobases in length, has the potential to capture dozens of proteins. This unique ability cannot be replicated by any protein molecules. For example, PNCTR arcRNA is a > 10 kb RNA transcript in HeLa cells and contains 565 non-overlapping PTBP binding motifs (YUCUYY/YYUCUY) (Yap et al. 2018). PNCs act as sponges to sequester approximately 10%–20% of cellular PTBP1 from the nucleoplasm, antagonizing the splicing regulatory function of PTBP1. HSATIII arcRNAs are mostly composed of GGAAU repeat sequences, which recruit large amounts of SRSF9 and N^6-methyladenosine (m^6A)-related factors to nSBs (Fig. 8.2a, b) (Ninomiya et al. 2021). nSBs act not only as sponges that sequester m^6A factors, but also as crucibles to facilitate enzymatic reactions by simultaneously accumulating the phosphatase CLK1 and its substrate SRSF9 (Fig. 8.2a, b) (Ninomiya et al. 2020). In addition, a single molecule of NORAD lncRNA can capture 18 molecules of PUM, and these multivalent NORAD-PUM RNA–protein interactions facilitate PUM-PUM protein–protein interactions, thereby recruiting 42 times more PUM to the NORAD-PUM bodies in the cytoplasm than the original RNA capacity (Elguindy and Mendell 2021). Thus, even RNAs expressed at low levels have the potential to induce phase separation by assembling numerous RBPs, which may have significant impacts on the surrounding nuclear environment. In some cases, RNA itself can promote protein-independent phase separation. For example, RNAs with ALS/FTD-associated G4C2 repeats form RNA gels independent of proteins in vitro, likely through the formation of G-quadruplex structures (Fig. 8.2b) (Jain and Vale 2017). NEAT1_2 arcRNAs are also predicted to form G-quadruplex structures, the inhibition of which by chemical compounds was reported to disrupt paraspeckle formation (Simko et al. 2020). In summary, a single RNA molecule is an ideal molecule to effectively capture proteins or RNAs to induce phase separation through both RNA–protein and RNA–RNA interactions.

Third, RNA can dictate the size and morphology of MLOs. RNA is a highly negatively charged biopolymer, and a high charge density can alter the properties of MLOs. Analyses using artificial RNP (RNA–protein) MLOs called ArtiG showed that phase-separated MLOs with a higher density of RNA on the surface tend to be smaller in size (Cochard et al. 2022; Garcia-Jove Navarro et al. 2019). This may be due to the high surface charge density of the MLOs, which causes electrostatic repulsion between them and prevents their coalescence. In a similar example, the Ki-67 proteins have been shown to function as surfactants by localizing to the mitotic chromosome surfaces and forming an electrostatic charge barrier (Cuylen et al. 2016). These results suggest that RNA can also act as a surfactant and is an important determinant of MLO size and number.

RNA is typically a much longer polymer than proteins, enabling a single RNA molecule to capture a wide variety of RBPs, even if their physical properties are quite different (e.g., hydrophilic and hydrophobic). In mammals, approximately 1500 genes encode RBPs having a wide variety of biological functions and physical properties (Castello et al. 2012; Gerstberger et al. 2014). Such RNP complexes act as block copolymers consisting of two or more chemically distinct polymers joined by covalent bonds and form MLOs with optimal size and shape through micellization as a type of phase separation (Yamamoto et al. 2022; Yamazaki et al. 2022, 2021). Thus, RNA is a long, flexible, and negatively charged molecule and forms RNP complexes with proteins possessing a wide variety of properties that can dictate the size and shape of MLOs.

In summary, RNA is an ideal biomolecule that functions as a scaffold of biomolecular condensates with a variety of structures in a spatiotemporally controlled, rapid, and reversible manner. Whereas arcRNAs play various roles in stress responses as described above, RNA-mediated phase separation can negatively affect cells and organisms in certain pathological conditions. In the next section, we will introduce aberrant RNA condensates formed in repeat expansion diseases.

8.2.3 Disadvantages of RNA-Mediated Phase Separation in Diseases

Expansions of short nucleotide repeats cause a subset of disorders, such as myotonic dystrophy type 1 (DM1), fragile X-associated tremor ataxia syndrome (FXTAS), and C9orf72 amyotrophic lateral sclerosis and/or frontotemporal dementia (C9 ALS/FTD). In some contexts, RNAs expressed from genomic sites possessing expanded repeats accumulate in the cell nucleus and form MLO-like RNA foci (Malik et al. 2021; Swinnen et al. 2020). For example, DM1 is caused by the abnormal expansion of a CTG-trinucleotide repeat in the gene encoding dystrophia myotonic protein kinase (DMPK), and CUG RNA foci are formed in the nucleus and sequester specific RBPs such as muscleblind-like (MBNL) family proteins: MBNL1, MBNL2, and MBNL3 (Fig. 8.2b) (Miller et al. 2000; Taneja et al. 1995). One of the proposed pathogenic mechanisms behind the effects of toxic RNAs is the sequestration of MBNL proteins from the nucleoplasm, leading to altered splicing of several mRNAs (Charlet et al. 2002; Fugier et al. 2011; Mankodi et al. 2001; Philips et al. 1998). Because these altered splicing events have been confirmed in patient tissues and correlate with clinical features, MBNL1 dysfunction is considered as the key mechanism involved in DM1 (Freyermuth et al. 2016; Fugier et al. 2011; Savkur et al. 2001). The depletion of MBNL1 proteins by RNAi reduces the formation of nuclear foci containing CUG-repeat RNA in C2C12 cells, suggesting that MBNL proteins promote the aggregation of CUG-repeat RNA foci (Querido et al. 2011). In addition, CUG-repeat RNAs can induce protein-independent RNA phase separation and form RNA gels in vitro (Jain and Vale 2017). These findings suggest that CUG-repeat RNA foci are formed through combinations of RNA–RNA and RNA–protein interactions. One therapeutic strategy for repeat expansion diseases is to develop small molecular compounds that specifically bind to repeat RNAs and disrupt the RNA–protein and/or RNA–RNA interactions. A recent study reported that naphthyridine carbamate dimer (NCD) targets disease-causing UGGAA repeat RNAs in spinocerebellar ataxia type 31 (SCA31) and inhibits the interaction of the UGGAA repeat RNAs with a set of RBPs and the formation of RNA foci (Shibata et al. 2021). In summary, it has been suggested that the RNA expression from the repeat expanded regions causes RNA toxicity and promotes the formation of aberrant phase-separated condensates or insoluble aggregates through RNA–RNA and RNA–protein interactions in repeat expansion diseases.

8.3 Specific Molecular Features of NEAT1_2 arcRNA

NEAT1_2 is a well-characterized arcRNA, which is an essential scaffold of paraspeckle nuclear bodies. This section discusses how NEAT1_2 arcRNAs construct paraspeckles through phase separation and perform their cellular and physiological functions.

8.3.1 Overview of NEAT1_2 arcRNA and Paraspeckles

Paraspeckles are discrete subnuclear MLOs with an average diameter of approximately 360 nm (Souquere et al. 2010). Prominent paraspeckles are found only in the small population of cells that express high levels of Neat1_2 in adult mouse tissues (Nakagawa et al. 2011), although paraspeckles are detectable in many cultured mammalian cell lines. Paraspeckles were originally identified as distinct nuclear bodies called interchromatin granule-associated zones (IGAZs), which were observed as electron-dense structures by electron microscopy (Visa et al. 1993). In 2002, paraspeckles were defined as nuclear bodies found in close proximity to nuclear speckles and enriched in characteristic DBHS (Drosophila Behavior Human Splicing) RBPs such as SFPQ and NONO (Fig. 8.3a) (Fox et al. 2002). Inhibition of RNA polymerase II transcription by actinomycin D or D-ribofuranosylbenzimidazole (DRB) causes rapid disintegration of paraspeckles along with NEAT1_2 degradation (Fox et al. 2005; Sasaki et al. 2009; Sunwoo et al. 2009), suggesting that paraspeckles require either RNA or RNA polymerase II transcription for their generation or maintenance. In 2009, four groups independently reported that NEAT1 lncRNA functions as an arcRNA of paraspeckles (Fig. 8.3b) (Chen and Carmichael 2009; Clemson et al. 2009; Sasaki et al. 2009; Sunwoo et al. 2009). NEAT1 has two isoforms: the longer NEAT1_2 (22.7 kb in humans) and the shorter NEAT1_1 (3.7 kb in humans), and both isoforms are transcribed from the same transcription start site but undergo different 3′ end processing (Fig. 8.3c) (Naganuma et al. 2012; Sunwoo et al. 2009). NEAT1_2 lacks a canonical poly(A) tail and possesses a characteristic triple-helix structure at its 3′ end (3′TH) (Brown et al. 2012; Sunwoo et al. 2009; Wilusz et al. 2012). In *Neat1*-knockout mouse embryonic fibroblasts, it was shown that transient expression

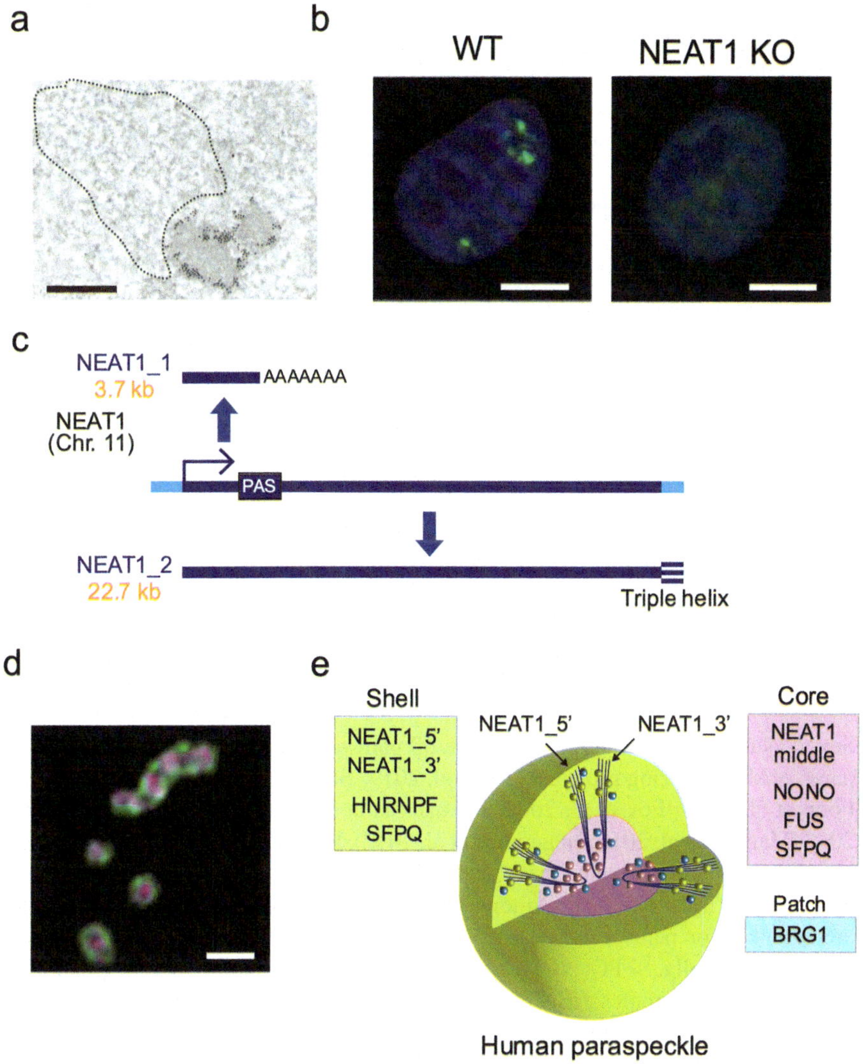

Fig. 8.3 The NEAT1_2 arcRNA in paraspeckle nuclear bodies. (**a**) An electron micrograph showing a paraspeckle (an electron-dense structure) formed in close proximity to a nuclear speckle (black dashed circle). Paraspeckles were detected using gold particles [probes against NEAT1_5′ regions] in human haploid cell lines treated with MG132 (5 μM) for 17 h. Scale bar, 500 nm. (**b**) Visualization of paraspeckles in wild-type (WT) and NEAT1-knockout (KO) human cells. The paraspeckles visualized by immunostaining of NONO (green signals in WT) are abolished in NEAT1 KO cells. Scale bar, 5 μm. (**c**) A schematic illustration of two isoforms of NEAT synthesized by alternative 3′ end processing. NEAT1_2 possesses a noncanonical 3′ triple-helix structure for its stability. (**d**) A super-resolution image of paraspeckles detected by RNA-FISH with NEAT1 antisense probes (NEAT1_5′ and NEAT1_3′: green; NEAT1_middle: magenta). (**e**) A schematic model of the core-shell arrangement of paraspeckle components and NEAT1_2 arcRNAs. (**a**, **b**, **d**) Reproduced from (Yamazaki et al. 2019), with permission

of NEAT1_2, but not NEAT1_1, can rescue paraspeckle formation, demonstrating an architectural role of NEAT1_2 in paraspeckle formation (Naganuma et al. 2012). Paraspeckles are massive RNP complexes containing NEAT1_2 arcRNAs and multiple RBPs, and both NEAT1_2 and RBPs are spatially arranged to form an ordered core-shell structure (Fig. 8.3d, e). A single paraspeckle is estimated to contain approximately 50 molecules of NEAT1_2 (Chujo et al. 2017). The transcription of NEAT1 is upregulated by various stress conditions, such as proteasome inhibition and viral infections, resulting in increases in the size and number of paraspeckles, and subsequent sequestration of specific RBPs from the nucleoplasm to control gene expression (Hirose et al. 2014; Imamura et al. 2014). Physiologically, paraspeckles are formed in the corpus luteal cells in the ovary and the luminal epithelial cells in the mammary gland, and NEAT1 is required for the establishment of pregnancy and mammary gland development in mice (Nakagawa et al. 2011, 2014; Standaert et al. 2014). NEAT1 is also involved in the progression of various cancers, fibrosis, and neurodegenerative diseases (Adriaens et al. 2016; Fukushima et al. 2020; Mello et al. 2017; Nishimoto et al. 2013), but the detailed molecular mechanisms underlying their physiological and pathological functions remain unclear.

8.3.2 The Paraspeckle Components

Several studies have reported more than 60 paraspeckle protein components (PSPs; paraspeckle-localizing proteins) (An et al. 2019; Barra et al. 2020; Chen et al. 2020; Fong et al. 2013; Fox et al. 2002; Fukushima et al. 2020; Kawaguchi and Hirose 2015; Mannen et al. 2016; Naganuma et al. 2012). Some proteomic studies identified NEAT1-interacting proteins that could be localized to paraspeckles as PSPs (West et al. 2014; Yap et al. 2022). Among them, seven PSPs and the SWI/SNF complex are essential for paraspeckle formation (Kawaguchi et al. 2015; Naganuma et al. 2012). Specifically, SFPQ, NONO, and RBM14 are essential for NEAT1_2 stability, and HNRNPK promotes NEAT1_2 synthesis by inhibiting polyadenylation of the NEAT1_1 isoform (Naganuma et al. 2012). Meanwhile, the remaining FUS, DAZAP1, HNRNPH3, and the SWI/SNF complex are required for paraspeckle assembly (Kawaguchi et al. 2015; Naganuma et al. 2012; Yamazaki et al. 2018). More than half of the PSPs contain IDRs, low-complexity domains (LCDs), and prion-like domains (PLDs) (Fox et al. 2018; Naganuma et al. 2012), suggesting that there is a network of IDR-mediated protein–protein interactions within paraspeckles. Indeed, the PLDs of FUS and RBM14 form hydrogels in vitro and are required for paraspeckle formation in vivo (Hennig et al. 2015).

We previously found that NEAT1_2 is poorly extracted by the conventional RNA extraction method using the acid guanidium thiocyanate-phenol-chloroform (AGPC) reagents such as TRIzol or TRI reagent (Chujo et al. 2017). Needle shearing or heating at 55 °C of cell lysate in RNA extraction reagent improved NEAT1_2 extraction by up to 20-fold, whereas using a conventional method, NEAT1_2 was trapped in the protein phase. This unusual semi-extractability of NEAT1_2 is partially dependent on FUS through its PLD, suggesting that tenacious RNA–protein and protein–protein interactions, which are required for nuclear body formation, are responsible for the semi-extractability of the arcRNAs (Chujo et al. 2017).

In addition to the protein components, it has also been proposed that multiple types of RNAs localize to the paraspeckles: mRNAs containing inverted repeats of Alu in their 3′ UTR (Chen and Carmichael 2009; Hu et al. 2015; Wang et al. 2018b), CTN-RNA (cationic amino acid transporter 2 RNA) (Prasanth et al. 2005), mRNAs and introns containing purine-rich sequences (Wang et al. 2018b; West et al. 2016), poorly processed and A-to-I-edited transcripts (Yap et al. 2022), and U1 snRNA (Visa et al. 1993). NEAT1 depletion affects mitochondrial dynamics and functions by altering the sequestration of mRNAs for mitochondrial proteins into the paraspeckles (Wang et al. 2018b). Further elucidation of the roles of RNAs localized to paraspeckles should provide important insights into gene regulation by the paraspeckles.

8.3.3 Functional RNA Domains of NEAT1_2

Functional elements/domains dictating lncRNA functions have been poorly investigated. Considering that each lncRNA associates with specific partner proteins to build functional RNP machinery, lncRNAs should have functional domain clustering protein-binding sites that consist of specific sequences, local structures, and/or chemical modifications. As a well-characterized example, multiple functional domains of NEAT1_2 that enable it to exert its architectural function have been identified by using CRISPR/Cas9-mediated extensive deletion analyses in human haploid cells (Fig. 8.4a) (Modic et al. 2019; Yamazaki et al. 2018, 2021). The 5′ end (0–1 kb) and 3′TH of NEAT1_2 are required for the stability of NEAT1_2 arcRNA (Yamazaki et al. 2018). Two regions upstream (2.1–2.8 kb) and downstream (4.0–5.0 kb) of the polyadenylation site (PAS) of NEAT1_1 promote the expression of NEAT1_2 by repressing NEAT1_1 polyadenylation (Yamazaki et al. 2018). The middle domain of NEAT1_2 (8–16.6 kb) is required and sufficient for paraspeckle assembly (Yamazaki et al. 2018). This NEAT1_2 middle domain can recruit essential PSPs, such as NONO/SFPQ and FUS (Fig. 8.4b) (Yamazaki et al. 2018). In particular, the NOPS domain of NONO, which contributes to dimerization and subsequent oligomerization with DBHS family proteins along the RNA molecule, is required for the paraspeckle assembly (Fig. 8.4b) (Yamazaki et al. 2018). Recently, we have identified the NEAT1_2 RNA domains required for the core-shell structuring of paraspeckles. The deletion of either 5′ (0–1.8 kb) or 3′ (16.6–22.6 kb) regions caused redistribution of the 5′ or 3′ terminal region of NEAT1_2 to the paraspeckle core, unlike in wild-type (WT) cells (Fig. 8.5) (Yamazaki et al. 2021). When both 5′ and 3′ terminal regions are simultaneously deleted, both ends of NEAT1_2 became randomly distributed within the paraspeckle (Fig. 8.5) (Yamazaki et al. 2021). These phenotypes due to the deletion of specific

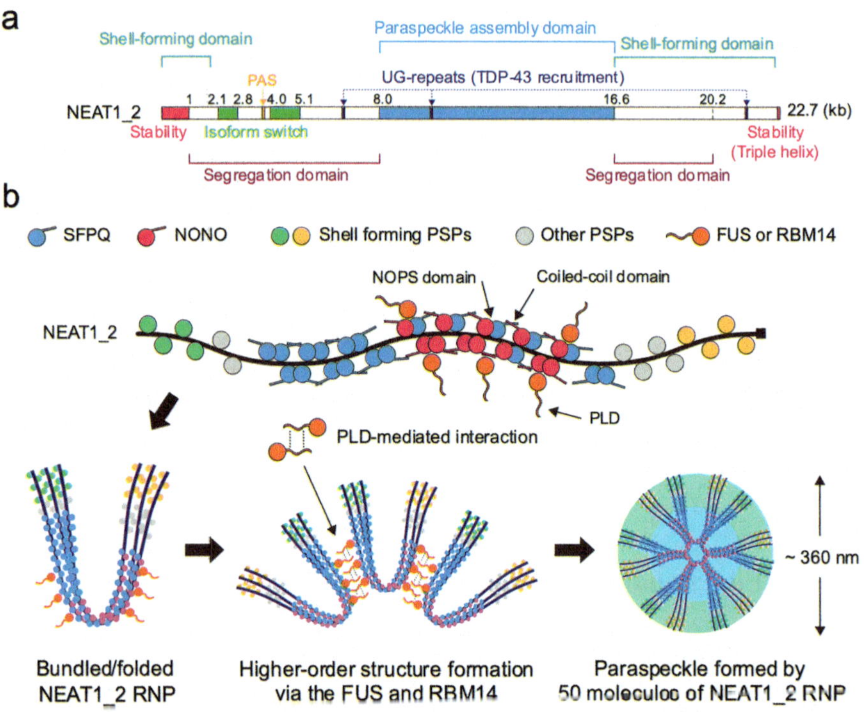

Fig. 8.4 The modular domain structure of the human NEAT1_2 arcRNA. (**a**) A schematic illustration shows the modular domains of human NEAT1_2 arcRNA required for paraspeckle formation and functions. These domains are involved in NEAT1_2 stability, isoform switching, polyadenylation signaling (PAS), sequestration of TDP-43 proteins by UG-repeats, paraspeckle assembly, core-shell organization, and paraspeckle segregation from nuclear speckles. The position of each domain relative to the 5′ end (kb, kilobases) is also shown. (**b**) A model of paraspeckle formation through micellization (top). Various PSPs bind to the high-affinity binding sites within functional domains of NEAT1_2. NONO and SFPQ are loaded onto paraspeckle assembly domains and spread through the NOPS and coiled-coil domains of NONO and SFPQ. Concurrently, FUS and RBM14 are recruited to the NEAT_2 RNPs, likely by NONO and SFPQ (bottom left). Several NEAT1_2 RNPs are bundled and folded into a V-shape. The bundled 5′ and 3′ domains may separately form specific RNP complexes that contribute to shell formation (bottom middle). Multivalent interactions between the PLDs of FUS and RBM14 assemble multiple units of bundled NEAT1_2 RNP and form higher-order structures.(bottom right). Finally, approximately 50 NEAT1_2 RNPs are assembled and form one spherical paraspeckle

RNA regions of NEAT1_2 support the idea that paraspeckles are formed through micellization, which will be discussed in the following section.

NEAT1_2 contains three UG repeat regions, which are evolutionarily conserved in humans and mice, and these repeat sequences are essential for the recruitment of TDP-43 protein that preferentially binds to UG-stretches (Fig. 8.4a) (Modic et al. 2019; Tollervey et al. 2011; Yamazaki et al. 2019). It has also been found that specific PSPs cannot be recruited to the paraspeckle in some NEAT1_2 mutants lacking specific RNA regions (Takakuwa et al. 2023; Yamazaki et al. 2018),

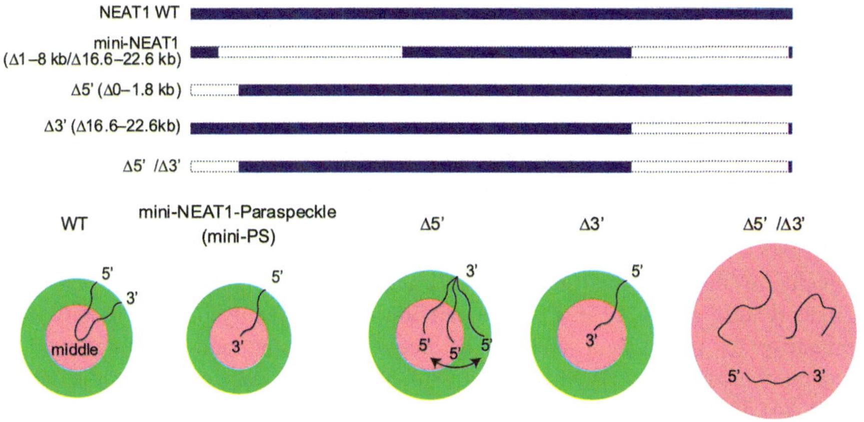

Fig. 8.5 Localization of NEAT1_2 within WT and mutant paraspeckles. Schematics of WT NEAT1_2 and mutants lacking the 5′ and/or 3′ domains and paraspeckles constructed by these mutants. The dashed boxes represent the deleted regions

suggesting that several PSPs are recruited to paraspeckles by binding to the specific RNA regions of NEAT1_2.

NEAT1_2 harbors a pseudo-miRNA (miR-612), which is not processed into the mature miRNA, at its 3′ terminal region. The pseudo-miRNA functions to attract microprocessors that process pri-miRNAs to pre-miRNAs in the nucleus (Jiang et al. 2017). The major PSPs, NONO/SFPQ, bind many expressed pri-miRNAs in HeLa cells and promote pri-miRNA processing by the microprocessors. These findings suggest that the paraspeckle acts as a crucible to facilitate pri-miRNA processing by concentrating specific processing factors.

The NEAT1_1 isoform is dispensable for paraspeckle formation, but its functional role is still unknown. Overexpression of NEAT1_1 in cells expressing NEAT1_2 increases the number of paraspeckles, suggesting that NEAT1_1 has a cooperative effect on paraspeckle formation (Clemson et al. 2009; Naganuma et al. 2012). Meanwhile, NEAT1_1 forms numerous non-paraspeckle foci termed "microspeckles," suggesting that NEAT1_1 may have paraspeckle-independent functions (Li et al. 2017a). Given that the amount of NEAT1_1 is ten-fold lower than that of NEAT1_2 in HeLa cells (Chujo et al. 2017), it is tempting to speculate that paraspeckles sequester NEAT1_1 RNPs in their shell and suppress the functions of NEAT1_1. Recent studies have shown that NEAT1_1-specific knockout mice do not exhibit obvious abnormal phenotypes under normal laboratory conditions (Adriaens et al. 2019; Isobe et al. 2020), but NEAT1_1-specific knockout mice may exhibit some phenotypes under certain experimental conditions.

8.3.4 Paraspeckles Are Formed Through Micellization

Several different types of analyses have revealed that paraspeckles possess features of phase-separated MLOs. First, fluorescence recovery after photobleaching (FRAP) experiments revealed that the major population of PSPs undergo rapid exchange between paraspeckles and the nucleoplasm (Audas et al. 2016; Mao et al. 2011; Wang et al. 2018b). Second, the PLDs of FUS and RBM14, which are essential for paraspeckle formation, can induce phase separation and form hydrogels in vitro, and tyrosine residues enriched in their PLDs are required for this function (Fox et al. 2018; Hennig et al. 2015). Phase separation of FUS family proteins is governed primarily by multivalent interactions between tyrosine and arginine residues (cation–π interactions) (Wang et al. 2018a). Third, 1,6-hexanediol, which is widely used to dissolve phase-separated condensates, can disrupt the formation of paraspeckles (Yamazaki et al. 2018). Fourth, paraspeckles undergo a fission and fusion process called "kiss-and-run/fusion" (Mao et al. 2011; Yang et al. 2019). These observations suggest that paraspeckles are phase-separated MLOs, but paraspeckles have several structural features that differ from typical MLOs formed by liquid–liquid phase separation (LLPS).

Although typical MLOs formed by LLPS show spherical shapes, paraspeckles have spherical or oblong structures with a short axis of constant length (~360 nm) and a long axis of variable length that become elongated from a spherical to a cylindrical shape when NEAT1_2 is upregulated (Hirose et al. 2014; Souquere et al. 2010). Notably, paraspeckles have highly organized internal core–shell structures, where 5′ and 3′ terminal regions of NEAT1_2 are localized in the shell, and the middle region is localized in the core, suggesting that NEAT1_2 forms a looped structure within paraspeckles (Fig. 8.3d, e) (Souquere et al. 2010; West et al. 2016). PSPs show three localization patterns: core, shell, and patch components (Fig. 8.3e) (Kawaguchi et al. 2015; West et al. 2016). As described above, NEAT1_2 lacking the 5′ and 3′ domains (Δ5′/Δ3′) forms paraspeckles without an internal ordered structure (Fig. 8.5) (Yamazaki et al. 2021). In this NEAT1_2 Δ5′/Δ3′ mutant, paraspeckles become larger and spherical (Yamazaki et al. 2021). These features are hallmarks of MLOs formed through arcRNA-driven LLPS (Yamamoto et al. 2020). A study using soft-matter physics theory showed that paraspeckles are formed by micellization, a new intracellular phase separation mechanism of MLOs distinct from LLPS, which controls NEAT1_2 internal organization, and the shape, number, and size of paraspeckles (Yamamoto et al. 2022; Yamazaki et al. 2021). In this model, NEAT1_2 RNPs are treated as block copolymers and paraspeckles as polymer micelles. For a more detailed discussion of the mechanism of paraspeckle formation through micellization and its significance, we refer the reader to the following review (Yamazaki et al. 2022). Although only paraspeckles have been reported to date as MLOs formed by micellization, it is possible that there are other RNP complexes that behave as block copolymers and that micellization is a molecular mechanism widely used by cells for MLO formation.

8.3.5 Significance of the Internal Core-Shell Structure of Paraspeckles

It has been suggested that the characteristic core-shell structure is crucial for facilitating the proper functioning of paraspeckles within the nucleus. First, the hydrophilic shell has the potential to serve as a platform for putative regulatory events that occur within paraspeckles. MLOs are spatially associated with specific chromosomal loci, enabling transcriptional regulation of their associated chromosomal loci as "chromatin hubs" (Fig. 8.2a) (Hirose et al. 2022; Shin and Brangwynne 2017). Several studies also reported the global association of NEAT1/paraspeckles with multiple chromosomal loci that tend to be transcriptionally active (Bonetti et al. 2020; Li et al. 2017b; Sridhar et al. 2017; West et al. 2014). The chromatin association likely occurs on the surface rather than the internal core of paraspeckles, suggesting that the hydrophilic shell serves as the platform for regulating the associated chromatin loci.

Second, the shell–core structure contributes to the independence of paraspeckles from other nuclear MLOs in nucleoplasmic space. Paraspeckles are usually observed adjacent to nuclear speckles that are enriched in splicing factors (Fox et al. 2002). Similarly, stress granules are often associated with the compositionally related p-bodies in the cytoplasm (Kedersha et al. 2005). In this case, a recent study has suggested that this multiphase coexistence is determined by protein networks that overlap between stress granules and p-bodies (Sanders et al. 2020). Therefore, it is possible that the protein-protein interactions between paraspeckles and nuclear speckles analogously contribute to the determination of the separate but attached arrangement. Indeed, we have recently reported that mini-NEAT1 paraspeckles (mini-PSs), formed by a NEAT1_2 deletion mutant possessing only the stability domains (0–1 kb and 3′TH) and the middle assembly domain (8–16.6 kb), are incorporated into nuclear speckles unlike WT-paraspeckles (WT-PSs) (Fig. 8.5 and 8.6a) (Takakuwa et al. 2023; Yamazaki et al. 2018). This observation suggests that specific NEAT1 regions and RBPs are required for the segregation of paraspeckles from nuclear speckles. The mini-PS maintained an ordered core-shell structure within the nuclear speckles (Fig. 8.5) (Takakuwa et al. 2023; Yamazaki et al. 2018). Artificial tethering of SFPQ, HNRNPF, or BRG1 to the shell of mini-PSs rescued the segregation of paraspeckles from nuclear speckles (Takakuwa et al. 2023). Electron microscopy and super-resolution microscopy observations showed that those proteins localize to the shell of WT-PSs (Kawaguchi et al. 2015; Takakuwa et al. 2023). Meanwhile, tethering of those proteins to the core of mini-PSs could not rescue the paraspeckle segregation, suggesting that proteins localized to the shell of the paraspeckle determine their positioning in the nucleus (Takakuwa et al. 2023).

SFPQ forms higher-order structures by spreading on the nucleic acids (Lee et al. 2015). Artificial tethering experiments with SFPQ mutant proteins revealed that oligomerization and PLD enriched with proline and glutamine residues are required for paraspeckle segregation (Takakuwa et al. 2023). Specifically, the SFPQ PLD

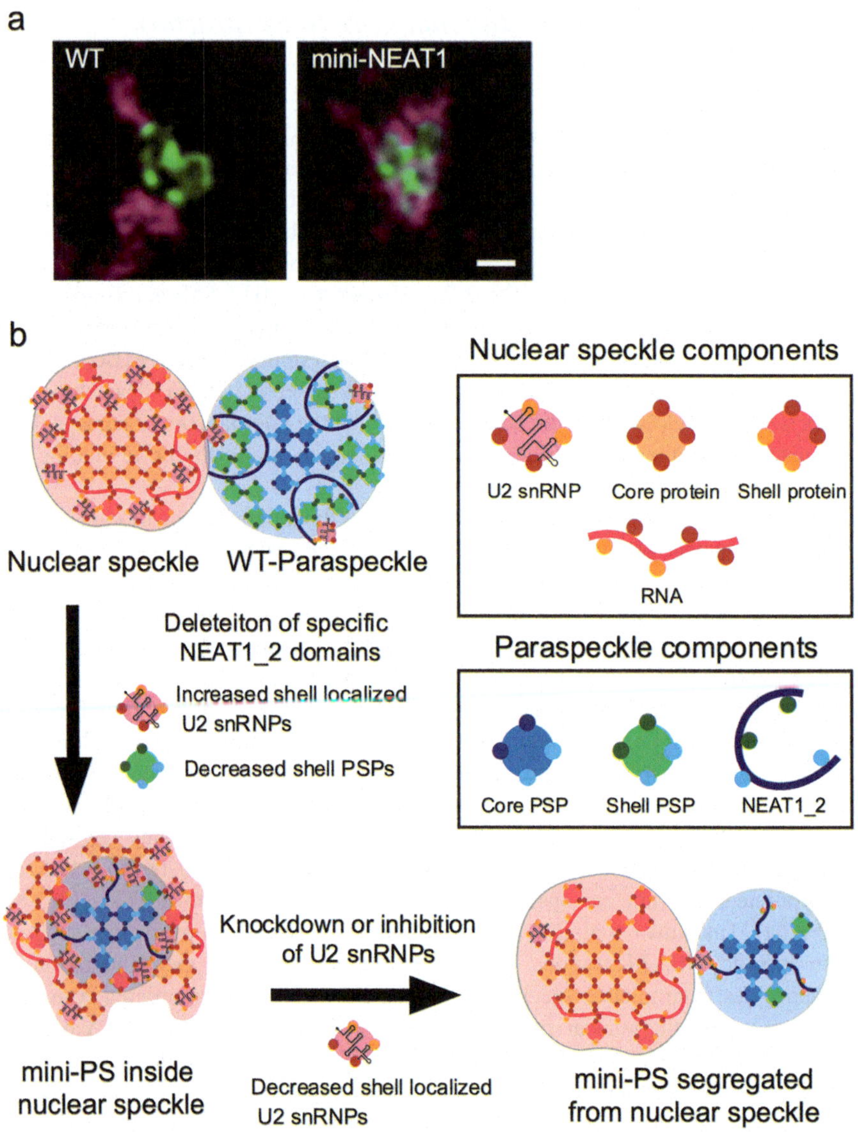

Fig. 8.6 Shell protein composition of paraspeckles dictates their independence from nuclear speckles. (**a**) 3D super-resolution images of paraspeckles and nuclear speckles detected by RNA-FISH with NEAT1 antisense probes (NEAT1_5′: green) and immunostaining of SRRM2 (magenta) in WT and mini-NEAT1 mutant cells. (**b**) Schematic illustrations show how protein networks between paraspeckles and nuclear speckles determine their localization in the nucleus. In paraspeckles with core-shell architecture, shell protein composition is a more important determinant of paraspeckle localization than core protein composition

mutant with glutamine substituted for glycine compromised the paraspeckle segregation function, although it retained the paraspeckle assembly function (Takakuwa et al. 2023). These findings demonstrate that SFPQ has two separable functions: paraspeckle assembly (in the core region) and segregation from nuclear speckles (in the shell region).

Mini-PSs are specifically incorporated into the nuclear speckles, but not into other MLOs such as promyelocytic leukemia protein (PML) bodies and histone locus bodies (HLBs) (Takakuwa et al. 2023), suggesting the existence of a molecular mechanism for the incorporation of mini-PSs into nuclear speckles. In vitro RNA pull-down experiments showed that U2 snRNP-related proteins bind to 8–9.8 kb regions, which localize to the shell of mini-PSs (Takakuwa et al. 2023). These results raise an intriguing possibility that U2 snRNP-related proteins localized to the shell of mini-PSs promote mini-PS incorporation into nuclear speckles. Indeed, artificial tethering of SF3B4 or SF3A3, components of U2 snRNP, to the shell of the WT-PSs induced paraspeckle incorporation into nuclear speckles like mini-PSs, whereas tethering of these proteins to the core of the paraspeckles did not (Takakuwa et al. 2023). Furthermore, knockdown or chemical inhibition of U2 snRNP components induced mini-PS segregation from nuclear speckles (Fig. 8.6b) (Takakuwa et al. 2023). Since U2 snRNP components interact with other nuclear speckle proteins and RNAs and are mainly present near the peripheral regions of nuclear speckles (Fei et al. 2017), these protein–protein and/or protein–RNA interactions occurring at the interface between MLOs would determine their multiphase structure (Fig. 8.6b). Although it has been proposed that interfacial tension between two MLOs plays a role in the formation of such multiphase structures (Feric et al. 2016; Gouveia et al. 2022), the biophysical determinants of this miscibility and/or immiscibility are still unknown. Further studies on the effects of these factors (e.g., SFPQ, BRG1, and U2 snRNP components) on the biophysical properties of MLOs should provide significant insights into the multiphase structures of MLOs.

8.4 Conclusion and Future Perspectives

This review focuses on the arcRNAs that act as architectural scaffolds of phase-separated MLOs. The identified arcRNAs play critical roles in stress responses, specific developmental stages, and pathological conditions. Therefore, the identification of novel arcRNAs and their cognate MLO components would provide important biological and pathological insights to evaluate the significance of phase separation in the cells.

Several methods have suggested the presence of numerous unidentified arcRNAs in the human transcriptome. Recently, we developed a method called semi-extractable RNA-seq to search for RNAs that exhibit semi-extractability, which may be a hallmark of arcRNAs (Chujo et al. 2017; Iwakiri et al. 2023). Using semi-extractable RNA-seq, we identified 45 semi-extractable RNAs in HeLa cells under normal conditions, of which the top ten most abundant RNAs form nuclear foci that are distinct from known nuclear bodies (Chujo et al. 2017). Furthermore, we searched for stress-inducible semi-extractable RNAs and found that hundreds of downstream-of-gene (DoG) transcripts, transcriptional readthrough from the upstream of protein-coding genes, induced by hyperosmotic stress or thermal stress exhibited semi-extractability (Iwakiri et al. 2023). Hyperosmotic stress-induced SRSF1 and RPS12 DoGs were found to form distinct nuclear foci that are sensitive to 1,6-HD (Iwakiri et al. 2023). A bioinformatic approach also showed that short tandem repeat-enriched RNAs (strRNAs) have strong RBP-interaction potential, and one such strRNA, the PNCTR arcRNA, is a core of PNC (Yap et al. 2018). It is thus important to establish and improve methods to search for arcRNAs in various stress conditions, cell lines, and species.

More than half of the PSPs have been identified by colocalization screening using a Venus-tagged cDNA library (Fong et al. 2013; Mannen et al. 2016; Naganuma et al. 2012). Meanwhile, comprehensive proteomic approaches are currently being used to determine the protein composition of MLOs. For example, our recent chromatin isolation by RNA purification followed by mass spectrometry (ChIRP-MS) analysis identified 141 proteins as nSB protein candidates (Ninomiya et al. 2020, 2021). In addition, proximity labeling methods using an engineered ascorbate peroxidase (APEX2) or BioID have been used to determine the protein composition of the phase-separated MLOs, such as stress granules (Markmiller et al. 2018; Marmor-Kollet et al. 2020; Padrón et al. 2019; Youn et al. 2018). A recent study using APEX2-based hybridization-proximity (HyPro) labeling technologies identified proteins and transcripts associated with RNAs including NEAT1_2 and PNCTR arcRNAs (Yap et al. 2022). Further identification of the protein and RNA components of the MLOs scaffolded by arcRNAs should provide important insights into their biological functions and formation mechanisms.

As a model for arcRNAs, we have studied the NEAT1_2 arcRNA to understand how arcRNAs form MLOs. However, many questions remain unanswered. While we have identified multiple functional NEAT1_2 domains that can recruit specific proteins to the appropriate area of paraspeckles, we have still not yet completely narrowed down which RNA elements are recognized by RBPs. Given the low cross-species sequence homology for NEAT1_2, it is possible that dispersed short sequence stretches, or conserved secondary or tertiary structures of NEAT1_2 are recognized by RBPs. Although paraspeckles are involved in developmental processes, stress responses, and various diseases such as cancer, viral infection, and neurodegenerative diseases, the molecular mechanisms underlying these physiological functions remain elusive. It is important to understand the links between the structural characteristics of paraspeckles and their physiological and pathological roles and ultimately understand how arcRNA sequence information dictates the function of MLOs.

Acknowledgments We thank all members of the Hirose laboratory for their encouragement and helpful discussions.

References

Adriaens C, Standaert L, Barra J et al (2016) P 53 induces formation of NEAT1 lncRNA-containing paraspeckles that modulate replication stress response and chemosensitivity. Nat Med 22:861–868. https://doi.org/10.1038/nm.4135

Adriaens C, Rambow F, Bervoets G et al (2019) The long noncoding RNA NEAT1_1 is seemingly dispensable for normal tissue homeostasis and cancer cell growth. RNA 25:1681. https://doi.org/10.1261/rna.071456.119

An H, Tan JT, Shelkovnikova TA (2019) Stress granules regulate stress-induced paraspeckle assembly. J Cell Biol 218:4127. https://doi.org/10.1083/JCB.201904098

Ashburner M (1970) Patterns of puffing activity in the salivary gland chromosomes of drosophila - V. Chromosoma 31:398. https://doi.org/10.1007/BF00321231

Audas TE, Jacob MD, Lee S (2012) Immobilization of proteins in the nucleolus by ribosomal intergenic spacer noncoding RNA. Mol Cell 45:147–157. https://doi.org/10.1016/j.molcel.2011.12.012

Audas TE, Audas DE, Jacob MD et al (2016) Adaptation to stressors by systemic protein amyloidogenesis. Dev Cell 39:155. https://doi.org/10.1016/j.devcel.2016.09.002

Banani SF, Lee HO, Hyman AA, Rosen MK (2017) Biomolecular condensates: organizers of cellular biochemistry. Nat Rev Mol Cell Biol 18:285–298. https://doi.org/10.1038/nrm.2017.7

Barra J, Gaidos GS, Blumenthal E et al (2020) Integrator restrains paraspeckles assembly by promoting isoform switching of the lncRNA NEAT1. Sci Adv 6:eaaz9072. https://doi.org/10.1126/sciadv.aaz9072

Bonetti A, Agostini F, Suzuki AM et al (2020) RADICL-seq identifies general and cell type–specific principles of genome-wide RNA-chromatin interactions. Nat Commun 11:1018. https://doi.org/10.1038/s41467-020-14337-6

Brown JA, Valenstein ML, Yario TA et al (2012) Formation of triple-helical structures by the 3′-end sequences of MALAT1 and MENβ noncoding RNAs. Proc Natl Acad Sci 109:19202–19207. https://doi.org/10.1073/pnas.1217338109

Castello A, Fischer B, Eichelbaum K et al (2012) Insights into RNA biology from an atlas of mammalian mRNA-binding proteins. Cell 149:1393. https://doi.org/10.1016/j.cell.2012.04.031

Charlet BN, Savkur RS, Singh G et al (2002) Loss of the muscle-specific Chloride Channel in type 1 myotonic dystrophy due to misregulated alternative splicing. Mol Cell 10:45–53. https://doi.org/10.1016/S1097-2765(02)00572-5

Chen L-L, Carmichael GG (2009) Altered nuclear retention of mRNAs containing inverted repeats in human embryonic stem cells: functional role of a nuclear noncoding RNA. Mol Cell 35:467–478. https://doi.org/10.1016/j.molcel.2009.06.027

Chen B, Deng S, Ge T et al (2020) Live cell imaging and proteomic profiling of endogenous NEAT1 lncRNA by CRISPR/Cas9-mediated knock-in. Protein Cell 11:641. https://doi.org/10.1007/s13238-020-00706-w

Chujo T, Yamazaki T, Hirose T (2016) Architectural RNAs (arcRNAs): a class of long noncoding RNAs that function as the scaffold of nuclear bodies. Biochim Biophys Acta 1859:139

Chujo T, Yamazaki T, Kawaguchi T et al (2017) Unusual semi-extractability as a hallmark of nuclear body-associated architectural noncoding RNAs. EMBO J 36:1447–1462. https://doi.org/10.15252/embj.201695848

Clark MB, Johnston RL, Inostroza-Ponta M et al (2012) Genome-wide analysis of long noncoding RNA stability. Genome Res 22:885. https://doi.org/10.1101/gr.131037.111

Clemson CM, Hutchinson JN, Sara SA et al (2009) An architectural role for a nuclear noncoding RNA: NEAT1 RNA is essential for the structure of Paraspeckles. Mol Cell 33:717–726. https://doi.org/10.1016/j.molcel.2009.01.026

Cochard A, Garcia-Jove Navarro M, Piroska L et al (2022) RNA at the surface of phase-separated condensates impacts their size and number. Biophys J 121:1675. https://doi.org/10.1016/j.bpj.2022.03.032

Cuylen S, Blaukopf C, Politi AZ et al (2016) Ki-67 acts as a biological surfactant to disperse mitotic chromosomes. Nature 535:308. https://doi.org/10.1038/nature18610

Dumbović G, Biayna J, Banús J et al (2018) A novel long non-coding RNA from NBL2 pericentromeric macrosatellite forms a perinucleolar aggregate structure in colon cancer. Nucleic Acids Res 46:5504. https://doi.org/10.1093/nar/gky263

Elguindy MM, Mendell JT (2021) NORAD-induced Pumilio phase separation is required for genome stability. Nature 595. https://doi.org/10.1038/s41586-021-03633-w

Fei J, Jadaliha M, Harmon TS et al (2017) Quantitative analysis of multilayer organization of proteins and RNA in nuclear speckles at super resolution. J Cell Sci 130:4180–4192. https://doi.org/10.1242/jcs.206854

Feric M, Vaidya N, Harmon TS et al (2016) Coexisting liquid phases underlie nucleolar subcompartments. Cell 165:1686. https://doi.org/10.1016/j.cell.2016.04.047

Fong KW, Li Y, Wang W et al (2013) Whole-genome screening identifies proteins localized to distinct nuclear bodies. J Cell Biol 203:149. https://doi.org/10.1083/jcb.201303145

Fox AH, Lam YW, Leung AKL et al (2002) Paraspeckles: a novel nuclear domain. Curr Biol 12: 13–25. https://doi.org/10.1016/S0960-9822(01)00632-7

Fox AH, Bond CS, Lamond AI (2005) P54nrb forms a heterodimer with PSP1 that localizes to Paraspeckles in an RNA-dependent manner. Mol Biol Cell 16:5304–5315. https://doi.org/10.1091/mbc.e05-06-0587

Fox AH, Nakagawa S, Hirose T, Bond CS (2018) Paraspeckles: where long noncoding RNA meets phase separation. Trends Biochem Sci 43:124

Freyermuth F, Rau F, Kokunai Y et al (2016) Splicing misregulation of SCN5A contributes to cardiac-conduction delay and heart arrhythmia in myotonic dystrophy. Nat Commun 7:11067. https://doi.org/10.1038/ncomms11067

Fugier C, Klein AF, Hammer C et al (2011) Misregulated alternative splicing of BIN1 is associated with T tubule alterations and muscle weakness in myotonic dystrophy. Nat Med 17:720. https://doi.org/10.1038/nm.2374

Fukushima K, Satoh T, Sugihara F et al (2020) Dysregulated expression of the nuclear exosome targeting complex component Rbm7 in nonhematopoietic cells licenses the development of fibrosis. Immunity 52:542. https://doi.org/10.1016/j.immuni.2020.02.007

Garbe JC, Pardue ML (1986) Heat shock locus 93D of Drosophila melanogaster: a spliced RNA most strongly conserved in the intron sequence. Proc Natl Acad Sci U S A 83:1812. https://doi.org/10.1073/pnas.83.6.1812

Garcia-Jove Navarro M, Kashida S, Chouaib R et al (2019) RNA is a critical element for the sizing and the composition of phase-separated RNA–protein condensates. Nat Commun 10:3230. https://doi.org/10.1038/s41467-019-11241-6

Gerstberger S, Hafner M, Tuschl T (2014) A census of human RNA-binding proteins. Nat Rev Genet 15:829. https://doi.org/10.1038/nrg3813

Gouveia B, Kim Y, Shaevitz JW et al (2022) Capillary forces generated by biomolecular condensates. Nature 609:255. https://doi.org/10.1038/s41586-022-05138-6

Hennig S, Kong G, Mannen T et al (2015) Prion-like domains in RNA binding proteins are essential for building subnuclear paraspeckles. J Cell Biol 210:529–539. https://doi.org/10.1083/jcb.201504117

Hirose T, Virnicchi G, Tanigawa A et al (2014) NEAT1 long noncoding RNA regulates transcription via protein sequestration within subnuclear bodies. Mol Biol Cell 25:169–183. https://doi.org/10.1091/mbc.e13-09-0558

Hirose T, Ninomiya K, Nakagawa S, Yamazaki T (2022) A guide to membraneless organelles and their various roles in gene regulation. Nat Rev Mol Cell Biol 24(4):288

Hogan NC, Traverse KL, Sullivan DE, Lou PM (1994) The nucleus-limited Hsr-omega-n transcript is a polyadenylated RNA with a regulated intranuclear turnover. J Cell Biol 125:21. https://doi.org/10.1083/jcb.125.1.21

Hu SB, Xiang JF, Li X et al (2015) Protein arginine methyltransferase CARM1 attenuates the paraspecklemediated nuclear retention of mRNAs containing IRAlus. Genes Dev 29:630. https://doi.org/10.1101/gad.257048.114

Imamura K, Imamachi N, Akizuki G et al (2014) Long noncoding RNA NEAT1-dependent SFPQ relocation from promoter region to paraspeckle mediates IL8 expression upon immune stimuli. Mol Cell 53:393–406. https://doi.org/10.1016/j.molcel.2014.01.009

Imamura K, Takaya A, Ishida Y et al (2018) Diminished nuclear RNA decay upon salmonella infection upregulates antibacterial noncoding RNA s. EMBO J 37:e97723. https://doi.org/10.15252/embj.201797723

Isobe M, Toya H, Mito M et al (2020) Forced isoform switching of Neat1_1 to Neat1_2 leads to the loss of Neat1_1 and the hyperformation of paraspeckles but does not affect the development and growth of mice. RNA 26:251. https://doi.org/10.1261/rna.072587.119

Iwakiri J, Tanaka K, Chujo T et al (2023) Remarkable improvement in detection of readthrough downstream-of-gene transcripts by semi-extractable RNA-sequencing. RNA 29:170. https://doi.org/10.1261/rna.079469.122

Jain A, Vale RD (2017) RNA phase transitions in repeat expansion disorders. Nature 546:243. https://doi.org/10.1038/nature22386

Jiang L, Shao C, Wu QJ et al (2017) NEAT1 scaffolds RNA-binding proteins and the microprocessor to globally enhance pri-miRNA processing. Nat Struct Mol Biol 24:816. https://doi.org/10.1038/nsmb.3455

Jolly C, Metz A, Govin J et al (2004) Stress-induced transcription of satellite III repeats. J Cell Biol 164:25. https://doi.org/10.1083/jcb.200306104

Kawaguchi T, Hirose T (2015) Chromatin remodeling complexes in the assembly of long noncoding RNA-dependent nuclear bodies. Nucleus 6:462–467. https://doi.org/10.1080/19491034.2015.1119353

Kawaguchi T, Tanigawa A, Naganuma T et al (2015) SWI/SNF chromatin-remodeling complexes function in noncoding RNA-dependent assembly of nuclear bodies. Proc Natl Acad Sci 112:4304–4309. https://doi.org/10.1073/pnas.1423819112

Kedersha N, Stoecklin G, Ayodele M et al (2005) Stress granules and processing bodies are dynamically linked sites of mRNP remodeling. J Cell Biol 169:871–884. https://doi.org/10.1083/jcb.200502088

Lee M, Sadowska A, Bekere I et al (2015) The structure of human SFPQ reveals a coiled-coil mediated polymer essential for functional aggregation in gene regulation. Nucleic Acids Res 43:3826–3840. https://doi.org/10.1093/nar/gkv156

Li R, Harvey AR, Hodgetts SI, Fox AH (2017a) Functional dissection of NEAT1 using genome editing reveals substantial localization of the NEAT1_1 isoform outside paraspeckles. RNA 23:872–881. https://doi.org/10.1261/rna.059477.116

Li X, Zhou B, Chen L et al (2017b) GRID-seq reveals the global RNA-chromatin interactome. Nat Biotechnol 35:940. https://doi.org/10.1038/nbt.3968

Lunde BM, Moore C, Varani G (2007) RNA-binding proteins: modular design for efficient function. Nat Rev Mol Cell Biol 8:479

Machitani M, Taniguchi I, Ohno M (2020) ARS2 regulates nuclear paraspeckle formation through 3′-end processing and stability of NEAT1 long noncoding RNA. Mol Cell Biol 40:e00269–e00219. https://doi.org/10.1128/mcb.00269-19

Maharana S, Wang J, Papadopoulos DK et al (2018) RNA buffers the phase separation behavior of prion-like RNA binding proteins. Science 360:918. https://doi.org/10.1126/science.aar7366

Malik I, Kelley CP, Wang ET, Todd PK (2021) Molecular mechanisms underlying nucleotide repeat expansion disorders. Nat Rev Mol Cell Biol 22:589

Mankodi A, Urbinati CR, Yuan QP et al (2001) Muscleblind localizes to nuclear foci of aberrant RNA in myotonic dystrophy types 1 and 2. Hum Mol Genet 10:2165. https://doi.org/10.1093/hmg/10.19.2165

Mannen T, Yamashita S, Tomita K et al (2016) The Sam68 nuclear body is composed of two RNasesensitive substructures joined by the adaptor HNR NPL. J Cell Biol 214:45. https://doi.org/10.1083/jcb.201601024

Mao YS, Sunwoo H, Zhang B, Spector DL (2011) Direct visualization of the co-transcriptional assembly of a nuclear body by noncoding RNAs. Nat Cell Biol 13:95–101. https://doi.org/10.1038/ncb2140

Markmiller S, Soltanieh S, Server KL et al (2018) Context-dependent and disease-specific diversity in protein interactions within stress granules. Cell 172:590. https://doi.org/10.1016/j.cell.2017.12.032

Marmor-Kollet H, Siany A, Kedersha N et al (2020) Spatiotemporal proteomic analysis of stress granule disassembly using APEX reveals regulation by SUMOylation and links to ALS pathogenesis. Mol Cell 80:876. https://doi.org/10.1016/j.molcel.2020.10.032

Mello SS, Sinow C, Raj N et al (2017) *Neat1* is a p53-inducible lincRNA essential for transformation suppression. Genes Dev 31:1095–1108. https://doi.org/10.1101/gad.284661.116

Miller JW, Urbinati CR, Teng-Umnuay P et al (2000) Recruitment of human muscleblind proteins to (CUG)(n) expansions associated with myotonic dystrophy. EMBO J 19:4439. https://doi.org/10.1093/emboj/19.17.4439

Modic M, Grosch M, Rot G et al (2019) Cross-regulation between TDP-43 and Paraspeckles promotes pluripotency-differentiation transition. Mol Cell 74:951–965.e13. https://doi.org/10.1016/j.molcel.2019.03.041

Mohammad Lellahi S, Rosenlund IA, Hedberg A et al (2018) The long noncoding RNA NEAT1 and nuclear paraspeckles are up-regulated by the transcription factor HSF1 in the heat shock response. J Biol Chem 293:18965. https://doi.org/10.1074/jbc.RA118.004473

Naganuma T, Nakagawa S, Tanigawa A et al (2012) Alternative 3′-end processing of long noncoding RNA initiates construction of nuclear paraspeckles. EMBO J 31:4020–4034. https://doi.org/10.1038/emboj.2012.251

Nakagawa S, Naganuma T, Shioi G, Hirose T (2011) Paraspeckles are subpopulation-specific nuclear bodies that are not essential in mice. J Cell Biol 193:31. https://doi.org/10.1083/jcb.201011110

Nakagawa S, Shimada M, Yanaka K et al (2014) The lncRNA *Neat1* is required for corpus luteum formation and the establishment of pregnancy in a subpopulation of mice. Development 141:4618–4627. https://doi.org/10.1242/dev.110544

Ninomiya K, Adachi S, Natsume T et al (2020) Lnc RNA -dependent nuclear stress bodies promote intron retention through SR protein phosphorylation. EMBO J 39:e102729. https://doi.org/10.15252/embj.2019102729

Ninomiya K, Iwakiri J, Aly MK et al (2021) M 6 a modification of HSATIII lncRNAs regulates temperature-dependent splicing. EMBO J 40:e107976. https://doi.org/10.15252/embj.2021107976

Nishimoto Y, Nakagawa S, Hirose T et al (2013) The long non-coding RNA nuclear-enriched abundant transcript 1-2 induces paraspeckle formation in the motor neuron during the early phase of amyotrophic lateral sclerosis. Mol Brain 6:1–8. https://doi.org/10.1186/1756-6606-6-31

Padrón A, Iwasaki S, Ingolia NT (2019) Proximity RNA labeling by APEX-Seq reveals the organization of translation initiation complexes and repressive RNA granules. Mol Cell 75:875. https://doi.org/10.1016/j.molcel.2019.07.030

Philips AV, Timchenko LT, Cooper TA (1998) Disruption of splicing regulated by a CUG-binding protein in myotonic dystrophy. Science 280:737. https://doi.org/10.1126/science.280.5364.737

Prasanth KV, Prasanth SG, Xuan Z et al (2005) Regulating gene expression through RNA nuclear retention. Cell 123:249. https://doi.org/10.1016/j.cell.2005.08.033

Prikryl J, Rojas M, Schuster G, Barkan A (2011) Mechanism of RNA stabilization and translational activation by a pentatricopeptide repeat protein. Proc Natl Acad Sci U S A 108:415. https://doi.org/10.1073/pnas.1012076108

Querido E, Gallardo F, Beaudoin M et al (2011) Stochastic and reversible aggregation of mRNA with expanded CUG-triplet repeats. J Cell Sci 124:1703. https://doi.org/10.1242/jcs.073270

Rizzi N, Denegri M, Chiodi I et al (2004) Transcriptional activation of a constitutive heterochromatic domain of the human genome in response to heat shock. Mol Biol Cell 15:543. https://doi.org/10.1091/mbc.E03-07-0487

Roden C, Gladfelter AS (2021) RNA contributions to the form and function of biomolecular condensates. Nat Rev Mol Cell Biol 22:183–195. https://doi.org/10.1038/s41580-020-0264-6

Sanders DW, Kedersha N, Lee DSW et al (2020) Competing protein-RNA interaction networks control multiphase intracellular organization. Cell 181:306–324.e28. https://doi.org/10.1016/j.cell.2020.03.050

Sasaki YTF, Ideue T, Sano M et al (2009) MENε/β noncoding RNAs are essential for structural integrity of nuclear paraspeckles. Proc Natl Acad Sci 106:2525–2530. https://doi.org/10.1073/pnas.0807899106

Savkur RS, Philips AV, Cooper TA (2001) Aberrant regulation of insulin receptor alternative splicing is associated with insulin resistance in myotonic dystrophy. Nat Genet 29:40. https://doi.org/10.1038/ng704

Shibata T, Nagano K, Ueyama M et al (2021) Small molecule targeting r(UGGAA)n disrupts RNA foci and alleviates disease phenotype in drosophila model. Nat Commun 12:236. https://doi.org/10.1038/s41467-020-20487-4

Shin Y, Brangwynne CP (2017) Liquid phase condensation in cell physiology and disease. Science 357:eaaf4382. https://doi.org/10.1126/science.aaf4382

Simko EAJ, Liu H, Zhang T et al (2020) G-quadruplexes offer a conserved structural motif for NONO recruitment to NEAT1 architectural lncRNA. Nucleic Acids Res 48:7421. https://doi.org/10.1093/nar/gkaa475

Souquere S, Beauclair G, Harper F et al (2010) Highly ordered spatial Organization of the Structural Long Noncoding NEAT1 RNAs within Paraspeckle nuclear bodies. Mol Biol Cell 21:4020–4027. https://doi.org/10.1091/mbc.e10-08-0690

Sridhar B, Rivas-Astroza M, Nguyen TC et al (2017) Systematic mapping of RNA-chromatin interactions in vivo. Curr Biol 27:602. https://doi.org/10.1016/j.cub.2017.01.011

Standaert L, Adriaens C, Radaelli E et al (2014) The long noncoding RNA Neat1 is required for mammary gland development and lactation. RNA 20:1844. https://doi.org/10.1261/rna.047332.114

Sunwoo H, Dinger ME, Wilusz JE et al (2009) *MEN ε/β* nuclear-retained non-coding RNAs are up-regulated upon muscle differentiation and are essential components of paraspeckles. Genome Res 19:347–359. https://doi.org/10.1101/gr.087775.108

Swinnen B, Robberecht W, Van Den Bosch L (2020) RNA toxicity in non-coding repeat expansion disorders. EMBO J 39:e101112. https://doi.org/10.15252/embj.2018101112

Takakuwa H, Yamazaki T, Souquere S, et al (2023) Shell protein composition specified by NEAT1 domains dictates the formation of paraspeckles as distinct membraneless organelles. bioRxiv. 2023.05.21.541661. https://doi.org/10.1101/2023.05.21.541661

Taneja KL, McCurrach M, Schalling M et al (1995) Foci of trinucleotide repeat transcripts in nuclei of myotonic dystrophy cells and tissues. J Cell Biol 128:995. https://doi.org/10.1083/jcb.128.6.995

Tanu T, Taniue K, Imamura K et al (2021) hnRNPH1-MTR4 complex-mediated regulation of NEAT1v2 stability is critical for IL8 expression. RNA Biol 18:537. https://doi.org/10.1080/15476286.2021.1971439

Tollervey JR, Curk T, Rogelj B et al (2011) Characterizing the RNA targets and position-dependent splicing regulation by TDP-43. Nat Neurosci 14:452. https://doi.org/10.1038/nn.2778

Van Treeck B, Parker R (2018) Emerging roles for intermolecular RNA-RNA interactions in RNP assemblies. Cell 174:791–802. https://doi.org/10.1016/j.cell.2018.07.023

Visa N, Puvion-Dutilleul F, Bachellerie JP, Puvion E (1993) Intranuclear distribution of U1 and U2 snRNAs visualized by high resolution in situ hybridization: revelation of a novel compartment containing U1 but not U2 snRNA in HeLa cells. Eur J Cell Biol 60:308–321

Wang J, Choi JM, Holehouse AS et al (2018a) A molecular grammar governing the driving forces for phase separation of prion-like RNA binding proteins. Cell 174:688. https://doi.org/10.1016/j.cell.2018.06.006

Wang Y, Bin HS, Wang MR et al (2018b) Genome-wide screening of NEAT1 regulators reveals cross-regulation between paraspeckles and mitochondria. Nat Cell Biol 20:1145. https://doi.org/10.1038/s41556-018-0204-2

West JA, Davis CP, Sunwoo H et al (2014) The long noncoding RNAs NEAT1 and MALAT1 bind active chromatin sites. Mol Cell 55:791. https://doi.org/10.1016/j.molcel.2014.07.012

West JA, Mito M, Kurosaka S et al (2016) Structural, super-resolution microscopy analysis of paraspeckle nuclear body organization. J Cell Biol 214:817–830. https://doi.org/10.1083/jcb.201601071

Wilusz JE, JnBaptiste CK, Lu LY et al (2012) A triple helix stabilizes the 3′ ends of long noncoding RNAs that lack poly(a) tails. Genes Dev 26:2392–2407. https://doi.org/10.1101/gad.204438.112

Yamamoto T, Yamazaki T, Hirose T (2020) Phase separation driven by production of architectural RNA transcripts. Soft Matter 16. https://doi.org/10.1039/c9sm02458a

Yamamoto T, Yamazaki T, Hirose T (2022) Triblock copolymer micelle model of spherical paraspeckles. Front Mol Biosci 9:925058. https://doi.org/10.3389/fmolb.2022.925058

Yamazaki T, Souquere S, Chujo T et al (2018) Functional domains of NEAT1 architectural lncRNA induce paraspeckle assembly through phase separation. Mol Cell 70:1038–1053.e7. https://doi.org/10.1016/j.molcel.2018.05.019

Yamazaki T, Nakagawa S, Hirose T (2019) Architectural RNAs for membraneless nuclear body formation. Cold Spring Harb Symp Quant Biol 84:227–237. https://doi.org/10.1101/sqb.2019.84.039404

Yamazaki T, Yamamoto T, Yoshino H et al (2021) Paraspeckles are constructed as block copolymer micelles. EMBO J 40:e107270. https://doi.org/10.15252/embj.2020107270

Yamazaki T, Yamamoto T, Hirose T (2022) Micellization: a new principle in the formation of biomolecular condensates. Front Mol Biosci 9:974772. https://doi.org/10.3389/fmolb.2022.974772

Yang LZ, Wang Y, Li SQ et al (2019) Dynamic imaging of RNA in living cells by CRISPR-Cas13 systems. Mol Cell 76:981. https://doi.org/10.1016/j.molcel.2019.10.024

Yap K, Mukhina S, Zhang G et al (2018) A short tandem repeat-enriched RNA assembles a nuclear compartment to control alternative splicing and promote cell survival. Mol Cell 72:525–540.e13. https://doi.org/10.1016/j.molcel.2018.08.041

Yap K, Chung TH, Makeyev EV (2022) Hybridization-proximity labeling reveals spatially ordered interactions of nuclear RNA compartments. Mol Cell 82:463–478.e11. https://doi.org/10.1016/j.molcel.2021.10.009

Youn JY, Dunham WH, Hong SJ et al (2018) High-density proximity mapping reveals the subcellular organization of mRNA-associated granules and bodies. Mol Cell 69:517. https://doi.org/10.1016/j.molcel.2017.12.020

Part III
Biology

Chapter 9
Force-Dependent Remodeling of Cell-to-Cell Adhesion Through the Regulation of Phase Separation

Noriyuki Kinoshita, Yutaka Hashimoto, and Naoto Ueno

Abstract During development and regeneration, cells vigorously migrate to achieve morphogenesis of tissues and organs. Upon these dynamic processes, cells are exposed to various forces such as compression force, stretching force, etc. Although the importance of physical forces in the regulation of morphogenesis has long been discussed, the molecular and cellular mechanism of the force-dependent processes remain largely unknown. We have addressed the problem using embryos of amphibian *Xenopus laevis* and mouse as well as cultured cells. This series of studies highlighted that physical forces induce remodeling of cell-to-cell adhesion which is in part controlled through the regulation of phase separation. The finding suggests that the cellular response to forces is a mechanochemical feedback system to ensure normal morphogenesis during animal development.

Keywords Early embryo · Force · Cell movements · Cell-to-cell adhesion · ZO-1 · Liquid–liquid phase separation

N. Kinoshita
National Institute for Basic Biology, Okazaki, Japan

SOKENDAI, The Graduate University of Advanced Studies, Okazaki, Japan
e-mail: nkinoshi@nibb.ac.jp

Y. Hashimoto
Graduate School of Medical Sciences, Nagoya City University, Nagoya, Japan
e-mail: hashimot@med.nagoya-cu.ac.jp

N. Ueno (✉)
National Institute for Basic Biology, Okazaki, Japan

SOKENDAI, The Graduate University of Advanced Studies, Okazaki, Japan

International Research Collaboration Center, National Institutes of Natural Sciences, Tokyo, Japan
e-mail: nueno@nibb.ac.jp

R. Kurokawa (ed.), *Phase Separation in Living Cells*,
https://doi.org/10.1007/978-981-99-4886-4_9

9.1 Introduction

It is well recognized that tissues and organs of animal body are constantly exposed to various types of physical forces generated by heart beating, blood flow, luminal pressure of body cavities, locomotion, etc. (Chan et al. 2019; Paolini and Abdelilah-Seyfried 2018). It is also known that cells and tissues undergo drastic shape changes by forces generated mainly by dynamic morphogenetic events during development, mainly driven by cell and tissue rearrangements through cells' own proliferation and migration (Godard and Heisenberg 2019; Bodor et al. 2020). However, it is less understood how the development of early embryos are influenced by these extracellular forces. To understand how physical forces contribute to early development, it is very important to address how embryonic cells sense physical forces generated during development and how the cells respond to the stimuli, at the molecular and cellular levels. In order to directly investigate potential physiological roles of physical forces in development, direct approach is to apply forces on whole embryo or embryonic tissues (Goddard et al. 2020).

We first focused on cell stretching of epithelial cells of *Xenopus laevis* embryo that occurs during gastrulation cell movements which is a critical tissue remodeling for organogenesis. In this event, common to many multicellular animal embryos, an oriented and synchronized cell movement termed "epiboly" which is driven by actomyosin of leading cells pulls following epithelial cells, and therefore, the following cell become significantly stretched and adopt elongated shape along the axis of the force. Taking this phenomenon as a model, we investigated how the epithelial cells sense the stretching force and respond to change their properties (Hashimoto et al. 2019).

Second, we investigated biochemical responses of cellular proteins to forces, particularly focusing on protein phosphorylation. The comprehensive proteomic analysis revealed that an array of proteins which are related to cell-to-cell as well as cell-to-substrate adhesion became phosphorylated upon force application. The result suggested that mechanical stimuli could affect cell behaviors through the phosphorylation of proteins (Hashimoto et al. 2019).

Further cellular analyses identified that ZO-1, one of tight junction (TJ) proteins change its behavior in the responding cells: its condensation in the cytoplasm diminished upon force and changes its localization to the membrane. ZO-1 links transmembrane proteins such as occludin and claudins and actin cytoskeletons (Otani and Furuse 2020), thereby mechanical forces loaded on the TJ change the conformation of ZO-1 (Spadaro et al. 2017; Haas et al. 2020). Therefore, we speculated that this remodeling of ZO-1 upon physical force may contribute to the enhancement of tight junction.

Liquid–liquid phase separation (LLPS) is a physical process that facilitates the demixing of proteins through protein condensation and forms membrane-less organelles (Brangwynne 2013; Shin and Brangwynne 2017; Banani et al. 2017). Compartmentalization of functional proteins by LLPS has been implicated in many cellular phenomena, such as transcription (Hnisz et al. 2017), stress response (Protter and Parker 2016), neuronal development (Wu et al. 2020) and cell division (Ong and Torres 2020). Recent studies revealed that ZO-1 dynamics is regulated by LLPS in mammalian culture cells (Beutel et al. 2019) and in zebrafish embryos (Schwayer et al. 2019). Beutel et al. (2019) demonstrated that ZO-1 forms membrane-attached compartments via phase separation, which is required to assemble functional TJs. Schwayer et al. (2019) found that non-junctional ZO-1 forms clusters and are transported toward the TJs with the actomyosin flow in gastrulating zebrafish embryos. These findings suggested that maintenance and enhancement of tight junction may partly be regulated by LLPS.

In this chapter, we describe the possible mechanism of force responses of cells and propose a model for the enhancement of cell-to-cell adhesion during early development in which phase separation of ZO-1 plays a critical role.

9.2 Cell-to-Cell Adhesion Becomes Enhanced by Forces

To mimic the stretching force of gastrulating embryos, we applied two mechanical stimuli which are centrifugation and compression, respectively, and loaded these forces on cells of intact *Xenopus* embryos of early gastrulating stage (stage 11)

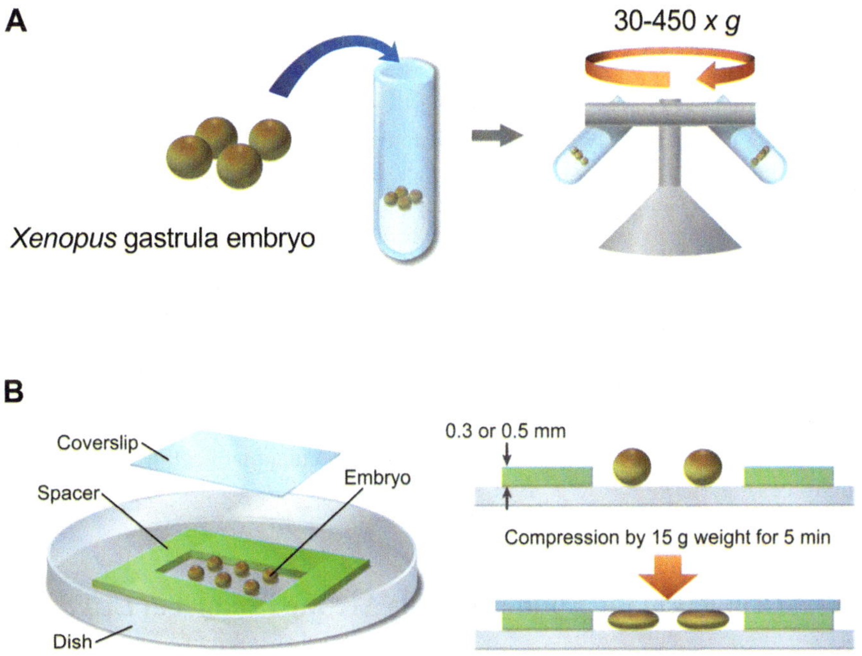

Fig. 9.1 Application of physical forces on *Xenopus* embryos. Early *Xenopus laevis* gastrula embryos were centrifuged at 30–450 × *g* for different time periods (**a**) or compressed by a cover slip and a weight of 15 g for 5 min (**b**) in the presence of a spacer

(Fig. 9.1), after which responses of embryonic cells were examined. Both types of forced mechanical stresses not only deformed and flattened the originally spherical embryos remarkably but also induced significant cell shape changes and elongated the cells to similar extents of early embryos undergoing gastrulation.

Strikingly, the deformed embryos recovered to their original shape within 2 h after the respective stimuli and most of them developed normally at least to the end of gastrulation/early tadpole stages. We then performed a series of phosphoproteomics analyses as we anticipated that phosphorylation is one of the earliest responses of cells to mechanical stresses. In fact, a number of proteins were found to be phosphorylated immediately upon the stimuli. It was noteworthy that among them, many cell-to-cell as well as cell-to-substrate adhesion-related proteins were listed as highly phosphorylated proteins. More interestingly, an enrichment analysis revealed that mesenchymal-to-epithelial transition (MET) proteins are upregulated in conjunction with the downregulation of the epithelial-to-mesenchymal transition (EMT) proteins upon the stimuli. This suggested that the force stimulation induces an MET-like phenotype, and thus the mechano-response was suspected to restrict cell motility by enhancing cell-to-cell adhesion (Hashimoto et al. 2019).

Next, we addressed how the phosphorylation, which is thought to occur in a cascade fashion, can influence cellular phenotype. Using immunostaining, we observed that cell-to-cell adhesion is remarkably reinforced as reveled by the enrichment of cadherin and ZO-1 proteins to adherens junction and tight junction, respectively (Fig. 9.2).

Detailed phosphoproteomic analyses of the force-loaded embryonic cells further identified that Erk2 and a set of its substrate candidates were highly phosphorylated, suggesting that receptor tyrosine kinases are involved in the sensing of the mechanical stimuli (Kinoshita et al. 2020). In fact, inhibitors of fibroblast growth factor receptor (FGFR) suppressed the force-dependent Erk2 phosphorylation as well as the accumulation of cadherin and ZO-1 to cellular junctions. This suggests that FGFR/Erk2 signaling contributes to the enhancement of cellular junctions by the force stimuli. In fact, FGFR is highly phosphorylated in intact embryos during gastrulation, while corresponding epithelial cells of dissected tissue (animal cap) which is free from cell stretching show little phosphorylation. Interestingly, however, there is no evidence so far that FGF ligands are necessary for the force-dependent activation of the receptor, shedding light on the previously unidentified mechanism for FGFR activation and the resulting regulation of cell-to-cell adhesion by force (Fig. 9.3).

9.3 Droplets in the Epithelial Cells Are the Product of LLPS

In sharp contrast to the accumulation of cadherin and ZO-1 to respective cell junctions, we noticed that fluorescent puncta of ZO-1, constantly observed in the cytoplasm of epithelial cells before gastrulation, disappeared from the cytoplasm

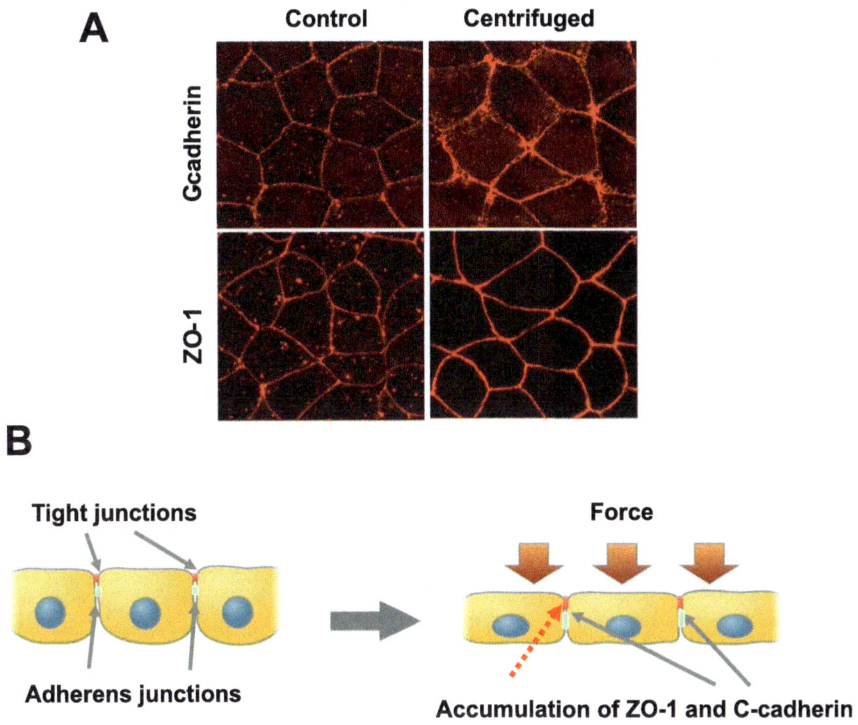

Fig. 9.2 Force-dependent enrichment of C-cadherin and ZO-1 to the junctions. (**a**) Immunostaining of centrifuged embryo with antibodies against C-cadherin and ZO-1, respectively, reveals that these junction proteins become enriched in respective junctions. From Kinoshita et al. 2020. (**b**) The illustration depicts the force-dependent accumulation of C-cadherin and ZO-1 to the junctions

upon gastrulation. This clearance of ZO-1 droplets from the cytoplasm seems to be controlled by cell stretching during gastrulation cell movements because the above-mentioned animal cap cells which are free from cell stretching retained the cytoplasmic puncta. Therefore, we speculated that the puncta possibly represent the ZO-1-containing condensates generated through LLPS, which in turn suggest that LLPS of ZO-1 might be negatively controlled by mechanical stresses.

To prove our hypothesis, we used cultured cells (*Xenopus*-derived A6 cells and canine MDCK cells) and GFP-fused ZO-1 was transiently expressed for microscopic observations at the subcellular level (Kinoshita et al. 2022). In the A6 cells, droplets with smooth surface are merely detected in the cytoplasm (Fig. 9.4a). Interestingly, however, when ZO-1's ability to bind F-actin was attenuated by expressing actin-binding peptide/protein such as Lifeact, utrophin, or the acting binding domain of moesin, or by inhibiting actin polymerization with latrunculin B, large-sized and round-shaped droplets which often adopt amorphous shape were observed by live-imaging, suggesting that in A6 cells, physical association of ZO-1 protein with

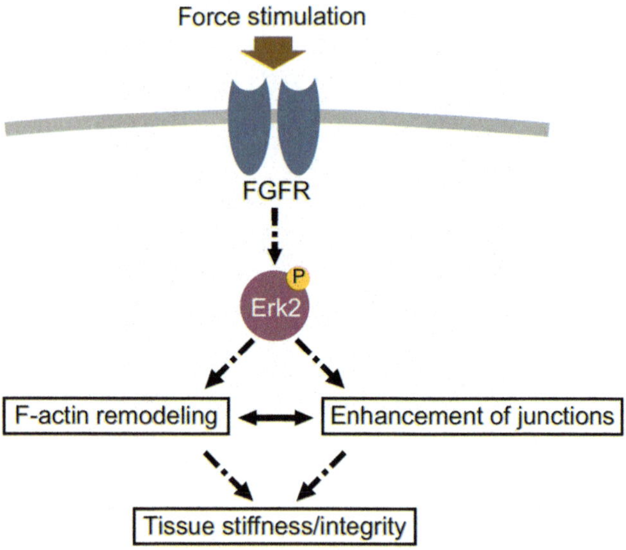

Fig. 9.3 Hypothesized mechano-signaling pathway. We hypothesize that the application of mechanical force to ectodermal cells in *Xenopus* embryos activates FGFR in a ligand-independent manner, which then activates the Erk2 signaling pathway. This mechanochemical signal transduction induces F-actin remodeling and enhancement of cell junctions, which in turn regulate tissue stiffness and integrity. From Kinoshita et al. 2020

F-actin restricts the formation of ZO-1 droplets (Fig. 9.4b). In fact, ZO-1 lacking actin-binding domain (ΔABD) formed large-sized droplets in A6 cells (Fig. 9.4c).

In contrast to A6 cells, MDCK cells stably forms large ZO-1 droplets when they were cultured in a sparse condition (Fig. 9.4d), indicating that ZO-1 protein behaves differently depending on cell types. Interestingly, however, when the cells reached confluent and cell-to-cell adhesion is fully established, ZO-1 droplets disappear from the cytoplasm. This can be interpreted that as the cells established F-actin network and adopted polygonal shape in the confluent sheet the droplets diminished. More interestingly, ZO-1 droplets emerge when an injury (a partial cell removal) was induced in the MDCK cell sheet and the edge cells started their migration toward the wound site extending cell protrusions. These finding also suggested the functional link between F-actin network and ZO-1 droplet formation and consistent with the fact that embryonic epithelial cells of *Xenopus* embryos being stretched remodel F-actin and localize it to the apical domain in conjunction with the loss of ZO-1 droplets.

We further addressed whether the ZO-1 droplets are indeed the product of LLPS. It was previously reported that ZO-1 protein undergoes LLPS and forms condensates. To confirm the possibility with *Xenopus*-derived A6 cells expressing the actin-binding domain of moesin (RFP-MoesinABD) to stimulate ZO-1 droplet formation (Fig. 9.4b), in addition to naïve MDCK cells, we treated the two types of cells with digitonin which permeabilize the cell membrane. This treatment is expected to

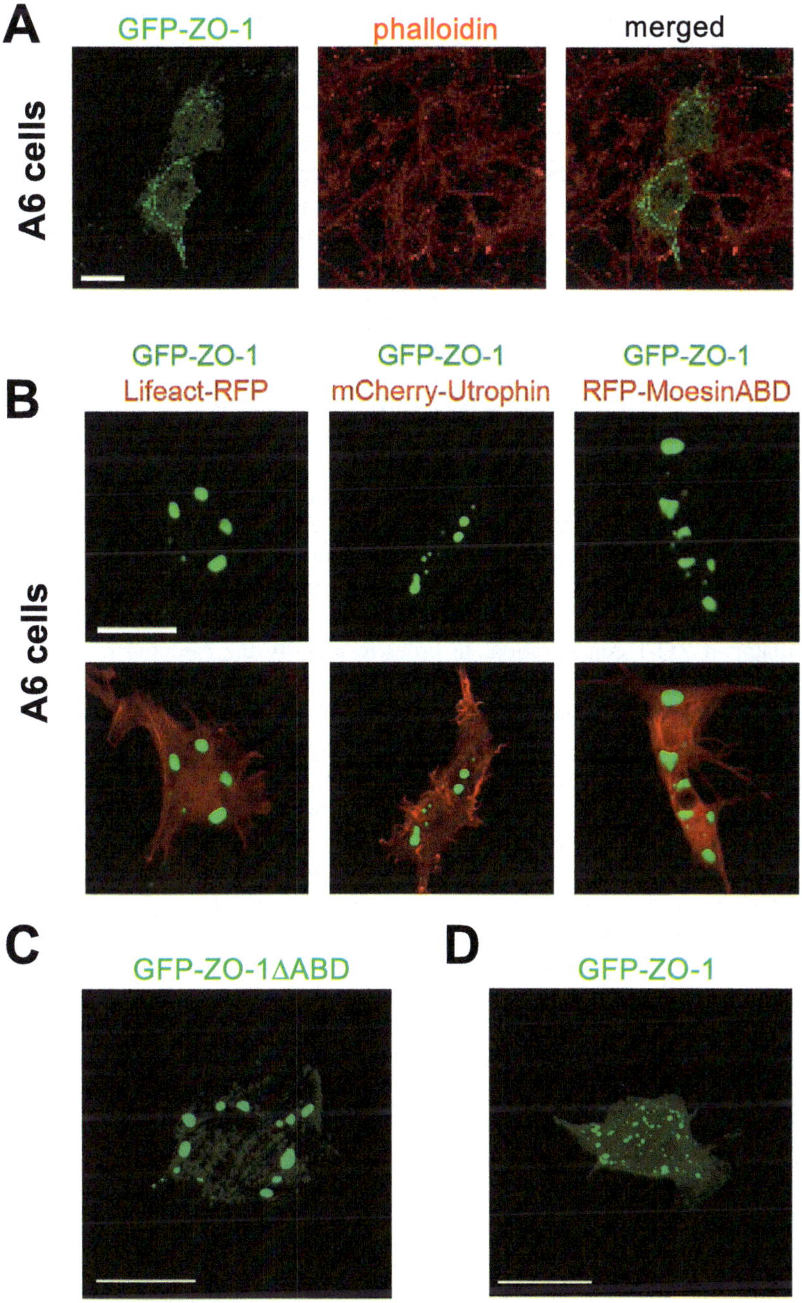

Fig. 9.4 ZO1 actin-binding domain inhibits the formation of ZO1 droplets in A6 cells. In A6 cells, droplet formation revealed by GFP fluorescence is not evident in the cytoplasm of normal cells (**a**), while round-shaped droplets can be observed when actin-binding peptides/proteins are expressed (**b**). (**c**) ZO-1 lacking the actin-binding domain (ZO-1-ΔABD) undergoes condensation and forms droplets. In contrast, naïve MDCK cells form ZO-1 droplets without inhibiting actin binding (**d**). Scale bars, 20 μm

disrupt the equilibrium of ZO-1 in the droplet and in the cytoplasm, and therefore diminishes the droplets. As we expected, ZO-1 droplets become fainted from the periphery and shrunk as the time proceeded. The results suggested that ZO-1 droplets are the condensates by LLPS. We also confirmed by fluorescence recovery after photobleaching (FRAP). In both cell conditions, ZO-1 fluorescence was recovered within 6 min after photobleaching, further supporting that the droplets are the ZO-1 condensates generated by LLPS and they are maintained in a dynamic equilibrium between the condensates and the cytoplasm. These findings made us to propose that the regulation of LLPS, at least of ZO-1, is one of the mechano-responses of embryonic cells.

9.4 Molecular Nature of ZO-1 Protein

ZO-1 is a relatively large protein whose molecular mass is approximately 200 k daltons. It has several PDZ domains and binding sites for F-actin, and tight junction proteins occludin, and claudin. Importantly, there are four stretches of intrinsically disordered region (IDR) which is thought to play critical roles in the protein condensation by LLPS. As described above, F-actin appears to be a negative regulator of ZO-1 condensation. In order to identify the essential regions of the protein for its condensation, we performed a structure-function relationship study by generating several deletion mutants using the GFP-ZO1-ΔABD as a positive control. Then, we deleted IDR1, IDR2, and IDR3, which reside in the N-terminal half, from GFP-ZO1-ΔABD (GFP-ZO1-ΔABD-ΔIDR123) (Fig. 9.5). We found that the mutant lacking the three IDRs still formed droplets, indicating that these regions are not required for phase separation of ZO-1 in A6 cells. In contrast, deletion of the C-terminal IDR4 (GFP-ZO1-ΔABD-ΔIDR4) failed to form droplets, and it rather dispersed in the cytoplasm. The result demonstrates that IDR4 of ZO-1 is critical for cytoplasmic liquid droplet formation by LLPS. Since the actin-binding domain overlaps with IDR4, it is possible that F-actin-binding may affect the LLPS-inducing activity of IDR4.

9.5 The Mechano-Response of Embryonic Cells Is Conserved across Species

It has been shown that in the leading cells of zebrafish during epiboly, droplets of ZO-1 vigorously flow to the junction where F-actin is enriched, which is suggesting the functional link between F-actin network and ZO-1 droplet dynamics. Our study suggested that formation and dissolution of ZO-1 condensates is a part of mechanochemical feedback system of embryonic and other types of cells to maintain tissue

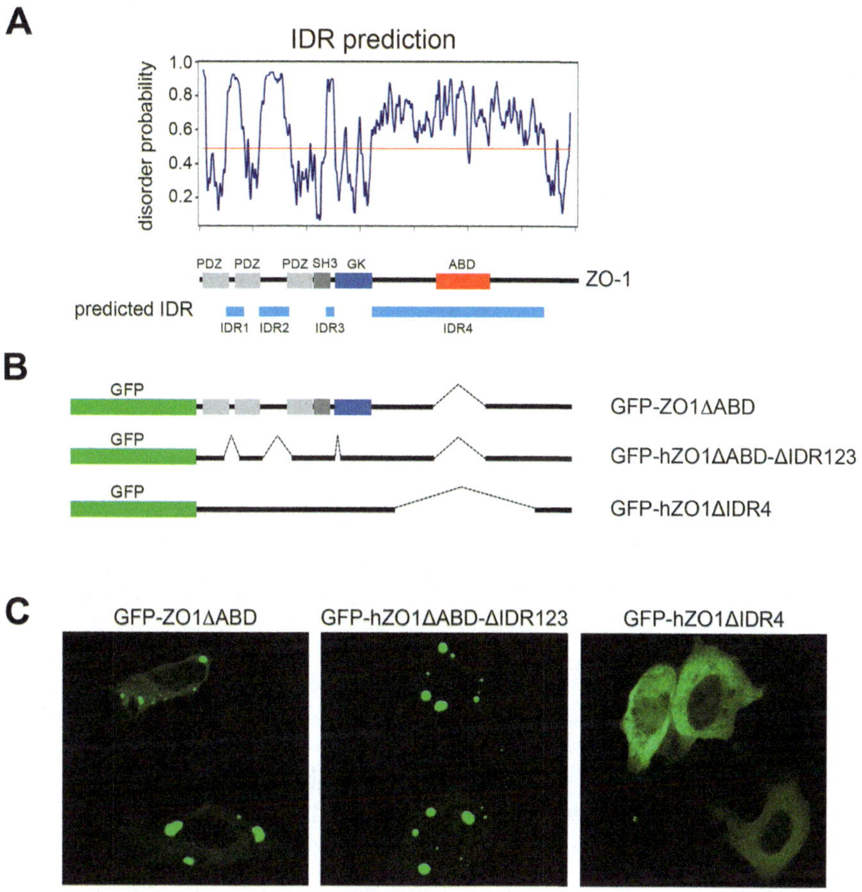

Fig. 9.5 Structure-functional analysis of ZO-1 for phase separation. (**a**) Four intrinsically disordered regions (IDRs) of ZO-1, IDR1–4 were predicted by the Protein Disorder Prediction System (http://prodos.hgc.jp/cgi-bin/top.cgi). (**b**) IDR deletion mutants were constructed based on the IDR prediction. (**c**) These mutants were expressed in A6 cells and GFP fluorescence was observed. Scale bars, 10 μm. From Kinoshita et al. 2022

integrity. We next examined how the feedback system is conserved across species using mouse embryos.

As early mouse embryo develops harboring an inner cavity called blastocoel cavity (BC) and its growth increases the inner pressure against embryonic cells. The effect is more evident on the trophectoderm (TE) cells of the molar side which consist a thinner layer being stretched by the luminal pressure compared to the polar side which includes inner cell mass (ICM) (Fig. 9.6a). Therefore, early mouse embryogenesis provides us an ideal experimental model to test whether the TE cells exposed to the luminal force dissolute ZO-1 condensates and instead enrich the ZO-1 protein to the cell junction. As expected, significant numbers of ZO-1

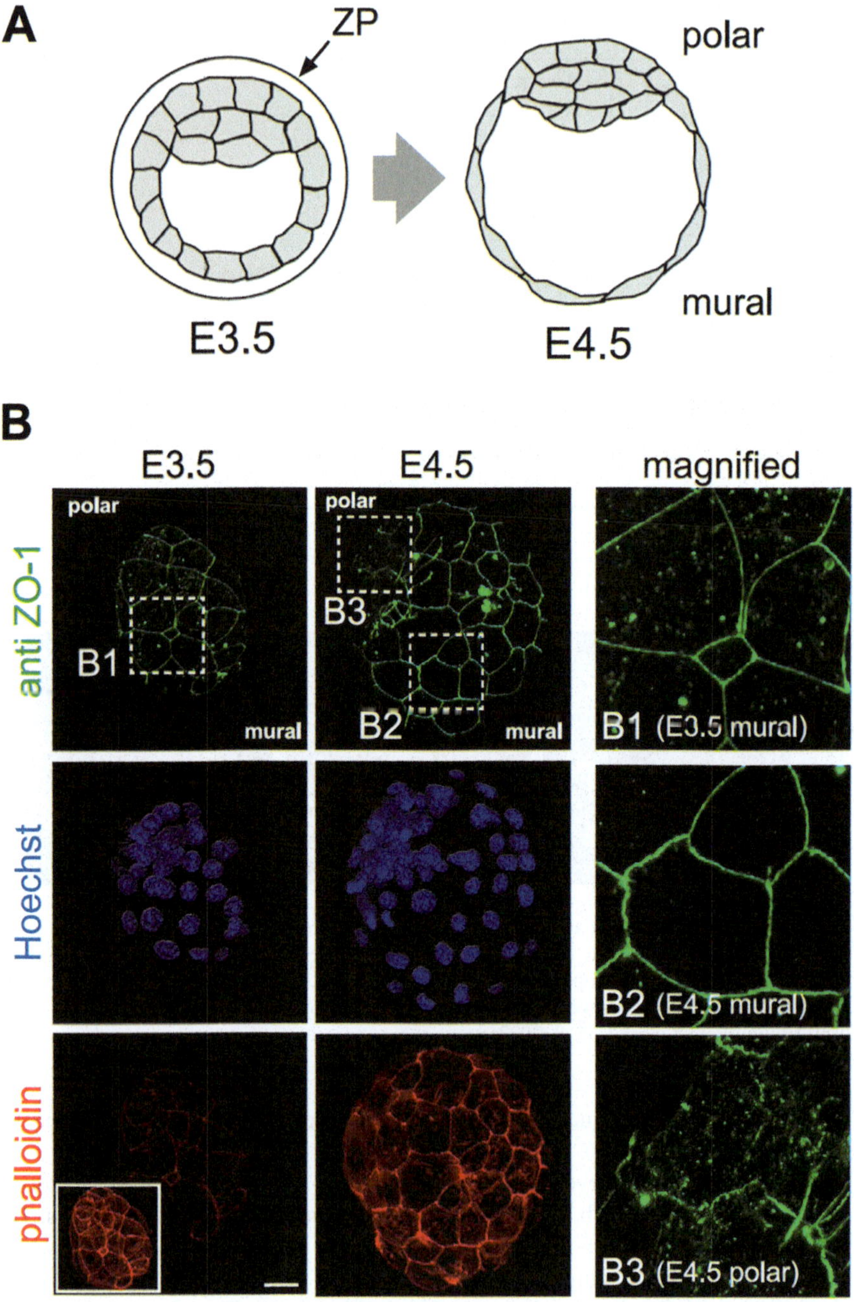

Fig. 9.6 ZO-1 condensation during mouse embryogenesis. (**a**) Schematic diagram of the hatching process of the mouse embryo. At E3.5, the embryo is covered with the zona pellucida (ZP). During hatching, the embryo is enlarged with an expansion of the trophectoderm (TE), and emerges from the ZP. (**b**) Immunofluorescence of mouse E3.5 and E4.5 embryos. Embryos were stained with an anti-ZO-1 antibody, Hoechst 33342 for nuclei, and Alexa Fluor 546 phalloidin for F-actin. Scale bar, 20 μm. The inset in the E3.5 phalloidin image demonstrates that the structure of cortical F-actin

Fig. 9.6 (continued) is formed at E3.5 even though the signal intensity was weaker than that in the E4.5 embryo. The right panels are enlarged images from the left panels indicated by the dashed squares B1, B2, and B3, and show that ZO-1 condensates disappear only in the mural side. From Kinoshita et al. 2022

droplet in E3.5 embryo disappeared as development proceeded to E4.5 and as the BC became larger (Fig. 9.6b), suggesting that the growing cavity's inner pressure diminished ZO-1 droplets. The recovery rate by FRAP analysis demonstrated that the mouse ZO-1 condensate has similar dynamics to those of A6 cells and MDCK cells (Kinoshita et al. 2022).

It was noteworthy that the release of the inner pressure by piercing E3.5 embryo or by treating the embryo with ouabain, a selective Na^+/K^+-ATPase inhibitor which inhibit water influx to the cavity, attenuated the reduction of the ZO-1 droplets in the cells of E4.5, suggesting that the ZO-1 condensates in mouse embryo is also force-sensitive. Taken together, these results strongly suggest that the mechanochemical feedback mechanism is highly conserved in mammals.

9.6 Conclusions and Perspectives

Our study demonstrated that embryonic epithelial cells have a unique feature that they remodel cellular junctions and enhance cell-to-cell adhesion responding to mechanical stimuli. This may be an essential mechano-response to maintain tissue integrity and ensure normal development resisting against the possible perturbation by physical forces generated both inside and outside of embryos. Important question remained may be whether this mechano-response is a property specific to the cells of epithelial origin. In this series of work, we focused on the behavior of embryonic epithelial cells of *Xenopus* and mouse from the early developmental point of view. In addition, both *Xenopus* A6 and MDCK cells are derived from kidney epithelial tissue including renal tube although the latter retains more typical natures of epithelialized cells. Therefore, it could be possible that this interesting property may be applied only to epithelial cells and ZO-1. Nevertheless, the mechanism in which LLPS is regulated by force may represent a common essential mechanism for securing the robustness of normal development and homeostasis under changing extracellular environments.

The clarification of the pathway for the mechano-feedback mechanism has just begun and we still need to address a number of questions, for example, how the initial cell response is triggered at the cell membrane, how the dissolution of ZO-1 condensate is biochemically controlled by the FGFR-ERK pathway, how protein phosphorylation and the other posttranscriptional modifications are involved in the condensate formation, whether any other proteins and/or RNAs are included in the condensate, and so on. We believe that investigation of the mechano-response pathway by addressing these questions in more detail would certainly contribute to the understanding, for example, the cellular mechanism of how cancer cells of epithelial origins, which represent a significant portion of cancer cases, can migrate out and transfer across tissues by EMT and settle in new places by MET.

Acknowledgments We thank Ileana Cristea for the collaboration for phosphoproteomics of *Xenopus* embryos,

Dr. Toshihiko Fujimori for collaboration for mouse embryo studies, Dr. Nobuyuki Shiina for technical advice and helpful comments, Dr. Michael Levine for useful discussions, and Dr. Takamasa S. Yamamoto and Ms. Naoko Yasue for technical assistance. We also thank NIBB core facilities for the use of microscopies and Dr. Hitoshi Morita for graphical illustrations. This research was supported by JSPS KAKENHI 20 K06663 to NK, JSPS KAKENHI 17H03689 and 16H06280 to TF, and MEXT KAKENHI 22127007, JSPS KAKENHI 15H05865 and 21H02493 to NU.

References

Banani SF, Lee HO, Hyman AA, Rosen MK (2017) Biomolecular condensates: organizers of cellular biochemistry. Nat Rev Mol Cell Biol 18:285–298

Beutel O, Maraspini R, Pombo-Garcia K, Martin-Lemaitre C, Honigmann A (2019) Phase separation of zonula Occludens proteins drives formation of tight junctions. Cell 179:923–936

Bodor DL, Ponisch W, Endres RG, Paluch EK (2020) Of cell shapes and motion: the physical basis of animal cell migration. Dev Cell 52:550–562

Brangwynne CP (2013) Phase transitions and size scaling of membrane-less organelles. J Cell Biol 203:875–881

Chan CJ, Costanzo M, Ruiz-Herrero T, Monke G, Petrie RJ, Bergert M, Diz-Munoz A, Mahadevan L, Hiiragi T (2019) Hydraulic control of mammalian embryo size and cell fate. Nature 571:112–116

Godard BG, Heisenberg CP (2019) Cell division and tissue mechanics. Curr Opin Cell Biol 60: 114–120

Goddard GK, Tarannum N, Woolner S (2020) Applying tensile and compressive force to Xenopus animal cap tissue. Cold Spring Harb Protoc 2020:105551

Haas AJ, Zihni C, Ruppel A, Hartmann C, Ebnet K, Tada M, Balda MS, Matter K (2020) Interplay between extracellular matrix stiffness and JAM-A regulates mechanical load on ZO-1 and tight junction assembly. Cell Rep 32:107924

Hashimoto Y, Kinoshita N, Greco TM, Federspiel JD, Jean Beltran PM, Ueno N, Cristea IM (2019) Mechanical force induces phosphorylation-mediated signaling that underlies tissue response and robustness in Xenopus embryos. Cell Syst 8:226–241

Hnisz D, Shrinivas K, Young RA, Chakraborty AK, Sharp PA (2017) A phase separation model for transcriptional control. Cell 169:13–23

Kinoshita N, Hashimoto Y, Yasue N, Suzuki M, Cristea IM, Ueno N (2020) Mechanical stress regulates epithelial tissue integrity and stiffness through the FGFR/Erk2 signaling pathway during embryogenesis. Cell Rep 30:3875–3888

Kinoshita N, Yamamoto TS, Yasue N, Takagi C, Fujimori T, Ueno N (2022) Force-dependent remodeling of cytoplasmic ZO-1 condensates contributes to cell-cell adhesion through enhancing tight junctions. iScience 25:103846

Ong JY, Torres JZ (2020) Phase separation in cell division. Mol Cell 80:9–20

Otani T, Furuse M (2020) Tight junction structure and function revisited. Trends Cell Biol 30:805–817

Paolini A, Abdelilah-Seyfried S (2018) The mechanobiology of zebrafish cardiac valve leaflet formation. Curr Opin Cell Biol 55:52–58

Protter DSW, Parker R (2016) Principles and properties of stress granules. Trends Cell Biol 26:668–679

Schwayer C, Shamipour S, Pranjic-Ferscha K, Schauer A, Balda M, Tada M, Matter K, Heisenberg CP (2019) Mechanosensation of tight junctions depends on ZO-1 phase separation and flow. Cell 179:937–952

Shin Y, Brangwynne CP (2017) Liquid phase condensation in cell physiology and disease. Science 357:eaaf4382

Spadaro D, Le S, Laroche T, Mean I, Jond L, Yan J, Citi S (2017) Tension-dependent stretching activates ZO-1 to control the junctional localization of its interactors. Curr Biol 27:3783–3795

Wu X, Cai Q, Feng Z, Zhang M (2020) Liquid-liquid phase separation in neuronal development and synaptic signaling. Dev Cell 55:18–29

Chapter 10
Regulation of Neuronal RNA Granule Dynamics Through Phase Separation in Memory Formation and Disease

Nobuyuki Shiina

Abstract Local translation in neuronal dendrites is an important basis for synaptic plasticity that underlies long-term memory formation. RNA granules, which are dynamic condensates consisting of mRNAs, ribosomes, and RNA-binding proteins, are essential for transporting mRNAs to dendrites and regulating local dendritic translation. Through coordinating and modulating the translation of specific mRNAs, these granules enable neurons to refine synaptic connections in response to synaptic inputs at the appropriate temporal and spatial scales. Recent studies have revealed that RNA granules form through liquid–liquid phase separation (LLPS), which allows them to adapt to changes in synaptic inputs and switch between different translational states. However, dysregulation of RNA granule dynamics, particularly the formation of aberrant aggregates, has been linked to neurodegenerative disorders such as amyotrophic lateral sclerosis (ALS) and frontotemporal lobar degeneration (FTLD). Thus, RNA granules play a pivotal role in maintaining synaptic plasticity and cognitive function in healthy neurons, while their dysregulation may contribute to neurodegeneration.

Keywords RNA granule · Local translation · mRNA transport · Synapse · Dendrite · Neuron · Phase separation · LLPS · Memory · Neurodegenerative disease

---

N. Shiina (✉)
Laboratory of Neuronal Cell Biology, National Institute for Basic Biology (NIBB), Okazaki, Aichi, Japan

Exploratory Research Center on Life and Living Systems (ExCELLS), National Institutes of Natural Sciences, Okazaki, Aichi, Japan

Department of Basic Biology, The Graduate University for Advanced Studies (SOKENDAI), Okazaki, Aichi, Japan
e-mail: nshiina@nibb.ac.jp

R. Kurokawa (ed.), *Phase Separation in Living Cells*,
https://doi.org/10.1007/978-981-99-4886-4_10

10.1 Introduction

Neurons receive input from a vast network of neurons via thousands of synaptic connections on their dendrites. In the context of learning and memory, only specific synaptic connections that receive input from activated neurons are selectively strengthened. This targeted approach may prevent a uniform enhancement of all synapses and the mixing of different memories. One mechanism that enables this selectivity is through local translation (Kang and Schuman 1996; Sutton and Schuman 2006; Costa-Mattioli et al. 2009).

Local translation is a process that produces proteins required for synapse formation and function, and delivers them to nearby postsynaptic sites (spines) on dendrites (Fig. 10.1). This mechanism involves a series of control mechanisms that place mRNAs, ribosomes, and RNA-binding proteins necessary for translation in a translation-repressed state near spines, and activate translation locally near the input-receiving synapses. These controls are mediated by "RNA granules," higher-order complexes composed of factors necessary for translation, which undergo discrete reorganization into translation complexes (polysomes) upon translational activation (Fig. 10.1) (Kiebler and Bassell 2006; Sudhakaran and Ramaswami 2017).

Recent evidence indicates that RNA granules form through LLPS of RNA-binding proteins and mRNAs, and changes in the LLPS status are believed to be responsible for the reorganization of RNA granules during the switching of local translational activity. While RNA granules play crucial physiological roles, the dysregulation of their dynamics, particularly their excessive aggregation, is associated with neurodegenerative diseases such as ALS and FTLD (Murakami et al. 2015; Bowden and Dormann 2016; Nedelsky and Taylor 2022).

10.2 Constituents of Neuronal RNA Granules

Neuronal RNA granules were initially discovered as assemblies that transport mRNA to dendrites in neurons (Knowles et al. 1996). These dendritic RNA granules contain various components, such as RNA-binding proteins, translational regulators, mRNAs, and ribosomes. Although mRNAs are also transported to axons, ribosomes have not been detected in axonal transport complexes, indicating that they are distinct from dendritic RNA granules (Holt et al. 2019).

The first identified RNA-binding protein in dendritic RNA granules was Staufen, which was later shown to participate in mRNA transport and translational activation (Kiebler et al. 1999; Tang et al. 2001). Since then, numerous constituent proteins have been identified, including cytoplasmic polyadenylation element-binding protein (CPEB) and germline development defective-2 (GLD-2), which are involved in mRNA polyA elongation; fragile X mental retardation protein (FMRP) and Pumilio2 (Pum2), which are involved in translational repression; RNA granule

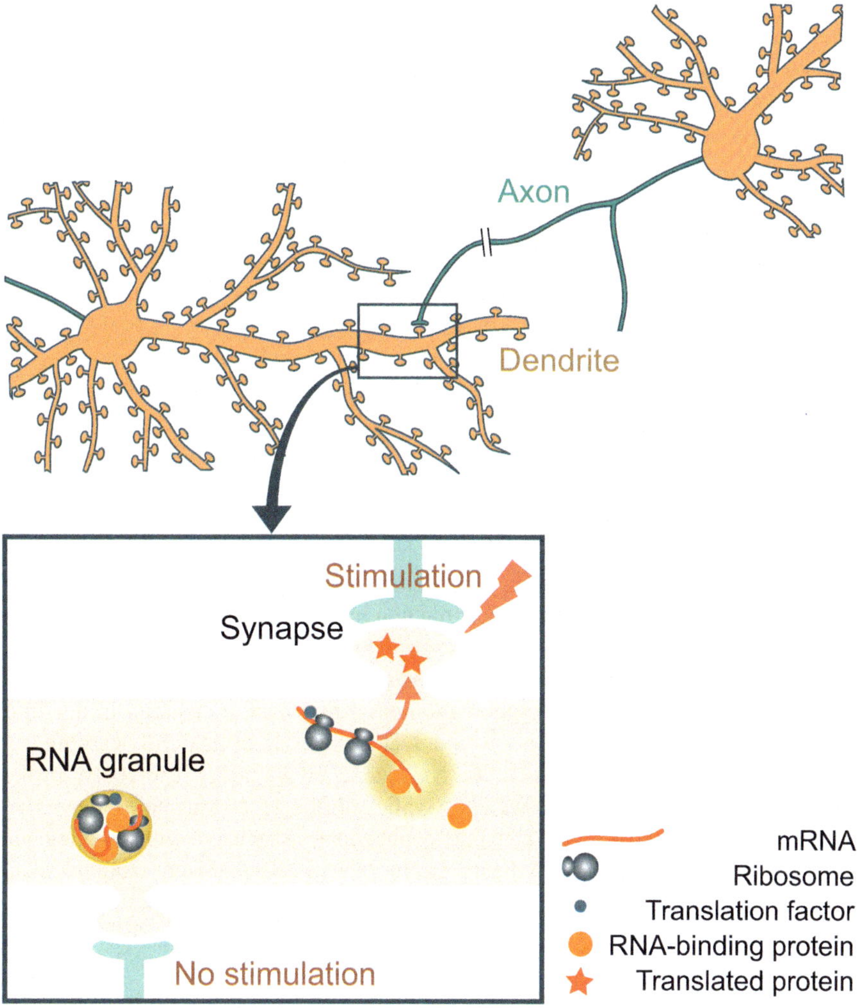

Fig. 10.1 Local control of translation by RNA granules. Neurons receive input from presynapses on the axons of other neurons through postsynapses (spines) on dendrites. In response to this synaptic input, RNA granules located near dendritic spines are reorganized into translation complexes (polysomes) to activate local translation. The locally translated proteins are then supplied to the input spine to strengthen synaptic connections

protein 105 (RNG105, also known as Caprin1), which is involved in mRNA transport; and fused in sarcoma (FUS) and TAR DNA-binding protein (Tardbp, also known as TDP-43), whose aggregation is associated with ALS and FLTD (Kiebler and Bassell 2006; Lagier-Tourenne et al. 2010; Roy et al. 2020). The comprehensive identification of RNA granule components has been facilitated by sucrose density gradient centrifugation-based RNA granule isolation and mass spectrometry, many of which overlap with the components of stress granules and germ granules that share similar structures and properties with neural RNA granules (El Fatimy et al. 2016). This overlap may contribute to shared characteristics among diverse granules, such as translational repression and rearrangement into polysomes.

Camk2a mRNA is a well-characterized mRNA that localizes to RNA granules. This mRNA has been observed to localize to the dendritic layer in the rat brain hippocampus, where dendrites are densely packed, and to be transported to dendrites in granular structures in cultured neurons (Rook et al. 2000). This localization is likely facilitated by its interaction with multiple RNA granule proteins, including FMRP, RNG105, CPEB, and Staufen (Darnell et al. 2011; Shiina et al. 2005; Udagawa et al. 2013; Heraud-Farlow et al. 2013). The translated product of *Camk2a* mRNA, Ca^{2+}/calmodulin-dependent protein kinase IIα (CaMKIIα) protein, is crucial for synaptic input-dependent CaMKII kinase activation and serves as a scaffold for postsynaptic densities, promoting synaptic long-term potentiation (LTP) (Zalcman et al. 2018).

Various methods have been employed to identify other mRNAs that localize to dendrites. These include in vitro binding experiments with RNA-binding proteins that constitute RNA granules, co-immunoprecipitation with RNA granule components from cell extracts, and a dendrite isolation method that isolates dendrite-localized mRNAs from the rodent hippocampus (Cajigas et al. 2012; Ainsley et al. 2014; Nakayama et al. 2017). The last method has revealed a significant number of mRNAs classified into gene ontology groups associated with synapses, spines, glutamate receptors, ribosomes, and translation elongation factors (Ohashi and Shiina 2020). These findings support the notion that local translation in dendrites is associated with spine formation and LTP, which is critical for learning and memory (Kang and Schuman 1996; Nakayama et al. 2017). However, further research is needed to identify which specific mRNA local translation events are responsible for strengthening synaptic connections associated with learning and memory, with *Camk2a* mRNA being the most well-characterized example so far (Miller et al. 2002).

10.3 Transport of RNA Granules to Dendrites in Neurons

The dynamics of RNA granules in neurons have been elucidated through fluorescence imaging of constituent proteins and mRNAs. RNA granules are transported bidirectionally to dendrites using microtubules, in both distal and proximal directions (Fig. 10.2). This bidirectional movement likely accounts for the even distribution of RNA granules throughout dendrites, as unidirectional transport tends to concentrate cargo at either end. Motor proteins, such as kinesin and dynein, are present in RNA granules and facilitate the transport of the granules along microtubules (Kiebler and Bassell 2006). FMRP, the component of RNA granules, interacts with kinesin, thereby providing a mechanism for recruiting motor proteins to RNA granules (Kanai et al. 2004; Dictenberg et al. 2008). Some RNA granules remain stationary within dendrites, often located near the bases of actin filament-rich spines,

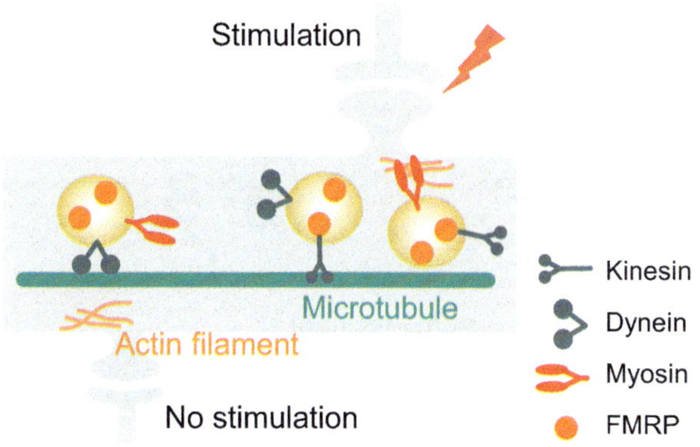

Fig. 10.2 Transport and anchoring of RNA granules. RNA granules are transported along microtubules by kinesin and dynein motor proteins and anchored to actin fibers at the synaptic base by myosin V. The interaction between kinesin and FMRP is enhanced in a neural activity-dependent manner, leading to increased transport of RNA granules to dendrites. Additionally, mRNA is anchored at the base of the spine where the synaptic input is received, depending on the specific input to the synapse

where myosin V, present within the granules, tethers them to actin filaments (El Fatimy et al. 2016) (Fig. 10.2).

The transport and tethering of RNA granules in neurons are influenced by neural activity and synaptic input. For example, *Camk2a* mRNA is transported to dendrites in a manner that depends on neural activity, resulting from the increased binding of FMRP and kinesin (Dictenberg et al. 2008). Tethering of β-actin (*Actb*) mRNA to actin at the spine base is also dependent on synaptic input (Yoon et al. 2016). However, it is uncertain whether all dendritic mRNAs are transported and tethered in the same manner in response to neural activity and synaptic input, given the wide range of RNA granules containing different constituent proteins and mRNAs among individual RNA granules (Mikl et al. 2011). Therefore, a comprehensive and detailed analysis of the transport and localization of individual constituent proteins and mRNAs remains a future challenge.

10.4 Synaptic Input-Dependent Regulation of Local Translation

The regulation of local translation is a finely-tuned process that involves repression during RNA granule transport and tethering, but activation in response to synaptic input. Several models have been proposed to explain these complex mechanisms. One model suggests that translation initiation is repressed in RNA granules due to the absence of translation initiation factor 4E (eIF4E) and the presence of cap-binding protein 80 (CBP80), which dissociates from mRNA before the steady-state translation by factors including eIF4E (Fritzsche et al. 2013) (Fig. 10.3). Even

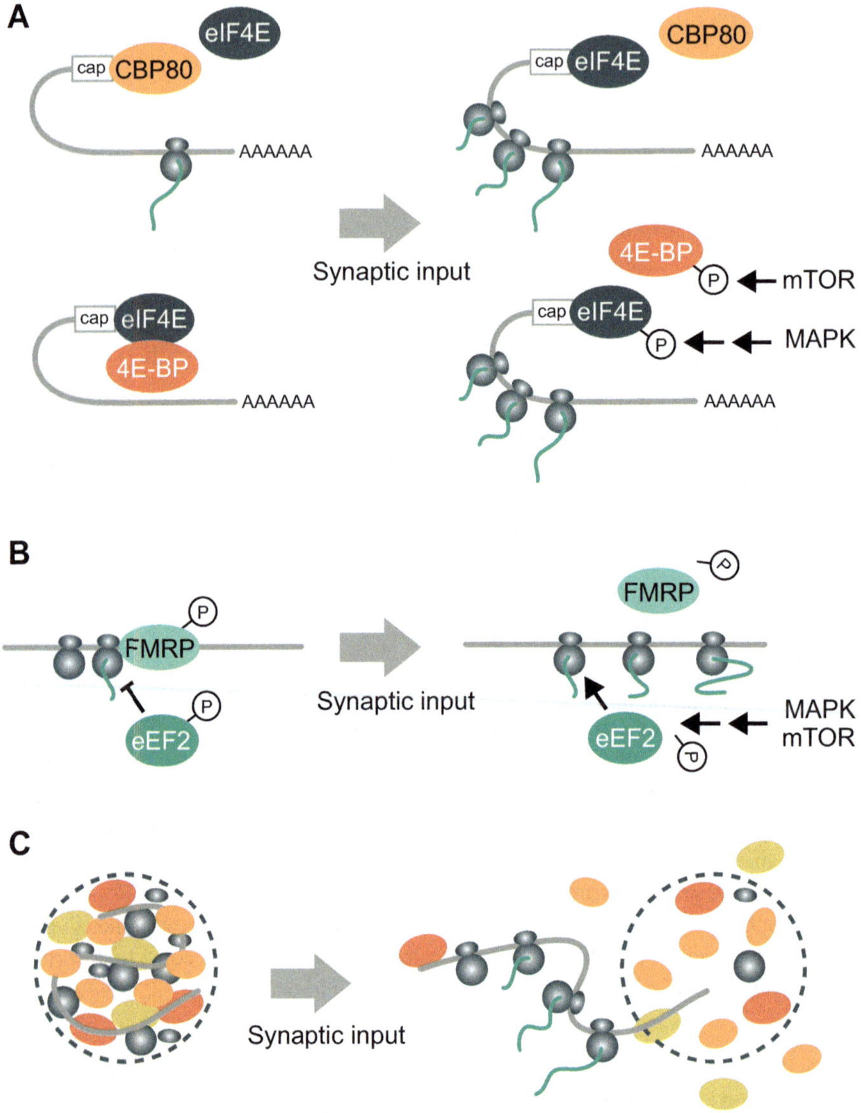

Fig. 10.3 Models of regulatory mechanisms of local translation repression and activation. (**a**) Regulation of translation initiation: CBP80 binds to the mRNA cap, halting steady-state translation initiation in the absence of synaptic input. Even if eIF4E binds to the cap, 4E-BP or Maskin can bind to it, inhibiting translation initiation. However, when synaptic input is received, CBP80 may dissociate from the cap, and 4E-BP and Maskin dissociate from eIF4E, leading to translation activation. Synaptic input activates MAP kinase (MAPK) and mTOR, leading to the phosphorylation of eIF4E and 4E-BP, respectively, which promotes translation initiation. (**b**) Regulation of translation elongation: Phosphorylated FMRP and eEF2 suppress translation elongation. However, when synaptic input is received, MAPK and mTOR are activated, leading to the phosphorylation and inactivation of eEF2 kinase, resulting in eEF2 dephosphorylation and lifting of elongation inhibition. FMRP is also dephosphorylated in a synaptic input-dependent manner, leading to the lifting of elongation inhibition. (**c**) In densely packed RNA granules, translation is repressed due to

in the presence of eIF4E, translation initiation can be inhibited by the binding of the CPEB complex and eIF4E-binding proteins (4E-BPs) to eIF4E. However, synaptic inputs can relieve this inhibition, leading to translation activation (Klann and Dever 2004) (Fig. 10.3). Alternatively, another model proposes that translation elongation is repressed in RNA granules but resumes in response to synaptic input (Na et al. 2016; Langille et al. 2019). This model suggests that mRNAs exist in a polysomal state where elongation is halted (Fig. 10.3). The activation of MAP kinase and mammalian target of rapamycin (mTOR) signaling pathways stimulates both translation initiation and elongation by phosphorylating eIF4E and 4E-BP and dephosphorylating FMRP and eukaryotic elongation factor 2 (eEF2) (Klann and Dever 2004; Narayanan et al. 2007) (Fig. 10.3).

In addition to the models discussed above, there is an intriguing hypothesis proposing that RNA granules are tightly packed structures that restrict molecular accessibility, thereby preventing mRNA translation. However, in response to input signals, the loosening of this packing triggers translation (Fig. 10.3). This notion is supported by electron microscopy studies (Krichevsky and Kosik 2001), as well as by the dissociation of RNG105 from RNA granules (Shiina et al. 2005) and the increased fluidity of *actb* mRNA and ribosomes due to synaptic input (Buxbaum et al. 2014; Park et al. 2014). Although the precise location of translation within or around these loosened granules remains unclear, recent studies using the SunTag method demonstrated that translation within stress granules, which share similar structures with neuronal RNA granules, is possible (Mateju et al. 2020). In contrast to loosening, CPEB has been observed to assemble into aggregates in response to input signals (Sudhakaran and Ramaswami 2017). While the dissociation and aggregation of RNA granule components appear to be opposing processes, both may reflect input-dependent reorganization processes of RNA granules into polysomes. The diverse models proposed for the repression/activation mechanisms of local translation by RNA granules may be due to the heterogeneity of RNA granules, or they may not be mutually exclusive, and further analyses are necessary to uncover the underlying regulatory mechanisms.

10.5 Control of RNA Granule Formation and Dynamics through LLPS

RNA granules are unique structures that lack membranes and are formed through intermolecular interactions, forming liquid-like condensates that separate from the cytoplasm via LLPS (Ryan and Fawzi 2019) (Fig. 10.4). The driving force behind the LLPS of RNA granules is weak protein–protein and protein–RNA interactions

Fig. 10.3 (continued) limited molecular accessibility to mRNA. However, in a synaptic input-dependent manner, the packing is loosened, leading to the activation of translation

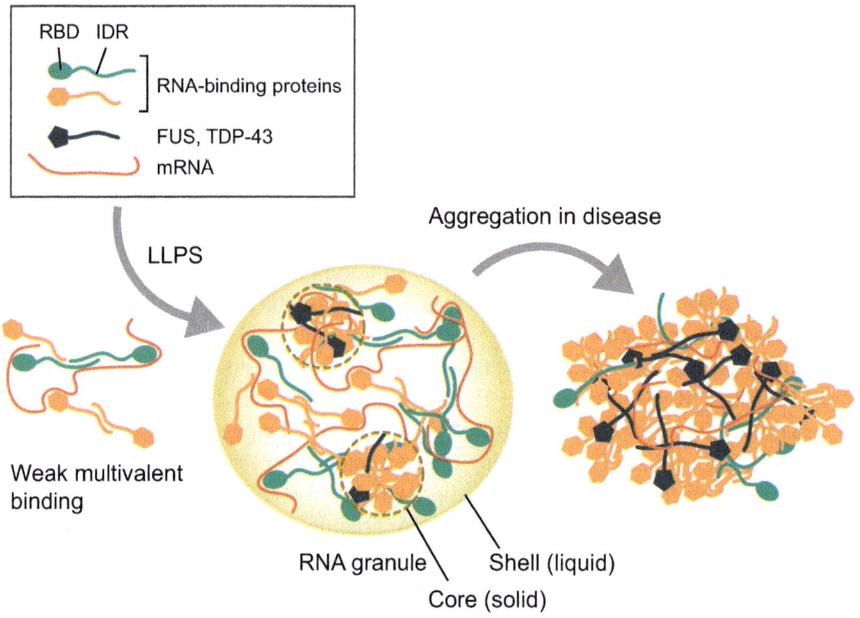

Fig. 10.4 Regulation of RNA granule formation through LLPS. RNA granules are formed through weak multivalent protein–protein and protein–mRNA interactions mediated by IDRs and RNA-binding domains (RBDs) of RNA-binding proteins, leading to LLPS. RNA granules have both liquid-like and solid-like subdomains, which are referred to as the shell and core, respectively. These subdomains differ in their constituent protein composition, with the shell being enriched in RNG105 and the core in FMRP. During learning and memory, changes in the LLPS state are believed to enhance the accessibility of translation factors to mRNAs and/or facilitate the release of mRNAs from the granules. FUS and TDP-43 are present in RNA granules in small amounts, contributing to their proper function. However, excessive accumulation of FUS and TDP-43 in neurodegenerative diseases can lead to the gelation and solidification of RNA granules, which is closely linked to the pathology of these diseases

mediated by RNA-binding sites and intrinsically disordered regions (IDRs) of RNA-binding proteins. IDRs, also known as low-complexity domains (LCDs) when composed of a limited set of amino acids, contain multiple interaction sites that enable the formation of multivalent and weak intermolecular interactions. These interactions can switch partners while preserving the fluidic liquid-phase assembly. RNA-binding proteins such as fused in sarcoma (FUS), TAR DNA-binding protein-43 (TDP-43), and heterogeneous nuclear ribonucleoprotein A1 (hnRNPA1) have been shown to undergo LLPS in vitro (Harrison and Shorter 2017), and their post-translational modifications play a key role in modulating LLPS status (Hofweber and Dormann 2017). For example, methylation and phosphorylation of the FUS IDR reduce its LLPS (Hofweber et al. 2018; Monahan et al. 2017), while phosphorylation of the FMRP IDR enhances its LLPS, and its methylation opposes it (Tsang et al. 2019). Additionally, the phosphorylation states of the FMRP and RNG105 IDRs regulate whether or not they undergo co-LLPS (Kim et al. 2019).

The recruitment of mRNAs into RNA granules is facilitated by their specific binding to RNA-binding proteins within these structures. However, an alternative mechanism exists whereby mRNAs can become part of LLPS through weak interactions with multiple molecules (Maharana et al. 2018; Guillén-Boixet et al. 2020) (Fig. 10.4). As a result, highly multivalent mRNAs are more likely to be incorporated into RNA granules. mRNA length is a key determinant of multivalency, with longer mRNAs having more interaction sites and thus a higher likelihood of being taken up into the granules, such as stress granules (Khong et al. 2017). Interestingly, it has been observed that mRNAs specifically bound by RNA-binding proteins are only slightly favored for uptake (Khong et al. 2017), likely due in part to changes in protein–mRNA-binding specificity upon LLPS (Hallegger et al. 2021). In neurons, the uptake of mRNAs is also influenced by alternative splicing, with isoforms containing longer 3′ untranslated regions (3′ UTRs) being preferentially localized to dendrites over those with shorter 3′ UTRs (Tushev et al. 2018). In addition, secondary structures and m^6A modifications of mRNAs can also modulate LLPS by altering the binding affinities between mRNAs and their binding proteins (Langdon et al. 2018; Ries et al. 2019).

One way to determine the formation of subcellular structures through LLPS is by observing whether they undergo fission and/or fusion. RNA granules in neurons display such events, providing evidence for their formation through LLPS (El Fatimy et al. 2016; Park et al. 2014). Fluorescence recovery after photobleaching (FRAP) is another useful method for studying granule dynamics. It has been demonstrated that the physical properties of RNA granules formed in fibroblasts depend on the type of RNA granule protein. For example, FUS and FMRP form near-solid RNA granules, whereas RNG105 forms highly fluid RNA granules (Shiina 2019) (Fig. 10.4). When heterologous proteins with different solid and liquid properties coexist within the same RNA granule, they form distinct subgranular structures, each retaining its unique dynamic properties. However, the presence of proteins with liquid properties in such granules biases solid-like proteins toward a loose liquid state (Shiina 2019). This coexistence of liquid and solid substructures, referred to as shell and core, respectively, has also been observed in stress granules and is likely a common structural feature of various granules (Jain et al. 2016). While the biological significance of such different subgranular structures with distinct dynamic properties is still unclear, the physical properties of RNA granules could tend toward a solid state if proteins with liquid properties in the shell are absent. Such a state may perturb translation and increase the risk of neurodegenerative diseases, as discussed later (Fig. 10.4).

As mentioned above, synaptic input induces the loosening of RNA granule packing, leading to a shift in the dynamics of RNG105, CPEB, *actb* mRNA, and ribosomes. These findings suggest that the modulation of LLPS in response to inputs plays a critical role in the transition from translational repression to activation and vice versa. Post-translational modifications of IDRs can alter the LLPS state, and such modifications in response to synaptic input could regulate the dynamics of granules and translational control. Alternatively, changes in ions, pH, and ATP concentration could also regulate the LLPS state (Jain et al. 2016; Krainer et al.

2021; Hong et al. 2020). Elucidating the contribution of these factors would be fundamental to unraveling the mechanisms underlying local translation regulation.

10.6 Long-Term Memory Formation Mediated by RNA Granule Components

Local translation is an essential process for synaptic LTP, which facilitates the input-dependent strengthening of synapses over prolonged periods. Conversely, local translation is dispensable for short-term potentiation (Fig. 10.5). Consistently, while de novo protein synthesis is a prerequisite for long-term memory, it is unnecessary for short-term memory. These observations have led to a significant question of whether RNA granules participate in long-term memory formation (Sudhakaran and Ramaswami 2017).

In *Drosophila*, studies have demonstrated that mutants or deletions of RNA granule proteins such as Staufen, CPEB (ORB), GLD-2, FMRP, Pumilio, and Ataxin-2 impair long-term memory. However, the knockout of their orthologues in mice has shown conflicting results (Roy et al. 2020; Ohashi and Shiina 2020). For example, FMRP (Fmr1) knockout mice exhibit inconsistent long-term memory impairments (The Dutch-Belgian Fragile X Consortium et al. 1994; Ding et al. 2014; Uutela et al. 2012; Baker et al. 2010; D'Hooge et al. 1997), while CPEB1 and GLD-2 knockout mice show normal long-term memory formation (Berger-Sweeney et al. 2006; Mansur et al. 2016). Staufen (Stau1)-deficient mice display impaired spine formation but intact long-term memory (Vessey et al. 2008), and

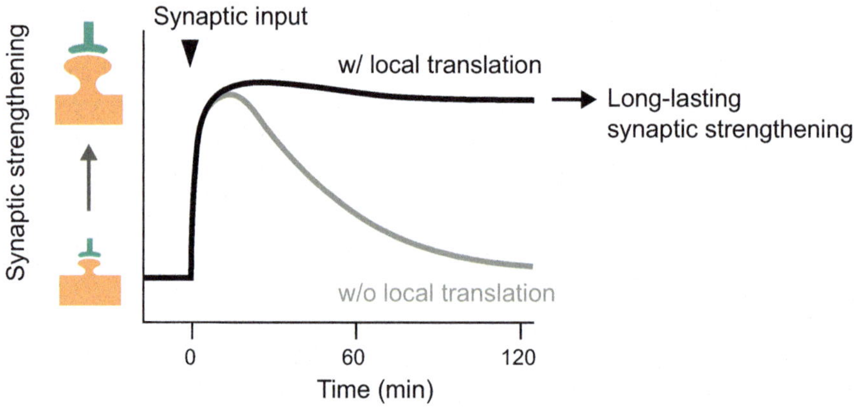

Fig. 10.5 Long-term synaptic potentiation through local translation. The strengthening of synaptic connections is attributed to synaptic enlargement and an increase in neurotransmitter receptors. While synaptic strengthening can occur without local translation for a short time after synaptic input, sustaining the strengthening for several hours or longer requires local translation. This long-lasting strengthening of synaptic connections forms the basis for long-term memory formation

Pum2 knockout mice even show improved long-term memory (Siemen et al. 2011). In Ataxin-2 knockout mice, conditioned fear memory is impaired, but spatial memory remains unaffected (Huynh et al. 2009). These varying effects could be due to functional redundancy or a negligible role of these proteins in the long-term memory formation process. Only RNG105 has been shown to impair long-term memory in both spatial and conditioned fear memory in mice when knocked out (Nakayama et al. 2017; Roy et al. 2020), indicating its unique role in the process. Although RNG105 may have no functional redundancy with other proteins, it is also plausible that RNG105 possesses unique properties that distinguish it from other RNA granule proteins, such as higher fluidity in RNA granules (Shiina 2019), which may facilitate mRNA incorporation into RNA granules and regulate local translation, crucial for long-term memory formation. Future research should investigate this possibility and the manipulation of the LLPS state at the individual level of mice.

In addition to investigating RNA granule proteins, manipulating mRNAs within RNA granules is another crucial area of study. One notable example is *Camk2a* mRNA, which is incorporated into RNA granules via its 3' UTR for dendritic transport (Rook et al. 2000; Mori et al. 2000). Deletion of this region reduces dendritic transport, decreases dendrite-translated CaMKIIα protein, and impairs long-term memory in mice (Miller et al. 2002). Although this finding is promising, it remains unclear whether incorporating various other mRNAs into RNA granules also holds biological significance.

10.7 Neurodegenerative Diseases Associated with Aggregation of FUS and TDP-43 in RNA Granules

ALS and FTLD are severe neurodegenerative diseases that progressively impair motor and cognitive function. Early in the course of these diseases, synapse formation and LTP decrease, while neuronal cell death occurs later on (Gelon et al. 2022; Ling et al. 2013). Histopathologically, the formation of protein aggregates in the brain is a characteristic hallmark of both diseases. FUS and TDP-43, the gene products responsible for ALS and FTLD, are key components of these aggregates. Normally, FUS and TDP-43 reside in the nucleus, with a fraction of these molecules localizing to cytoplasmic RNA granules, where they play a role in mRNA transport and local translation in dendrites (Ratti and Buratti 2016). However, in the disease state, excessive translocation of these proteins from the nucleus to the cytoplasm leads to the formation of cytoplasmic aggregates. Recent research has shown that RNA granule and stress granule proteins accumulate in these aggregates, suggesting a connection between neurodegenerative diseases and RNA granules (Kato et al. 2012; Ling et al. 2013; Murakami et al. 2015; Shin et al. 2017). Additionally, phase-separated condensates of these disease-causing proteins undergo gelation and eventually form amyloid fibrils (Harrison and Shorter 2017) (Fig. 10.4).

It is believed that gel and solid-phase aggregates of FUS and TDP-43 may limit the mobility of RNA granule proteins within them, possibly leading to RNA granule dysfunction (Kato et al. 2012; Shin et al. 2017; Murakami et al. 2015). However, only a small subset of RNA granule proteins, such as survival motor neuron (SMN) and Staufen, have been found to have reduced mobility within these aggregates (Murakami et al. 2015). The effects of aggregate formation on the localization and dynamics of many other proteins and mRNAs remain unknown. In contrast to previous hypotheses, it is also possible that RNA granule proteins and mRNAs are expelled from the aggregates, thereby preventing normal RNA granule function. In this regard, it has been observed that in patients with advanced disease, FUS and TDP-43 form fibrillated inclusions that do not co-localize with RNA granule components (Ryan and Fawzi 2019). The phase behavior of granules with excessive accumulations of FUS and TDP-43, and how it affects RNA granule proteins and mRNAs, are still unclear and require further investigation.

10.8 Conclusion

Over 100 proteins and several hundred mRNAs have been identified as constituents of RNA granules (El Fatimy et al. 2016; Ohashi and Shiina 2020). However, due to the nature of RNA granule formation through LLPS, they dissolve when removed from cells, leaving only the solid substructures of the granules for analysis. Innovative non-destructive techniques such as proximity labeling of proteins and RNAs (Padrón and Ingolia 2022) and photo isolation chemistry (PIC) (Honda et al. 2021), offer promising solutions to identify the proteins and mRNAs present within the liquid structure of RNA granules in cells.

While researchers have conducted knockouts of proteins or modifications of mRNAs to non-transportable forms in individual animals to unravel the role of RNA granules in learning and memory, a comprehensive understanding of this field remains elusive due to the limited number of studies. Additionally, a new challenge has arisen: elucidating the significance of LLPS in learning and memory. Addressing this issue requires modifying IDRs to prevent LLPS or altering the physical properties of phase-separated granules without causing loss of protein function, as seen in knockouts. This necessitates both expertise and innovative approaches to manipulate only the LLPS status. Moreover, understanding abnormalities in RNA granule dynamics, mRNA transport, and local translation in neurodegenerative diseases is crucial. Insights gained from these studies will improve our understanding of the underlying mechanisms of long-term memory formation and neurodegenerative diseases.

References

Ainsley JA, Drane L, Jacobs J, Kittelberger KA, Reijmers LG (2014) Functionally diverse dendritic mRNAs rapidly associate with ribosomes following a novel experience. Nat Commun 5:4510. https://doi.org/10.1038/ncomms5510

Baker KB, Wray SP, Ritter R, Mason S, Lanthorn TH, Savelieva KV (2010) Male and female Fmr1 knockout mice on C57 albino background exhibit spatial learning and memory impairments. Genes Brain Behav 9:562–574. https://doi.org/10.1111/j.1601-183X.2010.00585.x

Berger-Sweeney J, Zearfoss NR, Richter JD (2006) Reduced extinction of hippocampal-dependent memories in CPEB knockout mice. Learn Mem 13:4–7. https://doi.org/10.1101/lm.73706

Bowden HA, Dormann D (2016) Altered mRNP granule dynamics in FTLD pathogenesis. J Neurochem 138:112–133. https://doi.org/10.1111/jnc.13601

Buxbaum AR, Wu B, Singer RH (2014) Single β-actin mRNA detection in neurons reveals a mechanism for regulating its translatability. Science 343:419–422. https://doi.org/10.1126/science.1242939

Cajigas IJ, Tushev G, Will TJ, tom Dieck S, Fuerst N, Schuman EM (2012) The local transcriptome in the synaptic neuropil revealed by deep sequencing and high-resolution imaging. Neuron 74:453–466. https://doi.org/10.1016/j.neuron.2012.02.036

Costa-Mattioli M, Sossin WS, Klann E, Sonenberg N (2009) Translational control of long-lasting synaptic plasticity and memory. Neuron 61:10–26. https://doi.org/10.1016/j.neuron.2008.10.055

Darnell JC, Van Driesche SJ, Zhang C, Hung KY, Mele A, Fraser CE, Stone EF, Chen C, Fak JJ, Chi SW, Licatalosi DD, Richter JD, Darnell RB (2011) FMRP stalls ribosomal translocation on mRNAs linked to synaptic function and autism. Cell 146:247–261. https://doi.org/10.1016/j.cell.2011.06.013

D'Hooge R, Nagels G, Franck F, Bakker CE, Reyniers E, Storm K, Kooy RF, Oostra BA, Willems PJ, De Deyn PP (1997) Mildly impaired water maze performance in male Fmr1 knockout mice. Neuroscience 76:367–376. https://doi.org/10.1016/s0306-4522(96)00224-2

Dictenberg JB, Swanger SA, Antar LN, Singer RH, Bassell GJ (2008) A direct role for FMRP in activity-dependent dendritic mRNA transport links filopodial-spine morphogenesis to fragile X syndrome. Dev Cell 14:926–939. https://doi.org/10.1016/j.devcel.2008.04.003

Ding Q, Sethna F, Wang H (2014) Behavioral analysis of male and female Fmr1 knockout mice on C57BL/6 background. Behav Brain Res 271:72–78. https://doi.org/10.1016/j.bbr.2014.05.046

El Fatimy R, Davidovic L, Tremblay S, Jaglin X, Dury A, Robert C, De Koninck P, Khandjian EW (2016) Tracking the fragile X mental retardation protein in a highly ordered neuronal ribonucleoparticles population: a link between stalled polyribosomes and RNA granules. PLoS Genet 12:e1006192. https://doi.org/10.1371/journal.pgen.1006192

Fritzsche R, Karra D, Bennett KL, Ang FY, Heraud-Farlow JE, Tolino M, Doyle M, Bauer KE, Thomas S, Planyavsky M, Arn E, Bakosova A, Jungwirth K, Hörmann A, Palfi Z, Sandholzer J, Schwarz M, Macchi P, Colinge J, Superti-Furga G, Kiebler MA (2013) Interactome of two diverse RNA granules links mRNA localization to translational repression in neurons. Cell Rep 5:1749–1762. https://doi.org/10.1016/j.celrep.2013.11.023

Gelon PA, Dutchak PA, Sephton CF (2022) Synaptic dysfunction in ALS and FTD: anatomical and molecular changes provide insights into mechanisms of disease. Front Mol Neurosci 15:1000183. https://doi.org/10.3389/fnmol.2022.1000183

Guillén-Boixet J, Kopach A, Holehouse AS, Wittmann S, Jahnel M, Schlüßler R, Kim K, Trussina IREA, Wang J, Mateju D, Poser I, Maharana S, Ruer-Gruß M, Richter D, Zhang X, Chang YT, Guck J, Honigmann A, Mahamid J, Hyman AA, Pappu RV, Alberti S, Franzmann TM (2020) RNA-induced conformational switching and clustering of G3BP drive stress granule assembly by condensation. Cell 181:346–361.e17. https://doi.org/10.1016/j.cell.2020.03.049

Hallegger M, Chakrabarti AM, Lee FCY, Lee BL, Amalietti AG, Odeh HM, Copley KE, Rubien JD, Portz B, Kuret K, Huppertz I, Rau F, Patani R, Fawzi NL, Shorter J, Luscombe NM, Ule J

(2021) TDP-43 condensation properties specify its RNA-binding and regulatory repertoire. Cell 184:4680–4696.e22. https://doi.org/10.1016/j.cell.2021.07.018
Harrison AF, Shorter J (2017) RNA-binding proteins with prion-like domains in health and disease. Biochem J 474:1417–1438. https://doi.org/10.1042/BCJ20160499
Heraud-Farlow JE, Sharangdhar T, Li X, Pfeifer P, Tauber S, Orozco D, Hörmann A, Thomas S, Bakosova A, Farlow AR, Edbauer D, Lipshitz HD, Morris QD, Bilban M, Doyle M, Kiebler MA (2013) Staufen 2 regulates neuronal target RNAs. Cell Rep 5:1511–1518. https://doi.org/10.1016/j.celrep.2013.11.039
Hofweber M, Dormann D (2019) Friend or foe-post-translational modifications as regulators of phase separation and RNP granule dynamics. J Biol Chem 294:7137–7150. https://doi.org/10.1074/jbc.TM118.001189
Hofweber M, Hutten S, Bourgeois B, Spreitzer E, Niedner-Boblenz A, Schifferer M, Ruepp MD, Simons M, Niessing D, Madl T, Dormann D (2018) Phase separation of FUS is suppressed by its nuclear import receptor and arginine methylation. Cell 173:706–719.e13. https://doi.org/10.1016/j.cell.2018.03.004
Holt CE, Martin KC, Schuman EM (2019) Local translation in neurons: visualization and function. Nat Struct Mol Biol 26:557–566. https://doi.org/10.1038/s41594-019-0263-5
Honda M, Oki S, Kimura R, Harada A, Maehara K, Tanaka K, Meno C, Ohkawa Y (2021) High-depth spatial transcriptome analysis by photo-isolation chemistry. Nat Commun 12:4416. https://doi.org/10.1038/s41467-021-24691-8
Hong K, Song D, Jung Y (2020) Behavior control of membrane-less protein liquid condensates with metal ion-induced phase separation. Nat Commun 11:5554. https://doi.org/10.1038/s41467-020-19391-8
Huynh DP, Maalouf M, Silva AJ, Schweizer FE, Pulst SM (2009) Dissociated fear and spatial learning in mice with deficiency of ataxin-2. PLoS One 4:e 6235. https://doi.org/10.1371/journal.pone.0006235
Jain S, Wheeler JR, Walters RW, Agrawal A, Barsic A, Parker R (2016) ATPase-modulated stress granules contain a diverse proteome and substructure. Cell 164:487–498. https://doi.org/10.1016/j.cell.2015.12.038
Kanai Y, Dohmae N, Hirokawa N (2004) Kinesin transports RNA: isolation and characterization of an RNA-transporting granule. Neuron 43:513–525. https://doi.org/10.1016/j.neuron.2004.07.022
Kang H, Schuman EM (1996) A requirement for local protein synthesis in neurotrophin-induced hippocampal synaptic plasticity. Science 273:1402–1406. https://doi.org/10.1126/science.273.5280.1402
Kato M, Han TW, Xie S, Shi K, Du X, Wu LC, Mirzaei H, Goldsmith EJ, Longgood J, Pei J, Grishin NV, Frantz DE, Schneider JW, Chen S, Li L, Sawaya MR, Eisenberg D, Tycko R, McKnight SL (2012) Cell-free formation of RNA granules: low complexity sequence domains form dynamic fibers within hydrogels. Cell 149:753–767. https://doi.org/10.1016/j.cell.2012.04.017
Khong A, Matheny T, Jain S, Mitchell SF, Wheeler JR, Parker R (2017) The stress granule transcriptome reveals principles of mRNA accumulation in stress granules. Mol Cell 68:808–820.e5. https://doi.org/10.1016/j.molcel.2017.10.015
Kiebler MA, Bassell GJ (2006) Neuronal RNA granules: movers and makers. Neuron 51:685–690. https://doi.org/10.1016/j.neuron.2006.08.021
Kiebler MA, Hemraj I, Verkade P, Köhrmann M, Fortes P, Marión RM, Ortín J, Dotti CG (1999) The mammalian staufen protein localizes to the somatodendritic domain of cultured hippocampal neurons: implications for its involvement in mRNA transport. J Neurosci 19:288–297. https://doi.org/10.1523/JNEUROSCI.19-01-00288.1999
Kim TH, Tsang B, Vernon RM, Sonenberg N, Kay LE, Forman-Kay JD (2019) Phospho-dependent phase separation of FMRP and CAPRIN1 recapitulates regulation of translation and deadenylation. Science 365:825–829. https://doi.org/10.1126/science.aax4240

Klann E, Dever TE (2004) Biochemical mechanisms for translational regulation in synaptic plasticity. Nat Rev Neurosci 5:931–942. https://doi.org/10.1038/nrn1557

Knowles RB, Sabry JH, Martone ME, Deerinck TJ, Ellisman MH, Bassell GJ, Kosik KS (1996) Translocation of RNA granules in living neurons. J Neurosci 16:7812–7820. https://doi.org/10.1523/JNEUROSCI.16-24-07812.1996

Krainer G, Welsh TJ, Joseph JA, Espinosa JR, Wittmann S, de Csilléry E, Sridhar A, Toprakcioglu Z, Gudiškytė G, Czekalska MA, Arter WE, Guillén-Boixet J, Franzmann TM, Qamar S, George-Hyslop PS, Hyman AA, Collepardo-Guevara R, Alberti S, Knowles TPJ (2021) Reentrant liquid condensate phase of proteins is stabilized by hydrophobic and non-ionic interactions. Nat Commun 12:1085. https://doi.org/10.1038/s41467-021-21181-9

Krichevsky AM, Kosik KS (2001) Neuronal RNA granules: a link between RNA localization and stimulation-dependent translation. Neuron 32:683–696. https://doi.org/10.1016/s0896-6273(01)00508-6

Lagier-Tourenne C, Polymenidou M, Cleveland DW (2010) TDP-43 and FUS/TLS: emerging roles in RNA processing and neurodegeneration. Hum Mol Genet 19:R46–R64. https://doi.org/10.1093/hmg/ddq137

Langdon EM, Qiu Y, Ghanbari Niaki A, McLaughlin GA, Weidmann CA, Gerbich TM, Smith JA, Crutchley JM, Termini CM, Weeks KM, Myong S, Gladfelter AS (2018) mRNA structure determines specificity of a poly Q-driven phase separation. Science 360:922–927. https://doi.org/10.1126/science.aar7432

Langille JJ, Ginzberg K, Sossin WS (2019) Polysomes identified by live imaging of nascent peptides are stalled in hippocampal and cortical neurites. Learn Mem 26:351–362. https://doi.org/10.1101/lm.049965.119

Ling SC, Polymenidou M, Cleveland DW (2013) Converging mechanisms in ALS and FTD: disrupted RNA and protein homeostasis. Neuron 79:416–438. https://doi.org/10.1016/j.neuron.2013.07.033

Maharana S, Wang J, Papadopoulos DK, Richter D, Pozniakovsky A, Poser I, Bickle M, Rizk S, Guillén-Boixet J, Franzmann TM, Jahnel M, Marrone L, Chang YT, Sterneckert J, Tomancak P, Hyman AA, Alberti S (2018) RNA buffers the phase separation behavior of prion-like RNA binding proteins. Science 360:918–921. https://doi.org/10.1126/science.aar7366

Mansur F, Ivshina M, Gu W, Schaevitz L, Stackpole E, Gujja S, Edwards YJ, Richter JD (2016) Gld2-catalyzed 3′ monoadenylation of miRNAs in the hippocampus has no detectable effect on their stability or on animal behavior. RNA 22:1492–1499. https://doi.org/10.1261/rna.056937.116

Mateju D, Eichenberger B, Voigt F, Eglinger J, Roth G, Chao JA (2020) Single-molecule imaging reveals translation of mRNAs localized to stress granules. Cell 183:1801–1812. https://doi.org/10.1016/j.cell.2020.11.010

Mikl M, Vendra G, Kiebler MA (2011) Independent localization of MAP2, CaMKIIα and β-actin RNAs in low copy numbers. EMBO Rep 12:1077–1084. https://doi.org/10.1038/embor.2011.149

Miller S, Yasuda M, Coats JK, Jones Y, Martone ME, Mayford M (2002) Disruption of dendritic translation of CaMKIIalpha impairs stabilization of synaptic plasticity and memory consolidation. Neuron 36:507–519. https://doi.org/10.1016/s0896-6273(02)00978-9

Monahan Z, Ryan VH, Janke AM, Burke KA, Rhoads SN, Zerze GH, O'Meally R, Dignon GL, Conicella AE, Zheng W, Best RB, Cole RN, Mittal J, Shewmaker F, Fawzi NL (2017) Phosphorylation of the FUS low-complexity domain disrupts phase separation, aggregation, and toxicity. EMBO J 36:2951–2967. https://doi.org/10.15252/embj.201696394

Mori Y, Imaizumi K, Katayama T, Yoneda T, Tohyama M (2000) Two cis-acting elements in the 3′ untranslated region of alpha-CaMKII regulate its dendritic targeting. Nat Neurosci 3:1079–1084. https://doi.org/10.1038/80591

Murakami T, Qamar S, Lin JQ, Schierle GS, Rees E, Miyashita A, Costa AR, Dodd RB, Chan FT, Michel CH, Kronenberg-Versteeg D, Li Y, Yang SP, Wakutani Y, Meadows W, Ferry RR, Dong L, Tartaglia GG, Favrin G, Lin WL, Dickson DW, Zhen M, Ron D, Schmitt-Ulms G,

Fraser PE, Shneider NA, Holt C, Vendruscolo M, St KCF, George-Hyslop P (2015) ALS/FTD mutation-induced phase transition of FUS liquid droplets and reversible hydrogels into irreversible hydrogels impairs RNP granule function. Neuron 88:678–690. https://doi.org/10.1016/j.neuron.2015.10.030

Na Y, Park S, Lee C, Kim DK, Park JM, Sockanathan S, Huganir RL, Worley PF (2016) Real-time imaging reveals properties of glutamate-induced arc/Arg 3.1 translation in neuronal dendrites. Neuron 91:561–573. https://doi.org/10.1016/j.neuron.2016.06.017

Nakayama K, Ohashi R, Shinoda Y, Yamazaki M, Abe M, Fujikawa A, Shigenobu S, Futatsugi A, Noda M, Mikoshiba K, Furuichi T, Sakimura K, Shiina N (2017) RNG105/caprin 1, an RNA granule protein for dendritic mRNA localization, is essential for long-term memory formation. elife 6:e29677. https://doi.org/10.7554/eLife.29677

Narayanan U, Nalavadi V, Nakamoto M, Pallas DC, Ceman S, Bassell GJ, Warren ST (2007) FMRP phosphorylation reveals an immediate-early signaling pathway triggered by group I mGluR and mediated by PP2A. J Neurosci 27:14349–14357. https://doi.org/10.1523/JNEUROSCI.2969-07.2007

Nedelsky NB, Taylor JP (2022) Pathological phase transitions in ALS-FTD impair dynamic RNA-protein granules. RNA 28:97–113. https://doi.org/10.1261/rna.079001.121

Ohashi R, Shiina N (2020) Cataloguing and selection of mRNAs localized to dendrites in neurons and regulated by RNA-binding proteins in RNA granules. Biomol Ther 10:167. https://doi.org/10.3390/biom10020167

Padrón A, Ingolia N (2022) Analyzing the composition and organization of ribonucleoprotein complexes by APEX-Seq. Methods Mol Biol 2428:277–289. https://doi.org/10.1007/978-1-0716-1975-9_17

Park HY, Lim H, Yoon YJ, Follenzi A, Nwokafor C, Lopez-Jones M, Meng X, Singer RH (2014) Visualization of dynamics of single endogenous mRNA labeled in live mouse. Science 343:422–424. https://doi.org/10.1126/science.1239200

Ratti A, Buratti E (2016) Physiological functions and pathobiology of TDP-43 and FUS/TLS proteins. J Neurochem 138(Suppl 1):95–111. https://doi.org/10.1111/jnc.13625

Ries RJ, Zaccara S, Klein P, Olarerin-George A, Namkoong S, Pickering BF, Patil DP, Kwak H, Lee JH, Jaffrey SR (2019) m^6A enhances the phase separation potential of mRNA. Nature 571:424–428. https://doi.org/10.1038/s41586-019-1374-1

Rook MS, Lu M, Kosik KS (2000) CaMKIIalpha 3′ untranslated region-directed mRNA translocation in living neurons: visualization by GFP linkage. J Neurosci 20:6385–6393. https://doi.org/10.1523/JNEUROSCI.20-17-06385.2000

Roy R, Shiina N, Wang DO (2020) More dynamic, more quantitative, unexpectedly intricate: advanced understanding on synaptic RNA localization in learning and memory. Neurobiol Learn Mem 168:107149. https://doi.org/10.1016/j.nlm.2019.107149

Ryan VH, Fawzi NL (2019) Physiological, pathological, and targetable membraneless organelles in neurons. Trends Neurosci 42:693–708. https://doi.org/10.1016/j.tins.2019.08.005

Shiina N (2019) Liquid- and solid-like RNA granules form through specific scaffold proteins and combine into biphasic granules. J Biol Chem 294:3532–3548. https://doi.org/10.1074/jbc.RA118.005423

Shiina N, Shinkura K, Tokunaga M (2005) A novel RNA-binding protein in neuronal RNA granules: regulatory machinery for local translation. J Neurosci 25:4420–4434. https://doi.org/10.1523/JNEUROSCI.0382-05.2005

Shin Y, Berry J, Pannucci N, Haataja MP, Toettcher JE, Brangwynne CP (2017) Spatiotemporal control of intracellular phase transitions using light-activated opto droplets. Cell 168:159–171.e14. https://doi.org/10.1016/j.cell.2016.11.054

Siemen H, Colas D, Heller HC, Brüstle O, Pera RA (2011) Pumilio-2 function in the mouse nervous system. PLoS One 6:e25932. https://doi.org/10.1371/journal.pone.0025932

Sudhakaran IP, Ramaswami M (2017) Long-term memory consolidation: the role of RNA-binding proteins with prion-like domains. RNA Biol 14:568–586. https://doi.org/10.1080/15476286.2016.1244588

Sutton MA, Schuman EM (2006) Dendritic protein synthesis, synaptic plasticity, and memory. Cell 127:49–58. https://doi.org/10.1016/j.cell.2006.09.014

Tang SJ, Meulemans D, Vazquez L, Colaco N, Schuman E (2001) A role for a rat homolog of staufen in the transport of RNA to neuronal dendrites. Neuron 32:463–475. https://doi.org/10.1016/s0896-6273(01)00493-7

The Dutch-Belgian Fragile X Consortium, Bakker CE, Verheij C, Willemsen R, van der Helm R, Willems PJ (1994) Fmr1 knockout mice: a model to study fragile X mental retardation. Cell 78: 23–33. https://doi.org/10.1016/0092-8674(94)90569-X

Tsang B, Arsenault J, Vernon RM, Lin H, Sonenberg N, Wang LY, Bah A, Forman-Kay JD (2019) Phosphoregulated FMRP phase separation models activity-dependent translation through bidirectional control of mRNA granule formation. Proc Natl Acad Sci U S A 116:4218–4227. https://doi.org/10.1073/pnas.1814385116

Tushev G, Glock C, Heumüller M, Biever A, Jovanovic M, Schuman EM (2018) Alternative 3' UTRs modify the localization, regulatory potential, stability, and plasticity of mRNAs in neuronal compartments. Neuron 98:495–511.e6. https://doi.org/10.1016/j.neuron.2018.03.030

Udagawa T, Farny NG, Jakovcevski M, Kaphzan H, Alarcon JM, Anilkumar S, Ivshina M, Hurt JA, Nagaoka K, Nalavadi VC, Lorenz LJ, Bassell GJ, Akbarian S, Chattarji S, Klann E, Richter JD (2013) Genetic and acute CPEB1 depletion ameliorate fragile X pathophysiology. Nat Med 19: 1473–1477. https://doi.org/10.1038/nm.3353

Uutela M, Lindholm J, Louhivuori V, Wei H, Louhivuori LM, Pertovaara A, Akerman K, Castrén E, Castrén ML (2012) Reduction of BDNF expression in Fmr1 knockout mice worsens cognitive deficits but improves hyperactivity and sensorimotor deficits. Genes Brain Behav 11: 513–523. https://doi.org/10.1111/j.1601-183X.2012.00784.x

Vessey JP, Macchi P, Stein JM, Mikl M, Hawker KN, Vogelsang P, Wieczorek K, Vendra G, Riefler J, Tübing F, Aparicio SA, Abel T, Kiebler MA (2008) A loss of function allele for murine Staufen 1 leads to impairment of dendritic Staufen1-RNP delivery and dendritic spine morphogenesis. Proc Natl Acad Sci U S A 105:16374–16379. https://doi.org/10.1073/pnas.0804583105

Yoon YJ, Wu B, Buxbaum AR, Das S, Tsai A, English BP, Grimm JB, Lavis LD, Singer RH (2016) Glutamate-induced RNA localization and translation in neurons. Proc Natl Acad Sci U S A 113: E6877–E6886. https://doi.org/10.1073/pnas.1614267113

Zalcman G, Federman N, Romano A (2018) CaMKII isoforms in learning and memory: localization and function. Front Mol Neurosci 11:445. https://doi.org/10.3389/fnmol.2018.00445

Chapter 11
The Role of Liquid–Liquid Phase Separation in the Structure and Function of Nucleolus

Jing Wei and Shige H. Yoshimura

Abstract Nucleolus is the largest intracellular membrane-less organelle comprising of proteins, ribosomal RNAs, and small nucleolar RNAs. It had been described as a liquid-like organelle, which undergoes dynamic morphological changes during cell cycle, upon cellular stimuli, and under stress condition. Recent studies demonstrated that many nucleolar proteins contain large portions of disordered regions and high propensities of liquid–liquid phase separation in vitro. These intrinsic properties of nucleolar proteins, together with transcription and processing of ribosomal RNAs, contribute to a steady-state structure and function of the nucleolus. In this chapter, we focus on liquid-like properties of nucleolus and describe how liquid–liquid phase separation of nucleolar proteins and rRNAs play a role in structural and functional dynamics of nucleolus.

Keywords Nucleolus · Ribosome biogenesis · Liquid–liquid phase separation · Ribosomal RNA (rRNA) · Stress response

11.1 Overview

The nucleolus is the most prominent structure in the nucleus. It is not only responsible for ribosome synthesis, but also plays a role in various cellular events, such as stress responses and viral infection. Due to the absence of a surrounding membrane, biomolecules are dynamically exchanged within and around this organelle. The nucleoli display liquid-like behavior, and in mature *Xenopus laevis* germinal vesicle, the nucleoli could be fused by micromanipulation using a microneedle (Brangwynne et al. 2011). Many nucleolar proteins are also able to form liquid droplets in vitro (Mitrea et al. 2016; Yao et al. 2019; Feric et al. 2016). These active and dynamic features of the nucleoli are closely related to liquid–liquid phase separation (LLPS),

J. Wei · S. H. Yoshimura (✉)
Graduate School of Biostudies, Kyoto University, Kyoto, Japan
e-mail: yoshimura@lif.kyoto-u.ac.jp

R. Kurokawa (ed.), *Phase Separation in Living Cells*,
https://doi.org/10.1007/978-981-99-4886-4_11

which is referred to as a state in which biopolymers such as proteins, RNA, and DNA are separated into multiple phases with different polymer concentrations.

Increasing evidence has indicated that LLPS is ubiquitous in cells. It not only drives the formation of membrane-less organelles, such as nucleoli, stress granules, and ribonucleoprotein (RNP) bodies (Feric et al. 2016; Jain et al. 2016; Hubstenberger et al. 2013), but is also involved in many biological reactions, events, and homeostasis. In this review, we discuss the involvement of LLPS in the establishment of the nucleolar architecture, the largest membrane-less condensate, and explore how LLPS regulates the spatiotemporal behavior of signaling molecules involved in nucleolar functions, including ribosome biogenesis, nucleolar disassembly and reassembly during mitosis, and nucleolus-related disease progression.

11.2 Structure and Function of the Nucleolus

Nucleoli are liquid-like subnuclear organelles that undergo disassembly and reassembly during the cell cycle (Németh and Grummt 2018). Animal cells contain a few to a dozen nucleoli, and the number and morphology differ based on cell type and growth. Cells that proliferate rapidly tend to have a larger number and size of nucleoli owing to the large demand for protein synthesis (Derenzini et al. 2000), which is a major function of the nucleolus. The nucleolus is particularly enriched in proteins and RNA owing to an abundance of ribosomes. The ribosomal RNA (rRNA) in the nucleolus was found to constitute ~80% of the mass of RNA in the cells (Feng and Manley 2022). Moreover, their higher density compared with that of the surrounding nucleoplasm (Vincent 1955) makes it possible to isolate nucleoli from nuclei (Andersen et al. 2002). Additionally, the nucleolus plays an important role in sensing diverse stresses including genotoxic and oxidative stress, heat, starvation, oncogene activation, and viral infections (Boulon et al. 2010).

The nucleolus can be observed by light microscopy using phase contrast or differential interference contrast (Boisvert et al. 2007). It is composed of three sub-compartments: the fibrillar center (FC), dense fibrillar compartment (DFC), and granular component (GC) (Amoretti et al. 2002) (Fig. 11.1). Each sub-compartment has a different protein and RNA composition (see later sections) and can be distinguished by electron microscopy.

11.2.1 Nucleolar Sub-Compartments

The nucleolar sub-compartment structure is comparable to a plum pudding: several DFC plums are scattered within the GC pudding, and individual DFC plums contain an FC seed, although the plums are more dynamic and can be fused into large plums in the nucleoli (movie in (Feric et al. 2016)). GC consists of RNP complexes with a diameter of 15–20 nm that can be digested by proteinases and RNases.

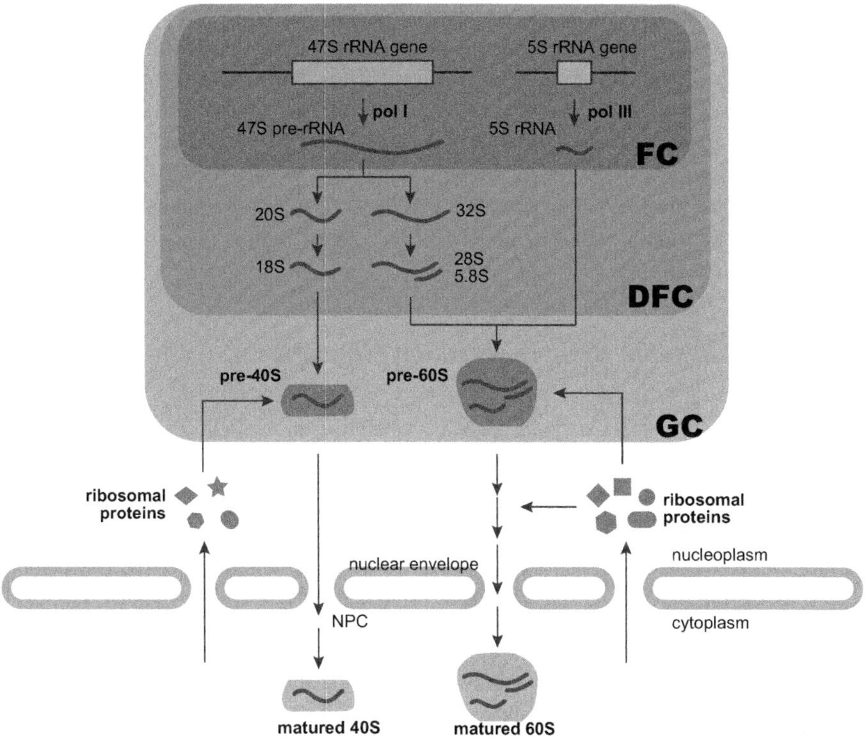

Fig. 11.1 Nucleolar sub-compartments and ribosomal biogenesis. Transcription of rRNA genes (47S and 5S) occur in the FC. The 47S pre-rRNA is separated from spacers (not shown) and fragmented into three rRNAs (5.8S, 18S, and 28S) in the DFC. The 18S rRNA is assembled with ribosomal proteins into 40S subunit, whereas 5S, 5.8S, and 28S form 60S subunit together with different sets of ribosomal proteins in the GC. After additional maturation processes in the nucleoplasm, both subunits are transported to the cytoplasm and matured

Nucleophosmin (NPM1) is a major protein component in GC and has been used as a marker of this sub-compartment. DFC is circular or half-moon-shaped (with the outer and inner diameters being ~630 and ~ 360 nm, respectively, in human cells (Yao et al. 2019)) and has the highest electron density (Boisvert et al. 2007). It is enriched in fibrillarin (FBL), rRNA, small nucleolar ribonucleoproteins (snoRNPs), and co-factors involved in pre-rRNA processing and modification (see a later section for ribosome biogenesis). The FC is a spherical structure with low electron density buried in the DFC. It contains RNA polymerase I (Pol I) subunits and FBL, which are involved in rDNA transcription and ribosome biogenesis (Shan et al. 2023).

In recent years, researchers have developed a systematic map of all human nucleolar proteins by combining immunofluorescence and confocal microscopy using the HPA Cell Atlas. They discovered 1318 nucleolar proteins, comprising approximately 7% of the human proteome; of these, 287 localize to the FC or DFC regions and 1031 localize to the whole nucleolus (Stenström et al. 2020).

In addition to these three sub-compartments, the nucleolar rim has recently been recognized as a new sub-compartment owing to its distinct proteomic composition. Nucleolar rim proteins tend to be more disordered than proteins in other sub-compartments (Stenström et al. 2020). The function of the nucleolar rim might be related to the mitotic perichromosomal region because some mitotic chromosome-associated proteins, such as Ki-67, are localized in the nucleolar rim. These proteins may have common molecular features and functions that require further investigation.

11.2.2 Ribosome Biogenesis in the Nucleolus

In the 1960s, the discovery of rRNA synthesis via nucleic acid hybridization (D. D. Brown and Gurdon 1964) opened a new chapter in the history of the nucleolus. Ribosomal biogenesis was subsequently identified as an important function of the nucleolus. There are three steps in ribosome biogenesis in the nucleolus.

11.2.2.1 Production of Precursor rRNAs

The eukaryotic 80S ribosome contains four rRNAs, three of which (18S, 5.8S, and 28S) are encoded by a single gene, transcribed into a single precursor RNA (47S pre-rRNA) by Pol I, and then processed (Fig. 11.1). The fourth rRNA, 5S, is encoded by a different gene and is transcribed by RNA polymerase III (Henras et al. 2015; Sørensen and Frederiksen 1991; Turi et al. 2019). Pol I and pre-rRNAs are localized at the edge of the FC (Yao et al. 2019), where the transcription of rRNA genes occurs. Transcription is initiated by a pre-initiation complex (PIC), which consists of an upstream binding factor (UBF), a selectivity factor (SL1, also known as TIF1-B), and transcription initiation factor 1A (TIF1-A or hRRN3). It begins with a UBF homodimer binding to the core promoter region and the upstream core element of the gene, creating a DNA loop. SL1, TIF1-A, and Pol I are then recruited to form a complete PIC (Turi et al. 2019). Transcription terminates when Pol I encounters DNA-bound transcription termination factor 1 (TTF-1). Pol I is stalled by TTF-1 and releases pre-rRNA with the help of PTRF (polymerase I and transcript release factor) (Jansa and Grummt 1999). Notably, as soon as the transcription of pre-rRNA initiates, the 5′ end of the nascent RNA is guided from the FC-DFC border to the DFC by FBL. The self-assembly of FBL into liquid droplets helps immobilize pre-rRNA in the DFC for further processing and modification (Yao et al. 2019).

As Pol I activity is a decisive factor in ribosomal production, cell growth, and proliferation, rRNA transcription must be tightly regulated. Recently, it was found that the size and liquidity of FC/DFC are important for Pol I-mediated transcription. In HeLa cells, FC/DFC units are coated with the RNA helicase DDX21. When DDX21 is in a closed conformation (mediated by the RNA regulator SLERT), it forms loose clusters that enable the FC/DFC units to be larger in size and sufficiently liquid-like, thus promoting rapid rDNA transcription and Pol I processivity (Wu et al. 2021).

11.2.2.2 rRNA Maturation (Processing and Modification)

In the 47S pre-rRNA transcript, 18S, 5.8S, and 28S rRNAs are intervened by internal transcribed spacers (ITS1 and ITS2) and flanked by external transcribed spacers (5′-ETS and 3′-ETS). During rRNA processing, the ITSs and ETSs are removed by the combined action of endonucleases and exonucleases (Henras et al. 2015) (Fig. 11.1).

Processing and modification of pre-rRNA transcripts occur mainly in the DFC, where box C/D and box H/ACA snoRNPs accumulate. In human cells, box C/D snoRNPs contain many processing factors, including methyltransferase FBL, accessory proteins Nop56 and Nop58, and small nucleolar RNAs (snoRNAs). Box H/ACA snoRNPs contain pseudouridine synthase dyskerin, accessory proteins Nhp2, Nop 10, GAR1, and snoRNAs. snoRNAs form base pairings with pre-rRNAs and help them fold properly to undergo methylation by FBL and pseudouridylation by dyskerin (Watkins and Bohnsack 2012). Interestingly, FBL is involved in many steps of ribosome biogenesis, including methylation, chaperoning, and exonucleolytic cleavages (Watkins and Bohnsack 2012). More than 100 snoRNAs have been reported to guide the extensive site-specific nucleotide modifications of rRNA during its processing (Bachellerie et al. 2002). Furthermore, ATPases, GTPases, and RNA helicases are also engaged in rRNA processing to facilitate the proper folding of rRNA (Turi et al. 2019).

11.2.2.3 Assembly of Precursor Ribosomal Subunits

Mature rRNAs (18S, 5.8S, and 28S) and 5S rRNA, which are transcribed from different gene loci, are then assembled in the GC with approximately 80 ribosomal proteins, which are translated in the cytoplasm and imported through the nuclear pore into the nucleus (Lafontaine et al. 2021; Turi et al. 2019) (Fig. 11.1). 18S rRNA is incorporated into the small ribosomal subunit 40S, whereas the other three rRNAs are incorporated into the large subunit 60S (Lafontaine et al. 2021, Turi et al. 2019). They are transported to the cytoplasm through the nuclear pore for the final assembly of 80S ribosomes to become fully functional protein-generating machines (Lafontaine et al. 2021, Turi et al. 2019).

In some cases, the ribosomal proteins are improperly folded or assembled into ribosomal subunits (Joazeiro 2019; Lafontaine 2010). In budding yeast, these unqualified products would be sensed under the surveillance of "No-body" (nucleolar body) and then degraded in the nucleoplasm. "No-body" is a specialized nucleolar condensate, containing the TRAMP-exosome system, in which TRAMP acts as an unqualified ribosomal subunit sensor while the exosome exerts decay activity (Lafontaine 2010).

11.2.3 Multiple Liquid Phases

The first documentation of the nucleolus was made independently by Wagner (Wagner 1835) and Valentin (Bennett and Malmfors 1975) in the 1830s (Pederson 2011; Lafontaine et al. 2021). Nucleolar morphology and materials were well documented in 1898 (Montgomery 1898). Ribosomal biogenesis was first discovered in the 1960s (Brown and Gurdon 1964). The liquid-like biophysical nature of the nucleolus was first reported in 1946 (Ehrenberg 2010). The nucleoli remained morphologically intact and highly dynamic when isolated from HeLa cells (Andersen et al. 2002). Since the finding that the formation of *Caenorhabditis elegans* germ granule (P granule) was driven by LLPS in 2009 (Brangwynne et al. 2009), the nucleolus has been studied in the context of LLPS. Recent studies have found that the nucleolus is a large, membrane-less, bimolecular condensate that contains various biopolymers and is formed by LLPS.

Although the mechanism of how nucleolar sub-compartments are established and maintained without biological membrane has not been fully understood, there have been several cues and evidence that show the establishment of the multiphase "core–shell" architecture by multi-component LLPS, in which protein immiscibility and hydrophobicity are key factors.

Purified NPM1 and FBL undergo LLPS in vitro (Feric et al. 2016). When the two proteins are mixed, the FBL-rich phase is always encapsulated by the NPM1-rich phase and shows a clear boundary similar to the nucleolar architecture (Feric et al. 2016). This is partly because homotypic interactions (NPM1–NPM1 and FBL–FBL) are more favorable than heterotypic interactions (NPM1–FBL) (Feric et al. 2016). In addition, the difference in hydrophobicity plays a role in the radial position of each protein phase. FBL is more hydrophobic than NPM1, and a droplet of a hydrophobic solute generally tends to be encapsulated by a more hydrophilic phase (Neeson et al. 2012). The multiphase core–shell architecture can also be found in stress granules and other RNP bodies (Jain et al. 2016; Hubstenberger et al. 2013).

RNA also plays a critical role in regulating nucleolar properties and structure. The addition of RNA affects the stability of FBL-rich droplets in vitro and their localization in DFC (Berry et al. 2015). Long noncoding RNA (lncRNA) SLERT affects the size and liquidity of FC and DFC by changing the conformation of the RNA helicase DDX21 (please also see Sect. 11.3.3).

11.2.4 Structural Dynamics of the Nucleolus during Cell Cycle

The nucleolus undergoes dynamic structural changes during the cell cycle. It begins to disassemble at the beginning of mitosis and reassembles in telophase. Recent studies have demonstrated that the mitotic dynamics of the nucleolus are closely related to its liquid-like properties. In particular, protein phosphorylation by mitotic kinases (e.g., Cdk1) and transcriptional suppression of rRNA genes play major roles.

Nucleolar disassembly begins in the early prophase. Cdk1 phosphorylates Pol I and suppresses rRNA transcription (Derenzini et al. 2005). The Pol I-dependent transcription decreases by approximately 30% during early prophase and stops entirely in late prophase (Gebrane-Younes et al. 1997; Prescott and Bender 1962), which results in a breakdown of FC and subsequently of DFC and GC (Danièle Hernandez-Verdun 2011, Derenzini et al. 2000). The mechanism underlying this orderly breakdown from the inner to the outer nucleolus is not well understood. This may be caused by sequential phosphorylation of protein components or the breakdown of FC affecting the LLPS of the bordering sub-compartment, DFC.

At the end of prophase, when the nuclear envelope is broken down and chromosomes are condensed, the nucleolus is almost completely disassembled. Most nucleolar proteins diffuse into the cytoplasm, while some are recruited to the chromosome surface (65 of 1318) (Stenström et al. 2020). In metaphase, some nuclear proteins (such as peripherin), nucleolar proteins (such as Ki-67 and FBL), and RNPs assemble on the mitotic chromosomal surface to form a liquid-like perichromosomal region (PR), which serves as an emulsifier to prevent chromosomes from coalescing (Hernandez-Verdun and Gautier 1994). Ki-67 localizes to the nucleolar rim of interphase cells and the PR of mitotic cells, and plays a major role in the PR formation; depletion of Ki-67 abolishes PR and the localization of other PR proteins on the surface of mitotic chromosomes (Remnant et al. 2021). Ki-67 is strongly phosphorylated by mitotic kinases. A recent study demonstrated that mitotic phosphorylation enhances the LLPS propensity of Ki-67 and plays a major role in PR formation (Yamazaki et al. 2022).

In early telophase, pre-nucleolar bodies (PNBs), which contain rRNA processing complexes and nucleolar proteins, such as FBL, NPM1, and Nop52, start to assemble on the chromosome surface (Hernandez-Verdun 2011; Leung et al. 2004). They are considered a liquid-like transient assembly of nucleolar components, and individual components have different protein compositions (Savino et al. 2001). PNBs interact with the nucleolar organizer regions (NORs), where 47S rRNA genes are tandemly repeated (Leung et al. 2004). NORs are formed around the short arms of the five acrocentric chromosomes (human chromosomes 13, 14, 15, 21, and 22) (McStay 2016; Lafontaine et al. 2021; Turi et al. 2019). In telophase, rRNA transcription resumes at the NORs (Hernandez-Verdun 2011). With the supply of nucleolar proteins and rRNA transcription processing complexes from the PNBs, the nucleolar structure is gradually formed from the FC toward the GC around individual NORs. After the G1 phase, when the nucleolar architecture is fully rebuilt, the PNBs subsequently disappear (Savino et al. 2001). PNB motion and nucleolar assembly require translocation of the processing complexes to NORs, involving dynamic long-distance interactions between spatially separated genomic regions. Owing to the liquidity of the PNB foci, LLPS may play a role in this process. However, the underlying mechanisms remain poorly understood.

11.3 Biochemical Properties of Nucleolar Proteins

Many nucleolar proteins undergo LLPS in vitro. As described in the previous section, FBL and NPM1 undergo strong LLPS, although their critical concentrations differ (Feric et al. 2016). Other nucleolar proteins such as Ki-67, NOC2 homolog, NEP1, BMS1, BRIX1, and RL1D1 also show strong LLPS propensities (Stenström et al. 2020). In this review, we examine the structural properties of these proteins.

11.3.1 Nucleolar Proteins Are Disordered

Proteins/protein regions containing no three-dimensional structures are called intrinsically-disordered proteins (IDPs)/regions (IDRs) and exhibit considerable conformational heterogeneity. They are also referred to as low-complexity sequence domains and are considered key drivers of phase separation. Nucleolar proteins contain higher fractions of IDRs than cytosolic proteins; 20% in nucleoli ($n = 848$) and 14% in cytosolic proteins ($n = 2054$) (Stenström et al. 2020). NPM1, for example, has an N-terminal oligomerization domain, followed by an IDR and a C-terminal substrate RNA-binding domain (Mitrea et al. 2016) (Fig. 11.2). In addition, FBL contains an IDR rich in glycine and arginine, also known as the glycine-arginine-rich (GAR) domain (Mitrea et al. 2016). Among the nucleolar proteins, the IDR content was found to be the highest in mitotic chromosome-associated proteins, nucleolar rim proteins, and FC proteins. The most disordered proteins are associated with PRs such as CCDC86, Ki-67, and RRP15 (Stenström et al. 2020). The high proportion of IDR-containing proteins in the nucleolus indicates that LLPS plays a prominent role in nucleolar signaling and function.

11.3.2 Charged Residues

IDRs are generally rich in both polar and charged residues. Charged residues are involved in various molecular interactions in LLPS and have multiple functions, many of which are based on electrostatic interactions. Long stretches of consecutive D and E are more prevalent in nucleic acid-related proteins, enabling DNA mimicry, mRNA processing, and transcription complex regulation (Chou and Wang 2015). Additionally, nucleolar methyltransferase TGS1 plays a significant role in the coalescence and steady morphology of the nucleolus, partly through its interaction with the conserved KKD/E domain of snoRNPs (Colau et al. 2004). Mutation of both basic and acidic residues of p26 (a protein of the virus PEMV2) leads to failure or a decrease in liquid droplet formation, thus hindering viral movement (Brown et al. 2021).

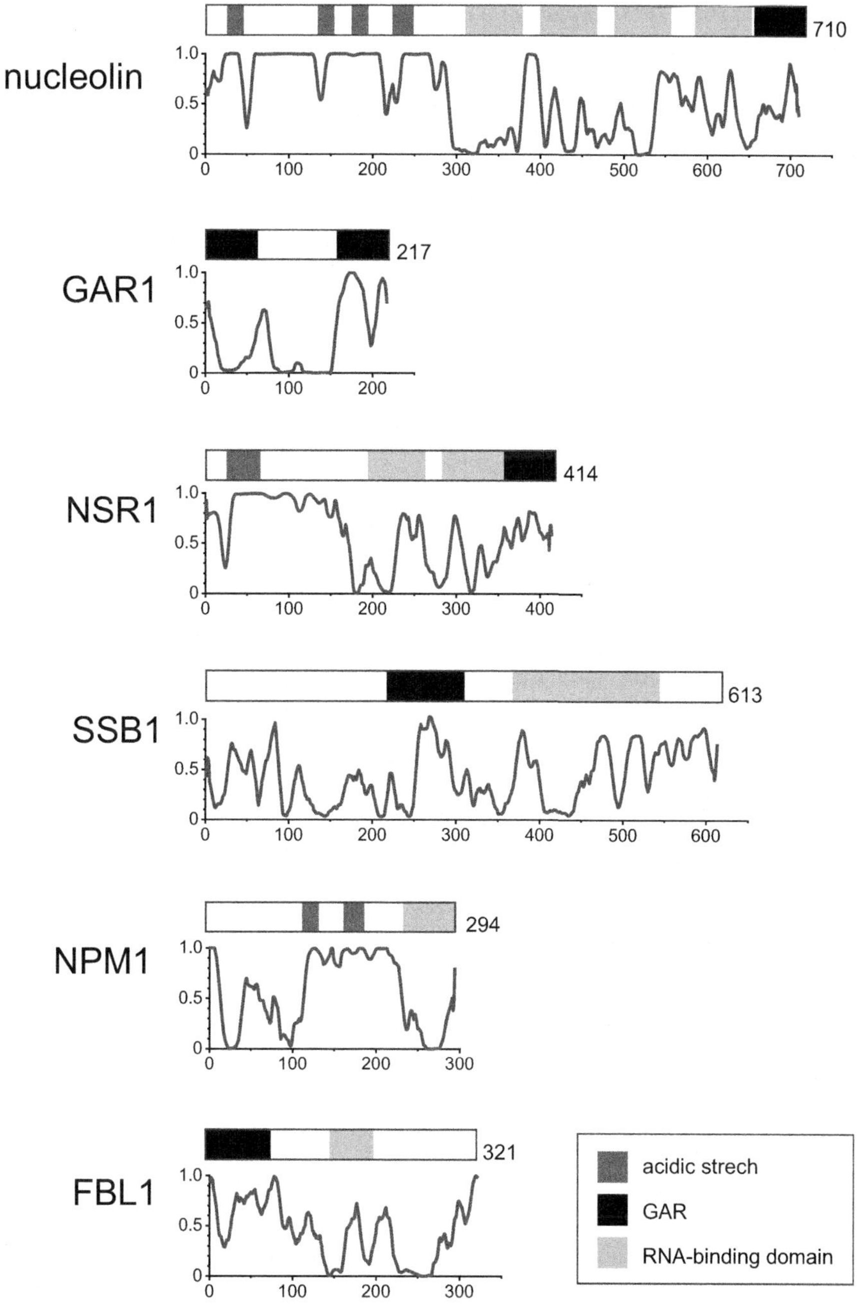

Fig. 11.2 Nucleolar proteins are disordered and charged. The domain structure and the disorder plot of several nucleolar proteins are shown. The disorder plot was made by PONDR algorithm

Arginine-rich IDRs play important roles in LLPS (Mitrea et al. 2016; Gallivan and Dougherty 1999). In addition to electrostatic interactions with negatively charged side chains or nucleic acids, arginine is involved in cation-π interactions with aromatic side chains (Gallivan and Dougherty 1999). Some nucleolar proteins [nucleolin, FBL, SSB1, NSR1, and GAR1 (Girard et al. 1992, Berry et al. 2015)] contain highly R-rich domains comprising several short stretches of 5–12 residues containing RGGXGGR or RGGXRGG sequences (X is usually a phenylalanine or, less frequently, serine, tyrosine or alanine) (Girard et al. 1992). These GAR sequences play roles in LLPS, RNA binding, and nucleolar localization. The N-terminal GAR domain of FBL is sufficient to form liquid-like droplets in vitro (Feric et al. 2016). GAR domain-mediated self-association of FBL promotes pre-rRNA sorting and facilitates DFC assembly (Yao et al. 2019). In *Xenopus* oocyte nucleoli, the C-terminal GAR domain of nucleolin plays a role in its efficient binding to nucleic acids and localization to the DFC (Heine et al. 1993). Although GAR domains also exist in nucleoplasmic [hnRNP A1, A2 (Watkins and Bohnsack 2012)] and cytoplasmic proteins [spectraplakins (Lane et al. 2017), GAS2L3 (Fackler et al. 2014)], the G/R residue content is higher in nucleolar proteins (Girard et al. 1992).

Extended stretches of charged residues in IDRs, such as consecutive K/R, or D/E repeats (Bigman et al. 2022) make up a "charge block." The distinctive patterning of the charge blocks can imbue proteins with transient attractive forces necessary to drive LLPS. Indeed, oppositely charged blocks of amino acid residues in the IDR of Ki67 regulate condensate formation and function via electrostatic interactions (Yamazaki et al. 2022). During mitotic phosphorylation, the randomly distributed charge blocks of nucleolar phosphoproteins Ki67 and NPM1 are tuned into alternating charge blocks, thus improving or weakening their LLPS tendencies, respectively (Yamazaki et al. 2022). Another example of an alternating charge block pattern is seen in MED1 (Lyons et al. 2023). Condensates composed of MED1 selectively partition RNA polymerase II together with its positive allosteric regulators, while excluding negative regulators. Disrupting or adding a charge pattern prevents or promotes partitioning, respectively, with functional consequences for gene activation (Lyons et al. 2023). A similar pattern of charged residues is required for the condensate formation of DDX4, NICD, and MeCP2 (Nott et al. 2015; Brown et al. 2010; Lin et al. 2015).

11.3.3 Role of rRNA in LLPS

NPM1 undergoes LLPS via polyampholytic IDRs in vitro (Mitrea et al. 2018); this type of LLPS is called homotypic. In addition, NPM1 undergoes LLPS with other nucleolar components such as rRNA and proteins containing multivalent Arg-rich residues (Mitrea et al. 2018); this type of LLPS is called heterotypic. Because the nucleolus is a liquid-like organelle with multiple components, heterotypic LLPS should be considered for a comprehensive understanding of the nucleolar structure and function.

In the nucleoli, where RNA-binding proteins (RBP) and rRNA are particularly enriched, the role of RNA in LLPS is significant. Both RNA–RNA and RNA–protein interactions affect condensate formation and properties. Studies on RNA in LLPS are especially complicated because of the generation of sequence complexity in RNA during transcription and processing. Many nucleolar proteins such as FBL, GAR1, NSR1, and SSB1 have an RNA recognition motif (RRM) that can interact with RNA, especially during pre-rRNA transcription (Watkins and Bohnsack 2012; Girard et al. 1992; Yao et al. 2019). Generally, RNA facilitate or accelerate phase separation. The LLPS propensity of FBL in vitro is enhanced in the presence of purified total RNA (Berry et al. 2015). Furthermore, RNA increases the formation rate, volume fraction, and coarsening rate of droplets (Berry et al. 2015). ncRNA SLERT regulates the size and liquidity of FC and DFC via a chaperone-like mechanism and affects the conformation of DDX21, an RNA helicase surrounding the FC/DFC (Wu et al. 2021).

The nucleotide sequence of RNA also affects LLPS. For example, a proline-arginine repeat peptide forms a gel-like network condensate with poly (G) RNA, whereas it forms a liquid-like condensate with other homotypic RNA species (A, U, or C). The condensates formed with poly(A) RNA are more viscous than those formed with poly(U) or poly(C) RNA (Boeynaems et al. 2019). The function−/property-encoding information in the RNA sequence and RNA specificity in the regulation of phase separation remain largely unknown.

11.4 Nucleolus as a Potential Stress Sensor and Biomarker for Disease and Aging

The nucleolus is closely associated with several diseases (ribosomopathies) and cancers. Recent research has shown that impaired rRNA synthesis and ribosome biogenesis under stress conditions such as heat shock, nutrient deprivation, oncogene activation, viral infection, and cancer may result in abnormal nucleolar structure and functions, and are related to a variety of signaling transductions, such as Mdm2-p53, NF-κB, and HIF-1α pathways (Hua et al. 2022). Here, we present some examples of the link between cellular stress, disease, and liquid-like properties of the nucleolus.

11.4.1 Heat Stress

In *Arabidopsis thaliana,* prolonged heat stress changes the sub-nucleolar localization of nucleolar proteins, causes a defect in DFC (DFC maker protein FBL localizes to GC), and disrupts ribosome biogenesis (Picart-Picolo et al. 2020). In this study, under heat stress, the circular or half-moon-shaped DFC regions were barely visible in the cells, and the abnormal distribution of DFC could be reversed by placing the cells under normal conditions (Picart-Picolo et al. 2020). When HeLa cells are exposed to prolonged heat stress, the nucleoli gradually show irregular shapes and harden, which may be due to the formation of amyloid-like foci in NPM1 (Frottin et al. 2019). Under relatively short-term heat stress (2 h), the liquid-like properties of nucleoli function as a protein quality control compartment (Frottin et al. 2019); the GC transiently stores metastable and misfolded nuclear proteins to prevent irreversible aggregation and maintain protein homeostasis (Frottin et al. 2019).

11.4.2 Viral Infection

Several reports have described the unique intracellular localization of viral proteins in membrane-less organelles via LLPS (McStay 2016). The nucleolus is also involved in many interactions with viral proteins. One example of a process where nucleolar protein is engaged in host-virus interactions is virus movement. The ORF protein of groundnut rosette virus (an umbravirus) interacts with FBL in the nucleoli to support viral movement through the phloem (systemic movement, in which the virus moves from an infection site to distant parts of the plant by hitching a ride on the plant's supply lines, such as the vein) (Kim et al. 2007). Consistent with this result, p26, a movement-related protein in PEMV2, interacts with FBL for systemic viral movement (Brown et al. 2021). Mutation in both the basic and acidic residues in p26 results in a failure to form or decreased formation of liquid droplets, hindering viral movement (Brown et al. 2021). Capsid proteins, which are responsible for packaging genomic RNA into viral particles of several Flaviviridae viruses, are targeted to the nucleolus (Saito et al. 2021). For instance, the capsid proteins of the dengue virus (DENV) are targeted to GC (Saito et al. 2021). In addition, the capsid protein of DENV-2 affects the recovery kinetics of NPM1-GFP in FRAP (Tiwary and Cecilia 2017). The West Nile virus (WNV) capsid protein translocates to the nucleus and interacts with the nucleolar protein DDX56 to assemble infectious WNV particles (Xu et al. 2011). Despite these findings, the functional role of the nucleolar localization of viral proteins is not fully understood, and further studies are required.

11.4.3 Cancer and Aging

In rapidly dividing cancer cells, the strong upregulation of ribosome assembly is an important molecular alteration owing to the high demand for ribosomes. The nuclear structure of cancer cells differs from that of normal cells in various aspects, including the size, shape, number of nucleoli, and "chromatin texture." These morphological changes in the nucleoli appear to be cancer type-specific (Zink et al. 2004). These changes are characteristic and distinctive among the given tumor types and stages, provide an important diagnostic feature, and may be related to changes in the functional properties of cancer cells (Zink et al. 2004). In fact, the quantification of interphase nucleoli by silver staining-mediated visualization of NORs (AgNORs) has been used as a reliable method to assist in the diagnosis of multiple cancer types (Pich et al. 2000). Nucleolar deformation is related to the biophysical state and the process of LLPS. Hence, the role of LLPS in the deformed nucleoli of cancer cells and potential abnormalities in ribosome biogenesis is worth investigating.

Regarding aging, several findings suggest that the rate of ribosome biogenesis is negatively correlated with lifespan, which has been proven in numerous living species, such as *C. elegans*, *Drosophila melanogaster*, yeast, mice, and humans (examples summarized in (Turi et al. 2019)). This theory is not only related to the ribosome production rate, but also to energy consumption and altered proteostasis, of which the latter is believed to be a hallmark of aging (López-Otín et al. 2013; Kaushik and Cuervo 2015).

However, this association remains controversial. For instance, in yeast, the instability of the chromosomal rDNA array per se is a cause of aging that shortens lifespan (Kasselimi et al. 2022). The instability of chromosomal rDNA is related to the decreased efficiency of rDNA replication and decreased rRNA levels and ribosomal proteins (Sanchez et al. 2017). Whether the rate of ribosome biogenesis, along with LLPS, is a key factor in aging remains to be investigated. How LLPS is engaged in these two processes remains largely unknown, yet worth investigating.

References

Amoretti M et al (2002) Production and detection of cold antihydrogen atoms. Nature 419(6906): 456–459

Andersen JS et al (2002) Directed proteomic analysis of the human nucleolus. Curr Biol 12(1):1–11

Bachellerie J-P, Cavaillé J, Hüttenhofer A (2002) The expanding SnoRNA world. Biochimie 84(8): 775–790

Bennett T, Malmfors T (1975) Repertorium Für Anatomie Und Physiologie. Comp Biochem Physiol C 52:47–49

Berry J et al (2015) RNA transcription modulates phase transition-driven nuclear body assembly. Proc Natl Acad Sci 112(38):E5237

Bigman LS, Iwahara J, Levy Y (2022) Negatively charged disordered regions are prevalent and functionally important across proteomes. J Mol Biol 434(14):167660

Boeynaems S et al (2019) Spontaneous driving forces give rise to protein−RNA condensates with coexisting phases and complex material properties. Proc Natl Acad Sci 116(16):7889–7898

Boisvert F-M, Van Koningsbruggen S, Navascués J, Lamond AI (2007) The multifunctional nucleolus. Nat Rev Mol Cell Biol 8(7):574–585

Boulon S et al (2010) The nucleolus under stress. Mol Cell 40(2):216–227

Brangwynne CP et al (2009) Germline P granules are liquid droplets that localize by controlled dissolution/condensation. Science 324(5935):1729–1732

Brangwynne CP, Mitchison TJ, Hyman AA (2011) Active liquid-like behavior of nucleoli determines their size and shape in *Xenopus Laevis* oocytes. Proc Natl Acad Sci 108(11):4334–4339

Brown DD, Gurdon JB (1964) Absence of ribosomal RNA synthesis in the anucleolate mutant of *Xenopus* levis. Proc Natl Acad Sci 51(1):139–146

Brown CJ, Johnson AK, Daughdrill GW (2010) Comparing models of evolution for ordered and disordered proteins. Mol Biol Evol 27(3):609–621

Brown SL, Garrison DJ, May JP (2021) Phase separation of a plant virus movement protein and cellular factors support virus-host interactions. PLOS Pathog 17(9):e1009622

Chou C-C, Wang AH-J (2015) Structural D/E-rich repeats play multiple roles especially in gene regulation through DNA/RNA mimicry. Mol BioSyst 11(8):2144–2151

Colau G et al (2004) The small Nucle(Ol)Ar RNA cap trimethyltransferase is required for ribosome synthesis and intact nucleolar morphology. Mol Cell Biol 24(18):7976–7986

Derenzini M et al (2000) Nucleolar size indicates the rapidity of cell proliferation in cancer tissues. J Pathol 191(2):181–186

Derenzini M et al (2005) Key role of the achievement of an appropriate ribosomal RNA complement for G1-S phase transition in H4–II-E-C3 rat hepatoma cells. J Cell Physiol 202(2):483–491

Ehrenberg L (2010) Influence of temperature on the nucleolus and its coacervative nature. Hereditas 32(3–4):407–418

Fackler M, Wolter P, Gaubatz S (2014) The GAR domain of GAS2L3 mediates binding to the chromosomal passenger complex and is required for localization of GAS2L3 to the constriction zone during abscission. FEBS J 281(9):2123–2135

Feng S, Manley JL (2022) Beyond rRNA: nucleolar transcription generates a complex network of RNAs with multiple roles in maintaining cellular homeostasis. Genes Dev 36(15–16):876–886

Feric M et al (2016) Coexisting liquid phases underlie nucleolar subcompartments. Cell 165(7): 1686–1697

Frottin F et al (2019) The nucleolus functions as a phase-separated protein quality control compartment. Science 365(6451):342–347

Gallivan JP, Dougherty DA (1999) Cation-π interactions in structural biology. Proc Natl Acad Sci 96(17):9459–9464

Gebrane-Younes J, Fomproix N, Hernandez-Verdun D (1997) When RDNA transcription is arrested during mitosis, UBF is still associated with non-condensed RDNA. J Cell Sci 110(19):2429–2440

Girard JP et al (1992) GAR1 is an essential small nucleolar RNP protein required for pre-RRNA processing in yeast. EMBO J 11(2):673–682

Heine MA, Rankin ML, DiMario PJ (1993) The Gly/Arg-rich (GAR) domain of Xenopus Nucleolin facilitates in vitro nucleic acid binding and in vivo nucleolar localization. Mol Biol Cell 4(11): 1189–1204

Henras AK et al (2015) An overview of pre-ribosomal RNA processing in eukaryotes: pre-ribosomal RNA processing in eukaryotes. Wiley Interdiscip Rev RNA 6(2):225–242

Hernandez-Verdun D (2011) Assembly and disassembly of the nucleolus during the cell cycle. Nucleus 2(3):189–194

Hernandez-Verdun D, Gautier T (1994) The chromosome periphery during mitosis. BioEssays 16(3):179–185

Hua L, Yan D, Wan C, Baoying H (2022) Nucleolus and nucleolar stress: from cell fate decision to disease development. Cell 11(19):3017

Hubstenberger A, Noble SL, Cameron C, Evans TC (2013) Translation repressors, an RNA helicase, and developmental cues control RNP phase transitions during early development. Dev Cell 27(2):161–173

Jain S et al (2016) ATPase-modulated stress granules contain a diverse proteome and substructure. Cell 164(3):487–498

Jansa P, Grummt I (1999) Mechanism of transcription termination: PTRF interacts with the largest subunit of RNA polymerase I and dissociates paused transcription complexes from yeast and mouse. Mol Gen Genet MGG 262(3):508–514

Joazeiro CAP (2019) Mechanisms and functions of ribosome-associated protein quality control. Nat Rev Mol Cell Biol 20(6):368–383

Kasselimi E, Pefani D-E, Taraviras S, Lygerou Z (2022) Ribosomal DNA and the nucleolus at the heart of aging. Trends Biochem Sci 47(4):328–341

Kaushik S, Cuervo AM (2015) Proteostasis and aging. Nat Med 21(12):1406–1415
Kim SH et al (2007) Interaction of a plant virus-encoded protein with the major nucleolar protein fibrillarin is required for systemic virus infection. Proc Natl Acad Sci 104(26):11115–11120
Lafontaine DLJ (2010) A "garbage can" for ribosomes: how eukaryotes degrade their ribosomes. Trends Biochem Sci 35(5):267–277
Lafontaine DLJ, Riback JA, Bascetin R, Brangwynne CP (2021) The nucleolus as a multiphase liquid condensate. Nat Rev Mol Cell Biol 22(3):165–182
Lane TR, Fuchs E, Slep KC (2017) Structure of the ACF7 EF-hand-GAR module and delineation of microtubule binding determinants. Structure 25(7):1130–1138
Leung, Lun AK et al (2004) Quantitative kinetic analysis of nucleolar breakdown and reassembly during mitosis in live human cells. J Cell Biol 166(6):787–800
Lin Y, Protter DSW, Rosen MK, Parker R (2015) Formation and maturation of phase-separated liquid droplets by RNA-binding proteins. Mol Cell 60(2):208–219
López-Otín C et al (2013) The hallmarks of aging. Cell 153(6):1194–1217
Lyons H et al (2023) Functional partitioning of transcriptional regulators by patterned charge blocks. Cell 186(2):327–345
McStay B (2016) Nucleolar organizer regions: genomic "dark matter" requiring illumination. Genes Dev 30(14):1598–1610
Mitrea DM et al (2016) Nucleophosmin integrates within the nucleolus via multi-modal interactions with proteins displaying R-rich linear motifs and RRNA. elife 5:e13571
Mitrea DM et al (2018) Self-interaction of NPM1 modulates multiple mechanisms of liquid–liquid phase separation. Nat Commun 9(1):842
Montgomery TSH (1898) Comparative cytological studies, with especial regard to the morphology of the nucleolus. J Morphol 15(2):265–582
Neeson MJ et al (2012) Compound sessile drops. Soft Matter 8(43):11042
Németh A, Grummt I (2018) Dynamic regulation of nucleolar architecture. Curr Opin Cell Biol 52: 105–111
Nott TJ et al (2015) Phase transition of a disordered Nuage protein generates environmentally responsive Membraneless organelles. Mol Cell 57(5):936–947
Pederson T (2011) The nucleolus. Cold Spring Harb Perspect Biol 3(3):a000638–a000638
Picart-Picolo A, Picart C, Picault N, Pontvianne F (2020) Nucleolus-associated chromatin domains are maintained under heat stress, despite nucleolar reorganization in Arabidopsis Thaliana. J Plant Res 133(4):463–470
Pich A, Chiusa L, Margaria E (2000) Prognostic relevance of AgNORs in tumor pathology. Micron 31(2):133–141
Prescott DM, Bender MA (1962) Synthesis of RNA and protein during mitosis in mammalian tissue culture cells. Exp Cell Res 26(2):260–268
Remnant L et al (2021) The intrinsically disorderly story of Ki-67. Open Biol 11(8):210120
Saito A et al (2021) How do Flaviviruses hijack host cell functions by phase separation? Viruses 13(8):1479
Sanchez JC et al (2017) Defective replication initiation results in locus specific chromosome breakage and a ribosomal RNA deficiency in yeast. PLOS Genet 13(10):e1007041
Savino TM et al (2001) Nucleolar assembly of the Rrna processing machinery in living cells. J Cell Biol 153(5):1097–1110
Shan L et al (2023) Nucleolar URB1 ensures 3′ ETS RRNA removal to prevent exosome surveillance. Nature 615(7952):526–534
Sørensen PD, Frederiksen S (1991) Characterization of human 5S RRNA genes. Nucleic Acids Res 19(15):4147–4151
Stenström L et al (2020) Mapping the nucleolar proteome reveals a spatiotemporal organization related to intrinsic protein disorder. Mol Syst Biol 16(8):e9469
Tiwary AK, Cecilia D (2017) Kinetics of the Association of Dengue virus capsid protein with the granular component of nucleolus. Virology 502:48–55

Turi Z, Lacey M, Mistrik M, Moudry P (2019) Impaired ribosome biogenesis: mechanisms and relevance to cancer and aging. Aging 11(8):2512–2540

Vincent WS (1955) Structure and chemistry of nucleoli. Int Rev Cytol 4:269–298

Wagner R (1835) Einige Bemerkungen und Fragen über das Keimbläschen (vesicular germinativa). In: Einige Bemerkungen und Fragen über die Keimbläschen (vesicula germinativa). AbeBooks, Victoria, pp 373–377

Watkins NJ, Bohnsack MT (2012) The box C/D and H/ACA SnoRNPs: key players in the modification, processing and the dynamic folding of ribosomal RNA: box C/D and H/ACA SnoRNPs. Wiley Interdiscip Rev RNA 3(3):397–414

Wu M et al (2021) LncRNA *SLERT* controls phase separation of FC/DFCs to facilitate pol I transcription. Science 373(6554):547–555

Xu Z, Anderson R, Hobman TC (2011) The capsid-binding nucleolar helicase DDX56 is important for infectivity of West Nile virus. J Virol 85(11):5571–5580

Yamazaki H et al (2022) Cell cycle-specific phase separation regulated by protein charge blockiness. Nat Cell Biol 24(5):625–632

Yao R-W et al (2019) Nascent pre-RRNA sorting via phase separation drives the assembly of dense Fibrillar components in the human nucleolus. Mol Cell 76(5):767–783

Zink D, Fischer AH, Nickerson JA (2004) Nuclear structure in cancer cells. Nat Rev Cancer 4(9): 677–687

Part IV
Medicine

Chapter 12
Phase Separation Orchestrates Cancer Signaling: Stress Granules as a Promising Target for Cancer Therapy

Reiko Sugiura, Ryosuke Satoh, Naofumi Tomimoto, and Teruaki Takasaki

Abstract Liquid–liquid phase separation (LLPS) has recently emerged as a new fundamental mechanism for the eukaryotic cellular organization via the formation of membrane-less intracellular organelles exemplified by nucleoli, P-bodies, and stress granules (SGs) (Mahboubi and Stochaj 2014; Luo et al. 2018; Molliex et al. 2015). These membrane-less condensates play important roles in various physiological and pathological processes, including regulation of gene expression, cellular stress responses, and signal transduction. Recent studies have highlighted the importance of these biomolecular condensates in tumorigenesis. Among them, SGs have attracted strong attention as a promising target for cancer treatment because of their involvement in various aspects of cancer progression, ranging from cancer formation to metastasis, as well as drug resistance. SGs regulate important cancer signaling pathways, such as mTOR and MAPK via spatial recruitment of signaling molecules thus indicating that SGs constitute signaling hubs that can rewire cancer signal transduction. Additionally, exciting discoveries in mammals as well as model organisms including yeasts have indicated that several molecules involved in cancer/proliferation signaling have been shown to upregulate SG formation. In this review, we summarize the fundamental aspects of SGs as a new paradigm for dynamic modulators of various cellular signal transduction systems and then focus on recent advances in our understanding of the role of SGs in cancer biology and its application for therapeutic strategies targeting SGs.

Keywords Phase separation · Cancer signaling · Cancer therapy · Ribonucleoprotein (RNP) granule · Stress granule · P-body · Condensate

---

The relationship between phase separation and cancer signaling, as well as recent advances in targeting stress granules for cancer therapy, will be summarized and discussed.

R. Sugiura (✉) · R. Satoh · N. Tomimoto · T. Takasaki
Laboratory of Molecular Pharmacogenomics, Department of Pharmaceutical Sciences, Faculty of Pharmacy, Kindai University, Higashiosaka City, Osaka, Japan
e-mail: sugiurar@phar.kindai.ac.jp

R. Kurokawa (ed.), *Phase Separation in Living Cells*,
https://doi.org/10.1007/978-981-99-4886-4_12

12.1 Introduction

Cells utilize various organelles to spatiotemporally compartmentalize and concentrate proteins and RNA molecules. Deterioration of this precise spatiotemporal regulation can cause aberrant regulation of basic physiological processes and thus lead to pathological consequences, as exemplified by cancer development.

Phase separation or liquid–liquid phase separation (LLPS) has attracted attention as a new layer of cellular compartmentalization mechanism by which proteins and other biomolecules form membrane-less organelles or droplet-like structures in the cytoplasm or nucleus. These biomolecular condensates range from the cytoplasmic stress granules (SGs) and P-bodies to nuclear speckles and nucleoli. The assembly of biomolecular condensates including SGs is largely driven by LLPS; however, the underlying molecular mechanisms are not fully understood. Among these biomolecular condensates, SGs are the most well-investigated, especially in terms of their functional relationship with proliferative and cancer signaling.

SGs are dynamic structures that can rapidly assemble and disassemble in response to changes in various stresses, including heat, hypoxia, oxidative stress, arsenite, and nutrient deprivation. During stress, general translation is inhibited, and the mRNAs and proteins involved in translation are sequestered into SGs. Thus, SG formation has major advantages for cell physiology to minimize energy expenditure, control protein, and ribostasis and improve cell survival under damaging conditions (Kedersha et al. 2002).

Beyond their role as mRNA triage centers as described above, the concept of "SGs as signaling hubs" has emerged because SGs have been shown to interact with a variety of signaling pathways that are often deregulated in cancer cells, including the mammalian target of rapamycin (mTOR) pathway and the mitogen-activated protein kinase (MAPK) pathway. This crosstalk between proliferation/cancer-associated signaling and SGs is mainly achieved by the compartmentalization of important signaling molecules, such as RACK, TRAF2, mTOR, and protein kinase C (PKC) into SGs. Many of these signaling enzymes or scaffold proteins function to promote pro-apoptotic signaling. As a consequence of the sequestration/compartmentalization of these signaling molecules, SGs can promote pro-survival effects.

Importantly, cancer cells are often exposed to various environmental stressors and utilize SGs to adapt to stressful conditions as exemplified by oxidative stress, hypoxia, ER stress, and nutrient deprivation. Cancer cells also use SGs to protect themselves from anticancer chemotherapeutics. Therefore, strategies modulating SG assembly/disassembly or targeting the expression/recruitment/activity of SG components, which can block SG formation could be a promising approach for cancer treatment. Recent exciting discoveries showed that several pro-tumorigenic signaling molecules as represented by the RAS oncogene have been shown to upregulate SG formation in cancer (Grabocka and Bar-Sagi 2016; Fonteneau et al. 2022). Thus, increasing evidence indicates that SGs are also implicated in the development and progression of cancer.

Notably, SGs are reported in numerous organisms and their assembly/disassembly is evolutionarily conserved stress responses. SGs have been studied in various model systems, ranging from yeasts, plants, flies, and worms, to animal tissues and patient samples (Mahboubi and Stochaj 2017). Among them, due to its genetic tractability, yeast is an excellent model system to understand the mechanisms of SG formation as well as SG biology. Despite the existence of some differences between the SGs of yeast and mammals in terms of regulation and components of SGs (Kedersha and Anderson 2009), yeasts are a valuable tool for addressing these issues. We have been studying the regulatory mechanisms of stress-responsive signaling pathways using fission yeast as a model organism (Sugiura et al. 1998; Takada et al. 2007). During the analysis, our molecular genetic screen identified multiple mRNA binding proteins and signaling proteins involved in the regulatory mechanisms of MAPK and calcineurin signaling pathways (Sugiura et al. 2003; Satoh et al. 2009, 2012, 2018; Kanda et al. 2016, 2021). Notably, many of them turned out to serve as components and/or regulators of SGs. More importantly, the compartmentalization of key signaling proteins into SGs plays a key role in signal transduction pathways.

In this chapter, we will introduce two aspects of the cross-talk between SGs and cancer. We first look at a glance of fundamental aspects of SGs and then, we introduce examples of the compartmentalization of important signaling proteins relevant to cancer/proliferation into SGs. In the last section, we will discuss the emerging roles of SGs in the regulation of tumor development and its application for cancer therapeutics.

12.2 Stress Granules at a Glance

SGs are dynamic structures that can rapidly assemble and disassemble in response to changes in various stresses, such as heat shock, oxidative stress, and viral infection, and regulate gene expression and cell survival during stress. SGs are composed of RNA-binding proteins, ribosomal 40S subunits, translation initiation factors, and mRNAs and are thought to function as a storage site for untranslated mRNAs during stress conditions (Fig. 12.1). SG components are highly conserved from yeast to humans, and the main SG components in humans are listed in Table 12.1. Many of the RNA-binding proteins found in SGs are involved in the regulation of translation and/or translation initiation, while others regulate transcription, splicing, and RNA stability (Table 12.1). SG proteins are characterized by an abundance of intrinsically disordered regions (IDRs) and low complexity (LC) regions, which are responsible for initiating liquid–liquid phase separation (LLPS) in vitro (Kato et al. 2012; Lin et al. 2012; Uversky 2017). SG formation is regulated by several different signaling pathways, including the eIF2α pathway, which is activated in response to various stresses and leads to the inhibition of translation initiation (Fig. 12.1). Other signaling pathways, such as the mTOR pathway, have also been implicated in stress granule formation (Fig. 12.1). During stress, general translation is inhibited, and

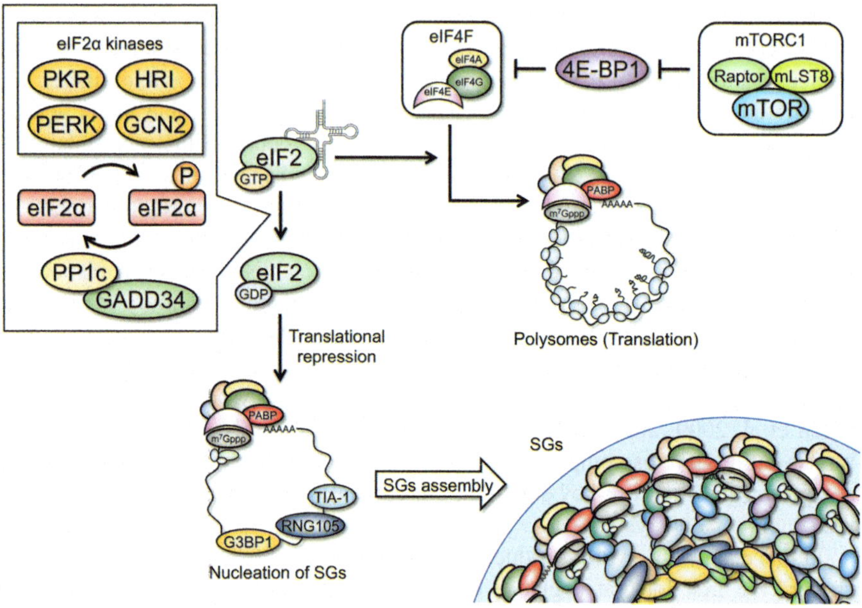

Fig. 12.1 Translational regulation and SG formation through eIF2α phosphorylation and mTORC1 signaling. mTORC1 is composed of mTOR, Raptor, and mLST8 and regulates translation and SG formation by phosphorylating 4E-BP1, which prevents it from binding and suppressing eIF4E. Conversely, eIF2 is a G protein consisting of three subunits (α, β, γ), and the γ subunit (eIF2γ) binds to GTP or GDP. eIF2 is activated by conversion from its GDP-bound form (inactive) to its GTP-bound form (active) by eIF2B and is involved in translation initiation. Actively translating mRNAs form macromolecular complexes called polysomes, which consist of multiple ribosomes that simultaneously translate a single mRNA into a polypeptide chain. eIF2α kinases (GCN2, HRI, PERK, PKR) phosphorylate eIF2α in response to distinct types of stress. The complex of PP1c and GADD34 specifically dephosphorylates eIF2α. Phosphorylation of eIF2α reduces the concentration of the eIF2-GTP-tRNA(Met) ternary complex, which is responsible for loading tRNA(Met) onto the 40S ribosome to initiate protein translation. Stress reduces the concentration of the ternary complex, which causes RNA-binding proteins such as TIA-1 and TIAR to facilitate the formation of a noncanonical preinitiation complex that does not include eIF2-GTP-tRNA(Met). Additionally, RNA-binding proteins such as G3BP and RNG105 are recruited to the translational repressed mRNA, and these RNA-binding proteins promote SG formation

the mRNAs and proteins involved in translation are sequestered into SGs. As a result, translation initiation is arrested while elongation continues, leading to the dissociation of ribosomes engaged in translation from mRNA and the disassembly of polysomes. This is thought to allow the cell to conserve energy and resources while also protecting these essential molecules from damage during stress. Interestingly, the overexpression of certain RNA-binding proteins (RBPs) can initiate the formation of SGs even in the absence of stress (Kedersha et al. 2005; Tourriere et al. 2003; Satoh et al. 2012, 2018). This highlights that LLPS is driven by the high local concentration of RBPs containing IDRs and/or LC regions, and this concentration

Table 12.1 List of main SG components in human

Component	Function	References
40S ribosome	Translation	Kedersha et al. (2002)
Ago1, 2	mRNA degradation	Leung et al. (2006), Pare et al. (2009)
Ataxin-2	Translation	Nonhoff et al. (2007)
CIRP	Translation	Rothe et al. (2006), De Leeuw et al. (2007)
CPEB1	Translation	Wilczynska et al. (2005)
DBPA	Splicing	Goodier et al. (2007)
DDX1	RNA helicase	Onishi et al. (2008)
DDX3	RNA helicase	Lai et al. (2008), Chalupnikova et al. (2008)
eIF2α	Translation initiation	Kedersha et al. (2002)
eIF3	Translation initiation	Kedersha et al. (2002)
eIF4A	Translation initiation	Low et al. (2005)
eIF4B	Translation initiation	Low et al. (2005)
eIF4E	Translation initiation	Kedersha et al. (2002)
eIF4G1	Translation initiation	Kedersha et al. (2002)
eIF4G2/DAP5	Translation initiation	Nousch et al. (2007)
FBP/KSRP	Splicing, mRNA degradation	Rothe et al. (2006)
FET proteins (FUS, EWS, and TAF15)	Splicing, transcription	Andersson et al. (2008)
FMRP/FXR1	Splicing, translation	Kim et al. (2006), Didiot et al. (2009), Mazroui et al. (2002)
G3BP	mRNA stability, translation	Tourriere et al. (2003)
hnRNP A1	Splicing	Guil et al. (2006)
hnRNP E1, 2	Splicing, mRNA stability, translation	Fujimura et al. (2009)
hnRNP K	Splicing, mRNA stability, translation	Fukuda et al. (2009)
hnRNP Q	Splicing	Quaresma et al. (2009)
HuD	RNA stability, translation	Burry and Smith (2006)
HuR	Splicing, RNA stability, translation	Gallouzi et al. (2000)
Lin28	miRNA processing	Balzer and Moss (2007)
MBNL1	Alternative splicing	Onishi et al. (2008)
MLN51	Splicing	Baguet et al. (2007)
Musashi-1	Translational repression	Kawahara et al. (2008)
p54/RCK/CGH-1	Translational repression	Wilczynska et al. (2005)
PABP	RNA stability, translation	Kedersha et al. (1999)
PACT	RNA silencing	Pare et al. (2009)
PAI-RBP1	mRNA stability	Goodier et al. (2007)

(continued)

Table 12.1 (continued)

Component	Function	References
PMR1	Endonuclease	Yang et al. (2006)
Pumilio 2	RNA stability, translation	Vessey et al. (2006)
RHAU helicase	RNA helicase	Chalupnikova et al. (2008)
RNG105/Caprin-1	mRNA transport, translation	Solomon et al. (2007)
Rpb4	Transcription	Lotan et al. (2005)
Rpp20	RNase P subunit	Hua and Zhou (2004b)
Smaug	Translational repression	Baez and Boccaccio (2005)
SMN	Splicing, snRNA assembly	Hua and Zhou (2004a)
SRC3	Transcription	Yu et al. (2007)
Staufen	dsRNA-binding	Thomas et al. (2005)
TIA-1	Transcription, translation, splicing	Kedersha et al. (1999)
TIAR	Transcription, translation, splicing	Kedersha et al. (1999)
TSN	Nuclease	Scadden (2007)
TTP/BRF-1	RNA stability	Stoecklin et al. (2004)
TUDOR-SN	Nuclease	Gao et al. (2010)
XRN1	mRNA degradation	Kedersha et al. (2005)
YB-1	mRNA chaperone	Yang and Bloch (2007)
ZBP1	mRNA localization	Stohr et al. (2006)

likely facilitates SG condensation by promoting protein–protein interactions (Molliex et al. 2015).

Speculated SG functions are described below. During stress, essential "housekeeping" mRNAs such as glyceraldehyde-3-phosphate dehydrogenase (GAPDH), β-actin, and insulin-like growth factor 2 (IGF2) (Stöhr et al. 2006) are sequestrated into SGs, leading to their temporary silencing. In contrast, mRNAs that encode proteins crucial for stress response, such as HSP70 (Kedersha and Anderson 2002) during heat shock, are excluded from SGs and are preferentially translated during stress. This enables rapid and reversible changes in the cellular proteome and facilitates physiological adaptations to environmental stress. SGs also play a role in both cell survival and apoptosis during stress, depending on the type and severity of the stress. SGs have also been implicated in the development and progression of cancer. Several studies have shown that stress granule formation can be dysregulated in cancer cells, leading to aberrant mRNA translation and altered gene expression. SG formation is increased in some cancer cells, which can protect the cancer cells from cell death and contribute to tumor survival. Furthermore, SGs are known to incorporate various signaling molecules and their regulators. Notably, SGs have been shown to interact with a variety of signaling pathways that are often deregulated in cancer cells, including the mammalian target of rapamycin (mTOR) pathway and the extracellular signal-regulated kinase (ERK) mitogen-activated protein kinase (MAPK) pathway. Additionally, SGs are highly dynamic structures that can rapidly assemble and disassemble in response to changes in stress levels. This suggests that SGs may act as a "stress sensor" in cells, allowing them to rapidly respond to stressful conditions and modulate their gene expression accordingly. Dysregulated SG formation may contribute to the activation of these pathways, which can promote cell proliferation and cell survival, and has been implicated in several diseases, including cancer (Buchan 2014; Asadi et al. 2021a), neurodegenerative disorders (Ash et al. 2014; Wolozin and Ivanov 2019; Asadi et al. 2021b), and autoimmune diseases (Johnson et al. 2016). SGs are important regulators of RNA metabolism and play critical roles in regulating cellular processes, including stress responses, mRNA translation, and germ cell development. Further research on RNP granules will be important for understanding their role in disease and identifying potential therapeutic targets.

12.3 Phase Separation of MAPK Signaling Regulators

MAPK signaling pathways play a crucial role in regulating a wide range of cellular processes, including cell proliferation, differentiation, inflammation, survival, and apoptosis (Zhang and Liu 2002). The MAPK signaling cascades are highly conserved in all eukaryotes and consist of three tiers of protein kinases (MAPKKK-MAPKK-MAPK). In mammals, there are more than a dozen MAPK enzymes that coordinately regulate cell proliferation, differentiation, and survival (Cargnello and Roux 2011). These MAPK enzymes are classified into at least three subfamilies of MAPKs, namely ERK, JNK, and p38. The ERK MAPK subfamily is activated by mitogenic stimuli, such as growth factors, and is associated with proliferative responses (Roux and Blenis 2004). In contrast, JNK and p38 are activated by various environmental stresses (e.g. DNA-damaging reagents, UV irradiation, or osmotic shock) and by cytokines (e.g. TNFα) and are associated with inflammation and/or apoptotic responses (Hammouda et al. 2020).

Dysregulation of MAPK signaling has been implicated in numerous diseases, as represented by cancers and neurodegenerative diseases (Kim and Choi 2010; Sugiura et al. 2021). In several types of cancer, mutations in upstream regulators of the ERK MAPK pathway, such as rat sarcoma virus (RAS) and rapidly accelerated fibrosarcoma (RAF), can lead to aberrant ERK activation even in the absence of external stimuli (Yang and Liu 2017). The activation of ERK MAPK in cancer cells can promote cell proliferation, survival, and angiogenesis, as well as inhibition of apoptosis and immune surveillance (Rubinfeld and Seger 2005; Baek et al. 2000). Thus, the ERK signaling pathway is an attractive target for cancer therapy and numerous inhibitors have been developed and some of them are in clinical trials/use as a therapeutic strategy for cancer treatment (Kohno and Pouyssegur 2006).

In this section, we will list the signaling molecules relevant to the regulation of MAPK signaling and/or cell fate (proliferation/apoptosis) that are reported as cargo

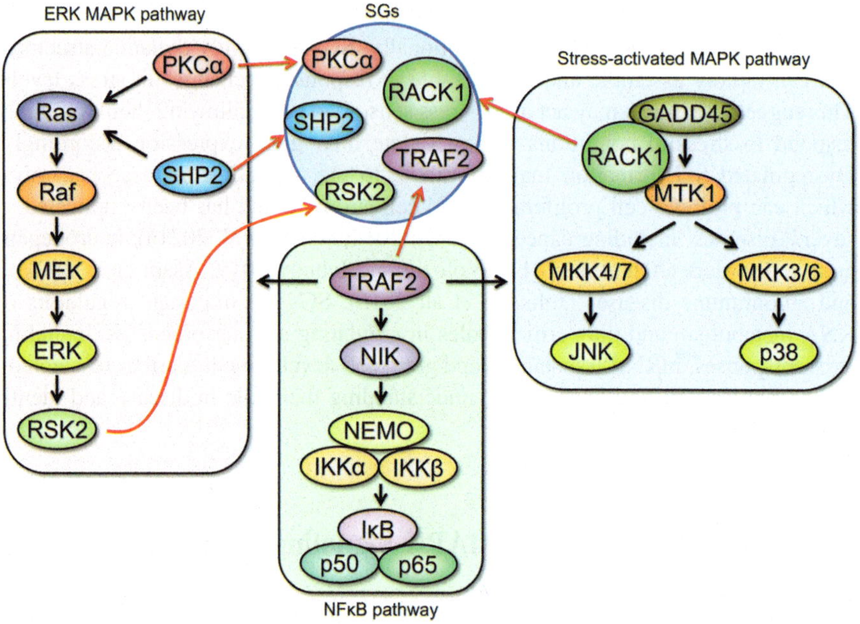

Fig. 12.2 MAPK signal and NFκB pathway regulators and SG. ERK MAPK signal regulators PKCα and SHP2 and ERK MAPK substrate RSK2 are sequestrated into SGs. RACK1 fails to activate the stress-activated MAPKKK MTK1 by being sequestrated into SGs. TRAF2, the upstream regulator of the NFκB pathway, regulates the ERK MAPK pathway and the stress-activated MAPK pathway and is sequestrated into SGs

for SGs and will discuss what are the functional consequences of the recruitment of the signaling proteins to SGs.

RACK1: Arimoto et al. reported pioneering work in the field of compartmentalization of signaling molecules relevant to MAPK signaling regulation by discovering the sequestration of the receptor for activated C kinase 1 (RACK1) into SGs and its relevance to apoptosis and cancer therapy (Arimoto et al. 2008).

RACK1 is a signaling scaffold protein that activates stress-responsive MTK1 MAPKKK by binding to MTK1 under stress conditions such as X-rays and genotoxic drugs. Activated MTK1 induces p38 and/or c-Jun N-terminal kinase (JNK) MAPK signaling and apoptosis (Gerwins et al. 1997; Takekawa et al. 1997) (Fig. 12.2). Conversely, RACK1 is sequestrated into SGs under stress conditions such as hypoxia, high temperature, and arsenite, and this sequestration inhibits the activation of p38 and JNK signaling and prevents apoptosis (Arimoto et al. 2008) (Fig. 12.2). This indicates that SG formation inhibits apoptosis by suppressing stress-responsive MAPK pathways through the compartmentalization of RACK1. This study suggests a potential mechanism for hypoxia-induced resistance to chemotherapy, as cells subjected to hypoxic shock form SGs, and are resistant to apoptosis induced by an anticancer compound etoposide (Arimoto et al. 2008).

Some anticancer therapies, like radiation and certain anticancer drugs, act by causing DNA damage and apoptosis of the cancer cells. Thus, the sequestration of RACK1 into SGs in such cancer cells could contribute to their resistance to anticancer therapies.

TRAF2: TRAF2 (*T*NF *r*eceptor-*a*ssociated *f*actor2) was also one of the first signaling molecules relevant to pro-apoptosis signaling that was shown to be targeted to SGs (Kim et al. 2005). TRAF2 is involved in various cancer-relevant cellular processes, such as the activation of transcription factors of the NF-κB family, stimulation of MAPK cascades, and the control of cell death programs (Siegmund et al. 2022). Importantly, the sequestration of TRAF2 into SGs was shown to interrupt TNF signaling to NF-κB activation under stress conditions (Kim et al. 2005) (Fig. 12.2). TRAF2 was shown to migrate to SGs under heat-stress conditions co-localizing with eIF4GI, which is a component of SGs and a scaffold protein interacting with many translation factors.

Upon recruitment to SGs, TRAF2 was sequestered and unable to relay signals to induce apoptosis downstream of TNF receptor activation through NF-κB. Intriguingly, following treatment with TNF-α, cells form a complex called the "TNFR1-mediated-signal transduction complex," which is composed of TNFR1, adaptor TRADD, kinase RIP1, and TRAF2 (Kim et al. 2005). However, TRAF2 is not associated with the TNFR1 complex under heat-stress conditions, instead, TRAF2 interacts with eIF4GI, which blocks TNF signaling. Thus, heat stress leads to reprogramming the macromolecular complex surrounding TNFα, which enables TRAF2 sequestration into SGs. Thus, SGs may promote cell survival during periods of stress by suppressing pro-apoptotic signaling cascades involving TNFα (Kim et al. 2005).

RSK2: RSK2 is a serine/threonine kinase and a member of the p90 ribosomal S6 kinase ($p90^{RSK}$; RSK; also known as MAPKAP-K) family, which regulates cell proliferation and transformation mediated by EGF, TPA, and UV radiation (Cho 2017). In signaling cascades, RSK2 is regulated under the control of ERK1/2 activities and functions upstream of transcription factors involved in cell proliferation and cancer development, such as CREB or YB2 (Cho 2017; Stratford et al. 2008). Thus, RSK2 plays a key role in mediating the proliferation signal downstream of MAP kinase cascades (Fig. 12.2). RSK2 is highly expressed in various cancer cells such as melanoma and the deletion of RSK2 inhibits cell proliferation and sensitizes various cancer cells to chemotherapy (Clark et al. 2005; Cho et al. 2012; Huynh et al. 2020; Wu et al. 2022).

An intriguing link between RSK2 and SGs was reported by Eisinger-Mathason et al. (Eisinger-Mathason et al. 2008). In stressed breast cells, endogenous RSK2 associates with and co-localizes with PABP1 and TIA-1, which regulates the assembly of SGs. RSK2 is a nucleocytoplasmic shuttling protein and in response to mitogen treatment, RSK2 accumulates in the nucleus. However, in stressed breast cells RSK2 is localized into SGs (Fig. 12.2). TIA-1 controls RSK2 localization and RSK2 controls TIA-1 recruitment into SGs, showing co-dependency of TIA-1 and

RSK2 localization. Importantly, this regulation is physiologically important because the loss of RSK2 decreases cell survival in response to stress. Moreover, the kinase activity or RSK2 is important for SG formation because they control both TIA-1 and PABP1 sequestration. Mitogen releases RSK2 from the SGs and the released RSK2 accumulates in the nucleus. Importantly, the nuclear localization of RSK2 is sufficient to enhance proliferation through the induction of cyclin D1. Loss of TIA-1 reduces RSK2 nuclear translocation and nuclear function to promote proliferation as evaluated by cyclin D1 expression. Thus, TIA-1-mediated SG assembly regulates RSK2 function by controlling its spatial localization, and RSK2 protects cells through its control of SG formation.

The involvement of another S6 kinase ($p70^{RSK}$) in the regulation of SGs and apoptosis will be described in the section on "mTOR signaling."

PKC: Protein kinase C (PKC) is an AGC kinase family member that is widely conserved in eukaryotes and regulates various cellular functions such as cell proliferation, cell death, differentiation, and angiogenesis through activating MAPK pathway and NF-κB pathway (Jacinto and Lorberg 2008). In mammalian systems, the PKC–Raf–MEK–ERK signaling pathway modulates gene expression and cell proliferation in various cell types, including endothelial cells, T-cells, and various cancer cell settings (Pintus et al. 2003; Hoyer et al. 2005; Li et al. 2012). In addition, PKC signaling is involved in the pathogenesis of various intractable diseases such as cancer, cardiovascular disease, and neurodegenerative disease (Isakov 2018; Marrocco et al. 2019; Kaleli et al. 2020). Recent studies from yeasts and mammals have implicated PKC in the regulation of SG formation and dynamics (Kobayashi et al. 2012; Kanda et al. 2021) (Fig. 12.2).

In human neuroblastoma cells SK-N-BE(2)C, PKCα directly binds to RAS GTPase-activating protein-binding protein 2 (G3BP2) via the regulatory domain of PKCα and the C-terminal RNA-binding domain of G3BP2 (Kobayashi et al. 2012). PKCα also phosphorylates G3BP2 in vitro and regulates SG formation following cellular stress. Downregulation of PKCα delays heat shock-induced phosphorylation of eIF2α and SG formation. Thus, PKCα is a component as well as a regulator of SGs in mammals. Importantly, the biological activity of PKC enzymes is tightly regulated by their subcellular localization (Newton and Johnson 1998; Gould and Newton 2008), as exemplified by PKC translocation to the plasma membrane in response to various stimuli. However, the physiological significance and/or the regulatory mechanisms of PKC localization at SGs remain unknown. More importantly, little is known whether or not PKC activity is regulated by compartmentalization to SGs.

In fission yeast, the PKC ortholog Pck2 activates Pmk1 MAPK (the fission yeast homolog of ERK1/2) and the Pck2/Pmk1 MAPK axis regulates cell proliferation and cell integrity (Sugiura et al. 1998; Arellano et al. 1999; Doi et al. 2015). A functional link between PKC and SGs was discovered through a molecular genetic screen aiming to identify genes that negatively regulate PKC/MAPK signaling (Kanda et al. 2021). Sugiura's group has identified negative regulators of MAPK signaling by isolating multi-copy suppressors of the MAPK hyperactivation-dependent cytotoxicity. These include dual-specificity phosphatase Pmp1, which dephosphorylates and inactivates Pmk1 MAPK; the SH3-adaptor protein Skb5, which spatially regulates Mkh1 MAPKKK (Sugiura et al. 1998; Takada et al. 2007; Kanda et al. 2016); and Ded1. Ded1 is similar to mammalian DDX3 (also known as DDX3X) and belongs to the highly conserved DEAD-box ATPases that are representative components of SGs (Grallert et al. 2000; Hilliker et al. 2011). The functional connection between the SG-resident protein Ded1 and PKC/MAPK signaling as revealed by a molecular genetic screen raised the hypothesis that SGs might be functionally relevant to regulate PKC/MAPK signaling activation.

Under unstressed conditions, Pck2 localizes to the cytoplasm and the growing cell tips (Madrid et al. 2017; Kanda et al. 2021). Upon high-heat stress (HHS), the localization of Pck2 dynamically changes from the plasma membrane to SGs through phase separation. Interestingly, SG localization of Pck2 was dependent on the Pmk1 MAPK pathway activation. This suggests that the downstream MAPK Pmk1 activation promotes the recruitment of an upstream regulator Pck2 into SGs, which intercepts MAPK hyperactivation and cell death upon HHS. Consistently, cells fail to activate MAPK signaling when Pck2 is sequestrated into SGs. Thus, SGs serve as a negative feedback circuit in controlling MAPK signaling.

Another unique feature of this study includes the heterogeneity of SGs, dependent on the nature/degree of extracellular stimuli. Intriguingly, HHS (45 °C), but not modest-heat stress (43 °C), targets Pck2 to SGs, independent of canonical SG machinery. Furthermore, Pck2 translocation to SGs is dependent on PKC kinase activity-dependent, indicating that HHS-induced SGs preferentially partitioned active Pck2 into SGs, thereby reducing Pck2 kinase activity and downstream MAPK signaling activation in the soluble fraction (Kanda et al. 2021). This finding, together with the dependency of the Pck2 SG translocation on MAPK hyperactivation, suggests the existence of a mechanism to dictate Pck2 phase separation when cells reach a certain threshold of MAPK activation detrimental to cell growth (Gutiérrez-Venegas et al. 2010). In this context, SGs may serve as a sensor to detect both extracellular and/or intracellular molecular alterations, thereby rewiring MAPK signaling and cell fate by selectively compartmentalizing specific signaling molecules including Pck2. Collectively, this study demonstrated that SGs safeguard against excess PKC/MAPK signaling hyperactivation induced by HHS via spatiotemporal PKC sequestration.

In budding yeast, SGs associate with eisosomes, filament-like structures, which are composed of specific peripheral membrane proteins and membrane lipids during glucose starvation (Amen and Kaganovich 2020). In the same condition, 5′-3′ exonuclease XRN1 accumulates at the eisosomes from P-bodies (Grousl et al. 2015). Further, the phosphoinositide-dependent protein kinases Pkh1/Pkh2 and Pkc1, the PKC ortholog in budding yeast as well as the effector kinase of Pkh1/Pkh2, regulate P-body assembly and mRNA metabolisms only under the nutrient-poor condition (Luo et al. 2011). Upon starvation, eisosomes clustered and promoted the formation of SGs. Pkc1 (the budding yeast isoform of protein kinase C) was required for starvation-induced eisosome reorganization and was recruited to SGs in its active form. In the absence of eisosomes, active Pkc1 was not retained in SGs, and recovery from starvation was delayed. Thus, this sequestration of Pkc1 into SGs is necessary to restart division after long-term starvation (Amen and Kaganovich 2020). These findings identify a mechanism whereby nutrient stress induces structural changes at the plasma membrane that promote the formation of SGs, thus protecting the cell during stress and enabling faster recovery.

Intriguingly, LLPS of the Par3/6 complex, key cell polarity-regulating proteins, was recently implicated in the apical redistribution of atypical protein kinase C (aPKC) activity during cell polarization (Liu et al. 2020). Conversely, PKCι activity may regulate LLPS of the Par complex. Correct polarization of aPKC promotes embryonic development and prevents tumor epithelial-mesenchymal transition (EMT) (Jung et al. 2019; Liu et al. 2020). Thus, the compartmentalization of PKC orthologues by LLPS may be a universal phenomenon, although the underlying mechanism and physiological significance can vary dependent on cell types and pathological contexts.

HOG1 (SAPK): High osmolarity glycerol 1 (HOG1) is a budding yeast stress-activated MAPK (SAPK) that plays a key role in the response to hyperosmotic stress, oxidative stress, high-temperature stress, and arsenite stress (Brewster and Gustin 2014). Hog1 translocates to the nucleus in response to hyperosmotic stress and phosphorylates downstream targets, including transcription factors such as Msn2/Msn4, Sko1, Hot1, and Smp1 (Hohmann 2002). Interestingly, the *Candida boidinii* (*C. boidinii*) Hog1, but not the *Saccharomyces cerevisiae* (*S. cerevisiae*) Hog1, can localize to SGs under high-temperature stress in a reversible manner (Shiraishi et al. 2018). β-sheet structure in the N-terminal region of the C. *boidinii* Hog1 is necessary and sufficient to localize to SGs. Intriguingly, Hog1 homologs in *Pichia pastoris* and *Schizosaccharomyces pombe*, but not the *S. cerevisiae* Hog1 exhibited similar dot formation under high-temperature stress. Under the high-temperature stress condition, activated Hog1 proteins are sequestrated in SGs, suggesting that the sequestration of activated Hog1 proteins in SGs was responsible for the downregulation of Hog1 activity under high-temperature stress.

Thus far, many of the cargo/clients of SGs in MAPK signaling were categorized as kinase or scaffold protein. Here, researchers uncover a link between phase separation and SHP2, a non-receptor protein tyrosine phosphatase upstream of the RAS-ERK MAPK signaling pathways, and clarify the underlying mechanism of human developmental disorders.

SHP2: Src homology region 2 domain-containing phosphatase-2 (SHP2), a non-receptor protein tyrosine phosphatase, encoded by PTPN11, activates several signaling pathways containing RAS-ERK MAPK signaling and Wnt signaling and suppresses PI3K/AKT pathway (Marin et al. 2011; Takahashi et al. 2011) (Fig. 12.2). Mutations in PTPN11 relate to various diseases such as leukemia and metachondromatosis (Bowen et al. 2011; Digilio et al. 2002; Tartaglia et al. 2003; Tartaglia and Gelb 2005). Especially, Noonan syndrome (NS) and Noonan syndrome with multiple lentigines (NS-ML) are rare human genetic disorders that prevent normal development in various parts of the body caused by mutations of the protein SHP2. The mechanism of NS and NS-ML pathogenesis has been unknown, and no effective treatment for the disorders has been reported.

Zhu et al. discovered that disease-associated SHP2 mutants acquire the capacity for liquid–liquid phase separation (LLPS) to boost enzymatic activity, leading to hyperactivation of the downstream cellular signaling pathway. Thus, SHP2 with disease-associated mutants recruits wild-type SHP2 in phase-separated condensates and this SHP2 condensation enhances SHP2 phosphatase activity and promotes RAS-ERK MAPK signaling (Zhu et al. 2020b). LLPS of SHP2 mutants is regulated by conformational changes in the SHP2 protein and is potently inhibited by the SHP2 allosteric inhibitor ET070. ET070 attenuates the phase separation of SHP2, providing a potential therapeutic strategy for targeting SHP2 LLPS to treat SHP2-associated developmental disorders as represented by the Noonan syndrome (Zhu et al. 2020b).

Rnc1: Rnc1, a K-homology (KH)-type RNA-binding protein, has been identified as a key regulator of posttranscriptional expression of the MAPK phosphatase Pmp1 in fission yeast through molecular genetic analysis (Sugiura et al. 2003). The overexpression of the *rnc1*$^{+}$ gene suppressed the phenotypes of Cl^{-} sensitivity associated with calcineurin deletion cells via stabilization of the mRNA encoding the MAPK phosphatase Pmp1, thereby inactivating Pmk1 MAPK. Thus, Rnc1 serves as a negative regulator of MAPK Pmk1 in fission yeast.

Rnc1 localizes to poly(A)+-binding protein (Pabp)-positive SGs in response to various stress stimuli, such as heat shock, and oxidative stresses (Satoh et al. 2018). Furthermore, Rnc1 overproduction induced massive aggregation of Pabp-positive SGs and eIF2α phosphorylation in the absence of stress. Conversely, overproduction of the KH domain-mutated Rnc1 failed to induce SG formation and eIF2α phosphorylation, suggesting that Rnc1 overproduction-induced SG assembly requires the RNA-binding activity of Rnc1. Importantly, SG formation in response to environmental stresses is significantly inhibited by the deletion of Rnc1, indicating that Rnc1 is a component and a regulator of SGs.

12.4 Phase Separation of mTOR Signaling

Accumulating evidence reveals complex crosstalk between the mammalian/mechanistic target of rapamycin (mTOR) signaling and SGs in various organisms. mTOR is a conserved serine/threonine kinase that regulates fundamental cell processes such as protein synthesis, cell metabolism, proliferation, migration, and autophagy (Brown et al. 1994; Laplante and Sabatini 2012). mTOR constitutes a central hub that integrates metabolic signals and adapts cellular processes to extrinsic and intrinsic changes and stressors in health, disease, and aging (Papadopoli et al. 2019; Liu and Sabatini 2020). Especially, oncogenic activation of the phosphatidylinositol-3-kinase (PI3K), protein kinase B (PKB/AKT), and mTOR

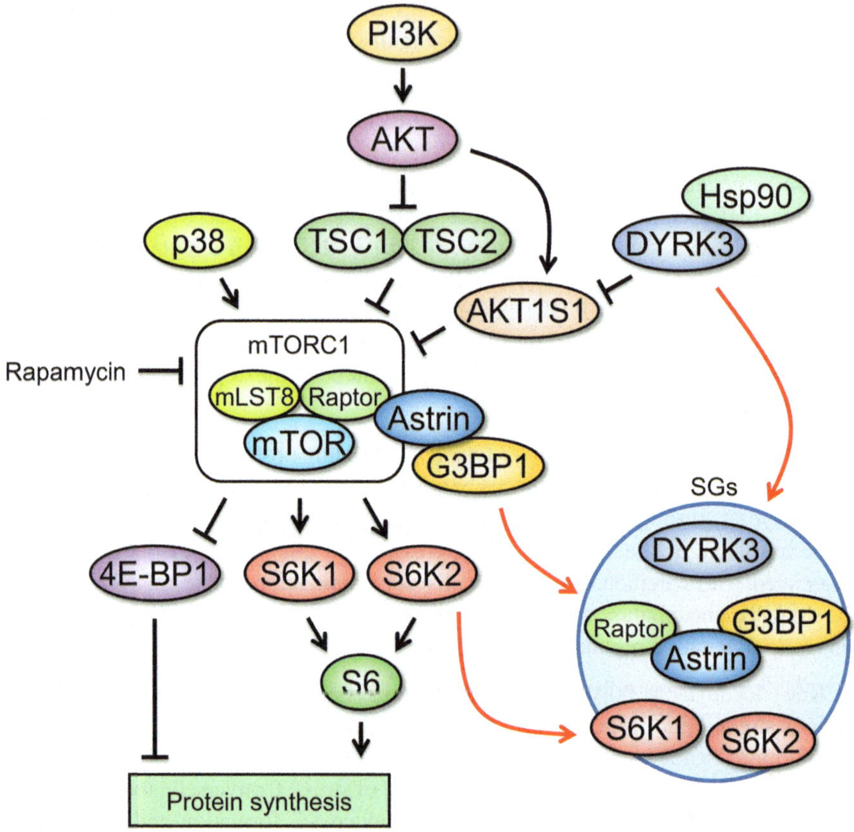

Fig. 12.3 mTORC1 and SGs. mTORC1 enhances protein synthesis by phosphorylating 4EBP1 and S6K1/S6K2. mTORC1 is regulated by the p38 pathway, the PI3K/AKT pathway, DYRK3/Hsp90, and Astrin. mTORC1 regulators DYRK3 and Astrin and the mTORC1 substrate S6K1/S6K2 are sequestrated into SGs

pathway has been shown to facilitate tumor formation, disease progression, and therapeutic resistance in various types of cancers.

mTOR exists in two separate complexes, mTOR complex 1 and 2 (mTORC1 and mTORC2). mTOR1 is responsible for regulating cell growth and metabolism, while mTOR2 is involved in the regulation of cell proliferation and survival (Wolfson and Sabatini 2017). mTORC1 is composed of mTOR, regulatory-associated protein of mTOR (Raptor), and mammalian lethal with SEC13 protein 8 (mLST8) and regulates protein synthesis by phosphorylating eucaryotic initiation factor 4E-binding protein 1 (4E-BP1) and p70 S6 kinase 1 (S6K1) (Burnett et al. 1998; Hara et al. 2002; Kim et al. 2002, 2003) (Fig. 12.3). mTORC2 is composed of mTOR, a rapamycin-insensitive companion of mTOR (Rictor), mLST8, and stress-activated MAPK-interacting protein 1 (mSin1) and regulates cell survival and proliferation by phosphorylating AKT. Dysregulation of the PI3K/AKT/mTOR signaling has been implicated in several diseases including cancer, cardiovascular disease, diabetes, and immune disease (Saxton and Sabatini 2017; Harwood et al. 2018).

The cross-talk between mTORC1 and SGs is two-fold, namely the control of SG formation/assembly by mTORC1 and the regulation of mTORC signaling by SGs.

In this section, the compartmentalization and regulation of the mTORC signaling network by SGs will be discussed.

12.4.1 Phase Separation Inhibits mTORC1 Signaling

Accumulating studies from yeast and mammals indicated the involvement of SGs to inhibit mTORC1 by several mechanisms (Takahara and Maeda 2012; Thedieck et al. 2013; Wippich et al. 2013; Mediani et al. 2021).

In budding yeast, genetic screening for negative regulators of TORC1 signaling using a hyperactivated TOR1 mutant identified Pbp1, a protein that is associated with Pab1. This study leads to the discovery that the TORC1 component Raptor (KOG1 in budding yeast) and TOR1 localize to SGs (Takahara and Maeda 2012). The overproduction of Pbp1 induces the formation of SGs in the absence of stress and sequestrates TORC1 component KOG1 into SGs, thereby suppressing TORC1 signaling in yeast. More importantly, KOG1 is sequestrated in SGs upon heat stress, a physiological SG-inducing condition. Importantly, TORC1 signaling is more rapidly reactivated during recovery after heat stress in the absence of SG formation. Thus, the sequestration of TORC1 delays the reactivation of TORC1 signaling after heat stress is removed. Moreover, the authors showed that SG formation prevents the heat stress-induced increase of mutation frequency via inhibition of TORC1. Collectively, SG disassembly is important for reactivating TORC1 during recovery from heat stress, and this reactivation serves as the protection of DNA and the reduction of heat stress-induced gene mutations.

In mammalian cells, SG recruitment of raptors is mediated by sperm-associated antigen 5 (SPAG5)/Astrin, a mTORC1 raptor interactor, which leads to the disassembly and inhibition of mTORC1 (Thedieck et al. 2013). Similar to budding yeast, mTOR in mammals also localizes to SGs (Wippich et al. 2013) (Fig. 12.3).

Reactive oxygen species (ROS) levels and redox condition is important to determine cell survival or death in cancer cells. High levels of ROS are common conditions in cancer, which is counteracted by an increased antioxidant capacity (Fruehauf and Meyskens Jr 2007; Sosa et al. 2013; Takasaki et al. 2018). SGs regulate oxidative stress-induced apoptosis via suppressing mTORC1 activity in cancer cells (Thedieck et al. 2013). Oncogenic kinase PI3K and stress-activated p38 MAPK signaling activate mTORC1 by affecting the raptor through the induction of SG assembly in the MCF-7 breast cancer cell line (Heberle et al. 2019). Further, hyperactivated mTORC1 and deficiency of SPAG5/Astrin enhance apoptosis (Shah et al. 2004; Lee et al. 2007; Gruber et al. 2002). Conversely, SG formation suppresses apoptosis by inhibiting the hyperactivation of mTORC1 (Wippich et al. 2013). SPAG5/Astrin inhibits the mTOR-raptor association and recruits raptor to

SGs to inhibit mTORC1 activity under oxidative stress conditions, and this inhibition mediates antiapoptotic SG functions in cancer cells such as HeLa cervical cancer strain and MCF-7 breast cancer strain (Thedieck et al. 2013). Conversely, tuberous sclerosis complex (TSC) 1-TSC2 regulates mTORC1 activity by controlling SPAG5/Astrin expression, which is important to enable the expression of stress response proteins (Thedieck et al. 2013). Importantly, highly expressed SPAG5/Astrin correlates with a negative prognosis in breast and lung cancer (Buechler 2009; Välk et al. 2010).

SGs also regulate mTORC1 via the kinase DYRK3 (dual-specificity tyrosine phosphorylation regulated kinase 3) (Wippich et al. 2013; Mediani et al. 2021). Under non-stressed conditions, cytosolic DYRK3 phosphorylates and represses the mTORC1's inhibitory subunit AKT1S1 at threonine 246, leading to mTORC1 hyperactivation (Wippich et al. 2013). In response to stress, SGs recruit inactive DYRK3, allowing active AKT1S1 to suppress mTORC1 (Wippich et al. 2013; Mediani et al. 2021) (Fig. 12.3). DYRK3 stabilizes SGs, enhancing the inhibitory effects of SGs on mTORC1. Because mTORC1 enhances SG formation by several mechanisms involving translation and autophagy as described in the next section, SG-mediated negative feedback of mTORC1 signaling may restrict its activity under stress, which contributes to the fine-tuning of cellular metabolism under stress.

Interestingly, HSP90, an essential chaperone that regulates the folding and stability of many clients, including stress response factors or cancer signaling-related kinases was shown to regulate DYRK3 and mTORC1 signaling via the SG assembly/disassembly process (Wippich et al. 2013; Mediani et al. 2021) (Fig. 12.3). Under non-stress conditions, HSP90 keeps DYRK3 in an active conformation. Upon stress or HSP90 inhibition (Wippich et al. 2013; Mediani et al. 2021), inactive DYRK3 is targeted to SGs via its IDR, resulting in AKT1S1 activation and mTORC1 inhibition. Hsp90 promotes the disassembly of SGs in part by binding and stabilizing DYRK3. Inhibition or loss of Hsp90 results in DYRK3 destabilization, SG persistence, and failure to restore mTORC1 signaling and translation. Thus, Hsp90 couples SG disassembly and mTORC1-dependent cell growth in part via DYRK3.

12.5 Phase Separation of PKA Signaling

Finally in this section, we will introduce an exciting example of the role of LLPS in the regulation of cAMP/PKA signaling and its implication in tumorigenesis.

PKA's regulatory subunit PKA RIα: cAMP signaling pathway is evolutionarily conserved and plays key roles in major physiological and pathological processes. cAMP controls a diverse set of cellular processes by binding and regulating several enzymes such as protein kinase A (PKA), a tetrameric holoenzyme composed of a pair of regulatory (R) subunits bound to a pair of catalytic (C) subunits (Johnson et al. 2001). At basal levels, the PKA holoenzyme is inhibited by the interaction of the R subunit inhibitory sequence (IS) with the C subunit active site (Johnson et al. 2001; Taylor et al. 2013). The binding of cAMP to the PKA regulatory subunits releases this inhibitory interaction, freeing the C subunits to phosphorylate a wide variety of downstream targets (Sassone-Corsi 2012; Taylor et al. 2013). Thus, cAMP/PKA signaling is a compelling regulator of cell growth and survival. Although the functions of cAMP signaling in tumors depend on cell types, the cAMP–PKA–CREB pathway plays a tumor-promoting role in many types of human tumors.

Recently, the type I regulatory subunit of PKA (PKARIα) was shown to undergo LLPS in response to cAMP dynamics (Zhang et al. 2020a). Interestingly, Riα biomolecular condensates sequester cAMP, thereby maintaining high PKA activity, which is critical for effective cAMP compartmentation. By directly binding and trapping cAMP, RIα condensates significantly restrict cytosolic free cAMP and thus lower its effective diffusion, which proves to be essential for compartmentalizing cAMP (Zhang et al. 2020a). Importantly, disruption of the phase-separated RIa condensates leads to oncogenic effects as a PKA fusion oncoprotein associated with an atypical liver cancer potently blocks the condensation of Ria, which leads to aberrant cAMP signaling, leading to "tumorigenic phenotypes," such as increased cell proliferation and induction of cell transformation. The authors proposed that Riα biomolecular condensates act as a dynamic "sponge" in recruiting and retaining cAMP and active PKA. This study discovered biomolecular condensates that regulate PKA signaling and tumorigenesis by showing how the cAMP/PKA pathway is spatially organized.

In addition to the signaling proteins discussed above, major signaling proteins reported to localize at SGs are listed in Table 12.2 (humans) and Table 12.3 (fission yeast and budding yeast), for readers' reference.

12.6 SGs Regulate Cancer

In the second section, we characterize the growing examples of key signaling pathways as a cargo/client of SGs and their functional consequences in terms of cell proliferation and cancer growth. These studies suggest that SGs function to regulate the switch between survival and cell death.

In this section, we will introduce exciting discoveries revealing an intimate link between SGs and every aspect of cancer progression, including cancer cell proliferation, invasion, metastasis, and tolerance to chemotherapy. Based on these studies, SGs emerge as key cancer signaling hubs and promising therapeutic targets.

Table 12.2 List of signal transduction proteins that translocate into SGs in human

Protein name	Description	References
ACBD5	Acyl-CoA binding domain containing 5	Markmiller et al. (2018)
AKAP13	A-Kinase anchoring protein 13	Marmor-Kollet et al. (2020)
AMOTL2	Angiomotin like 2	Marmor-Kollet et al. (2020)
AMPK-a2	Protein kinase AMP-activated catalytic subunit alpha 2	Mahboubi et al. (2015)
ANXA1	Annexin A1	Jain et al. (2016)
Argonaute 1	Argonaute 1, RISC catalytic component	Markmiller et al. (2018), Gallois-Montbrun et al. (2007)
Argonaute 2	Argonaute 2, RISC catalytic component	Detzer et al. (2011)
ARPC1B	Actin-related protein 2/3 complex subunit 1B	Jain et al. (2016)
ATP6V1G1	ATPase H+ transporting V1 subunit G1	Markmiller et al. (2018)
BIRC2	Baculoviral IAP repeat-containing 2	Marmor-Kollet et al. (2020)
CAPZB	Capping actin protein of muscle Z-line subunit beta	Marmor-Kollet et al. (2020)
CBFB	Core-binding factor subunit beta	Jain et al. (2016)
CCT6A	T-complex protein 1 subunit zeta	Jain et al. (2016)
Cdc20	Cell division cycle 20	Marmor-Kollet et al. (2020)
Cdc37	Cell division cycle 37	Pare et al. (2009)
Cdc73	Parafibromin	Jain et al. (2016)
CDK1	Cyclin-dependent kinase 1	Jain et al. (2016)
CDK2	Cyclin-dependent kinase 2	Moujalled et al. (2015)
CenPF	Centromere protein F	Marmor-Kollet et al. (2020)
CIT	Citron Rho-interacting kinase	Jain et al. (2016)
CK2α	Casein kinase 2 alpha 1	Reineke et al. (2017)
CNOT6L	CCR4-Not transcription complex subunit 6L	Youn et al. (2018)
Cofilin-1	Cofilin-1	Jain et al. (2016)
CPSF7	Cleavage and polyadenylation specificity factor subunit 7	Jain et al. (2016)
CRKL	CRK like proto-oncogene, adaptor protein	Markmiller et al. (2018)
CTNND1	Catenin delta-1	Jain et al. (2016)
Dcp2	Decapping mRNA 2	Youn et al. (2018)
DNCH1	Dynein cytoplasmic 1 heavy chain 1	Loschi et al. (2009)
DNCL1	Dynein light chain LC8-type 1	Markmiller et al. (2018), Tsai et al. (2009)
Dock4	Dedicator of cytokinesis 4	Marmor-Kollet et al. (2020)
DSP	Desmoplakin	Jain et al. (2016), Markmiller et al. (2018)
DST	Dystonin	Jain et al. (2016)
DYNLL2	Dynein light chain 2, cytoplasmic	Jain et al. (2016)
DZIP1	DAZ interacting zinc finger protein 1	Shigunov et al. (2014)

(continued)

Table 12.2 (continued)

Protein name	Description	References
eIF4B	Eukaryotic translation initiation factor 4B	Jain et al. (2016), Markmiller et al. (2018), Marmor-Kollet et al. (2020)
eIF4E	Eukaryotic translation initiation factor 4E	Kedersha et al. (2002)
eIF4G1	Eukaryotic translation initiation factor 4G1	Kedersha et al. (2002)
FABP5	Fatty acid binding protein 5	Markmiller et al. (2018)
FAK	Protein tyrosine kinase 2	Tsai et al. (2008)
FKBP1A	FKBP prolyl isomerase 1A	Marmor-Kollet et al. (2020)
GNB2	Guanine nucleotide-binding protein G(I)/G (S)/G(T) subunit beta-2	Jain et al. (2016)
GRB2	Growth factor receptor bound protein 2	Markmiller et al. (2018)
GRB7	Growth factor receptor bound protein 7	Tsai et al. (2008)
H2AFV	Histone H2A.V	Jain et al. (2016)
HDAC6	Histone deacetylase 6	Kwon et al. (2007)
HNRNPA1	Heterogeneous nuclear ribonucleoprotein A1	Guil et al. (2006)
HNRNPF	Heterogeneous nuclear ribonucleoprotein F	Markmiller et al. (2018)
HNRNPH1	Heterogeneous nuclear ribonucleoprotein H1	Markmiller et al. (2018)
HSP27	Heat shock protein family B (small) member 1	Jain et al. (2016), Kedersha et al. (1999)
HSP90	Heat shock protein HSP 90-alpha	Jain et al. (2016)
IGF2BP1	Insulin-like growth factor 2 mRNA-binding protein 1	Jain et al. (2016), Markmiller et al. (2018), Youn et al. (2018)
Importin-α1	Karyopherin subunit alpha 2	Fujimura et al. (2010)
ITGB1	Integrin beta-1	Jain et al. (2016)
JNK1	Mitogen-activated protein kinase 8	Wasserman et al. (2010)
KHDRBS3	KH domain-containing, RNA-binding, signal transduction-associated protein 3	Jain et al. (2016)
KIF2	Kinesin family member 2A	Loschi et al. (2009)
KNS2A	Kinesin light chain 1	Loschi et al. (2009)
LBR	Lamin-B receptor	Jain et al. (2016)
LEMD3	Inner nuclear membrane protein Man1	Jain et al. (2016)
MAGED1	Melanoma-associated antigen D1	Jain et al. (2016), Markmiller et al. (2018), Youn et al. (2018)
MAPRE1	Microtubule-associated protein RP/EB family member 1	Jain et al. (2016)
MOV-10	Mov10 RISC Complex RNA helicase	Gallois-Montbrun et al. (2007)
MSI2	RNA-binding protein Musashi homolog 2	Jain et al. (2016), Markmiller et al. (2018)
mTOR	Mechanistic target of rapamycin	Wippich et al. (2013), Sfakianos et al. (2018)
MYO6	Unconventional myosin-VI	Jain et al. (2016)

(continued)

Table 12.2 (continued)

Protein name	Description	References
Neuregulin-2	Neuregulin-2	Kim et al. (2016)
NUDC	Nuclear migration protein nudC	Jain et al. (2016)
NUDEL	NudE neurodevelopment protein 1 like 1	Markmiller et al. (2018)
NUP85	Nucleoporin 85	Zhang et al. (2018)
NUP98	Nuclear pore complex protein Nup98-Nup96	Jain et al. (2016), Zhang et al. (2018), Yang et al. (2020)
P4Hβ	Prolyl 4-hydroxylase subunit beta	Markmiller et al. (2018)
PAI-1	Serpine family E member 1	Omer et al. (2018)
PAK4	Serine/threonine-protein kinase PAK 4	Jain et al. (2016), Markmiller et al. (2018)
PARP-5a	Tankyrase	Leung et al. (2011)
PKCα	Protein kinase C alpha	Kobayashi et al. (2012)
POLR2B	DNA-directed RNA polymerase	Jain et al. (2016), Yang et al. (2020)
PPP2R1A	Serine/threonine-protein phosphatase 2A 65 kDa regulatory subunit A alpha isoform	Jain et al. (2016), Yang et al. (2020)
PPP2R1B	Serine/threonine-protein phosphatase 2A 65 kDa regulatory subunit A beta isoform	Markmiller et al. (2018)
PRMT1	Protein arginine N-methyltransferase 1	Jain et al. (2016)
Profilin 1	Profilin 1	Figley et al. (2014)
Profilin 2	Profilin 2	Figley et al. (2014)
PSF	Splicing factor proline and glutamine rich	Jain et al. (2016)
PSMD2	26S proteasome non-ATPase regulatory subunit 2	Jain et al. (2016), Turakhiya et al. (2018)
PTBP1	Polypyrimidine tract-binding protein 1	Markmiller et al. (2018)
PTGES3	Prostaglandin E synthase 3	Jain et al. (2016)
RACGAP1	Rac GTPase-activating protein 1	Jain et al. (2016)
RACK1	Receptor for activated C kinase 1	Arimoto et al. (2008)
RAD21	Double-strand-break repair protein rad21 homolog	Jain et al. (2016)
RANBP2	RAN binding protein 2	Jain et al. (2016)
RAPTOR	Regulatory-associated protein of mTOR complex 1	Wippich et al. (2013)
RBBP4	Histone-binding protein RBBP4	Jain et al. (2016)
RBFOX2	RNA-binding protein fox-1 homolog 2	Park et al. (2017)
RBMX	RNA-binding motif protein, X-linked	Youn et al. (2018)
RCC2	Protein RCC2	Jain et al. (2016)
RhoA	Ras homolog family member A	Tsai and Wei (2010)
Ribosomal Protein S6	Ribosomal protein S6	Sfakianos et al. (2018)
ROCK1	Rho associated coiled-coil containing protein kinase 1	Tsai and Wei (2010)
RSK2	Ribosomal protein S6 kinase A3	Eisinger-Mathason et al. (2008)

(continued)

Table 12.2 (continued)

Protein name	Description	References
S100A9	Protein S100-A9	Jain et al. (2016)
S6K1	Ribosomal protein S6 kinase B1	Sfakianos et al. (2018)
S6K2	Ribosomal protein S6 kinase B2	Sfakianos et al. (2018)
Sam68	KH RNA-binding domain containing, signal transduction-associated 1	Henao-Mejia and He (2009)
SFN	14-3-3 protein sigma	Jain et al. (2016)
Sirtuin 6	Sirtuin 6	Jedrusik-Bode et al. (2013)
SOX3	SRY-Box 3	Markmiller et al. (2018)
SQSTM1	Sequestosome 1	Chitiprolu et al. (2018)
SRC-3	Nuclear receptor coactivator 3	Yu et al. (2007)
STAT1	Signal transducer and activator of transcription 1-alpha/beta	Jain et al. (2016)
STIP1	Stress-induced-phosphoprotein 1	Jain et al. (2016)
STRAP	Serine-threonine kinase receptor-associated protein	Jain et al. (2016), Markmiller et al. (2018)
SYK	Spleen associated tyrosine kinase	Krisenko et al. (2015)
TERT	Telomerase reverse transcriptase	Iannilli et al. (2013)
TIA-1	TIA1 cytotoxic granule Associated RNA-binding protein	Kedersha et al. (1999)
TIAR	TIA1 cytotoxic granule associated RNA-binding Protein like 1	Kedersha et al. (1999)
TMOD3	Tropomodulin-3	Jain et al. (2016)
TNRC6A	Trinucleotide repeat-containing gene 6A protein	Markmiller et al. (2018), Youn et al. (2018)
TNRC6B	Trinucleotide repeat-containing gene 6B protein	Jain et al. (2016), Markmiller et al. (2018), Youn et al. (2018)
TNRC6C	Trinucleotide repeat-containing gene 6C protein	Markmiller et al. (2018), Youn et al. (2018)
TRA2B	Transformer 2 beta homolog	Youn et al. (2018)
TRAF2	TNF receptor-associated factor 2	Kim et al. (2005)
TUBA1C	Tubulin alpha-1C chain	Jain et al. (2016)
TUBA3C	Tubulin alpha-3C/D chain	Jain et al. (2016)
TUBA4A	Tubulin alpha-4A chain	Jain et al. (2016)
TUBB3	Tubulin beta-3 chain	Jain et al. (2016)
TUBB8	Tubulin beta-8 chain	Jain et al. (2016)
Ubiquitin	Ubiquitin	Meyerowitz et al. (2011)
ULK1	Unc-51 like autophagy activating kinase 1	Wang et al. (2019a)
USP13	Ubiquitin specific peptidase 13	Xie et al. (2018)
USP9X	Ubiquitin specific peptidase 9, X-linked	Narayanan et al. (2017)
VASP	Vasodilator-stimulated phosphoprotein	Jain et al. (2016)
VCP	Valosin containing protein	Buchan et al. (2013)
XPO1	Exportin 1	Jain et al. (2016)
YB-1	Y-Box binding protein 1	Yang and Bloch (2007)

(continued)

Table 12.2 (continued)

Protein name	Description	References
YES1	Tyrosine-protein kinase Yes	Jain et al. (2016)
ZBP1	Z-DNA binding protein 1	Stohr et al. (2006), Deigendesch et al. (2006)
14-3-3 α/β	Tyrosine 3-monooxygenase/tryptophan 5-Monooxygenase activation protein beta	Courchet et al. (2008)
14-3-3 η	14-3-3 protein eta	Jain et al. (2016)
14-3-3 θ	14-3-3 protein theta	Jain et al. (2016)

12.6.1 SG Assembly Mediated by Cancer Signaling Pathways

Representative pro-tumorigenic signaling including RAS and mTOR signaling has been shown to upregulate SG formation, which highlighted SGs as molecular structures that can integrate oncogenic signaling and tumor-associated stressors to enhance cancer cell progression.

12.6.1.1 TORC1 Signaling Regulates SG Assembly

The interplay between the mTOR pathway and SGs is important for regulating cell fate and cancer cell proliferation (Song and Grabocka 2023). As described in the second section, SGs and mTORC1 play a role in bilateral regulation. Namely, the mTORC1 signaling pathway is not only targeted to SGs, it contributes to inducing the formation of SGs. Canonical SG formation depends on the phosphorylation of eukaryotic translation initiation factor 2α (eIF2α), which blocks the assembly of productive translation preinitiation complexes (Kedersha et al. 1999). Heat shock, oxidative stress, hypoxia, and viral infections are triggers of SG formation and translational arrest (Anderson and Kedersha 2002). As mentioned earlier, mTORC1 components are sequestered into SGs in response to acute oxidative stress, which blocks mTORC1 activity (Thedieck et al. 2013; Sfakianos et al. 2018). Notably, however, the mTORC1 pathway regulates eIF2α phosphorylation and translation initiation in response to mild oxidative stress in Hela cells. In response to mild oxidative stress, mTORC1 activates downstream effector kinases S6K1 and S6K2, which leads to the inhibition of translation and the assembly of SGs. Importantly, S6K1 and S6K2 are required for SG formation as the treatment with an inhibitor of S6 kinase activity or silencing of either S6K1 or S6k2 decreased the number of cells displaying SGs in response to mild arsenite stress, although the contribution of S6K1 is larger than that of S6K2. Furthermore, S6 kinases promote SG assembly dependent upon their kinase activity. Importantly, the *C. elegans* S6 kinase orthologue, RSKS-1, was shown to promote the assembly of SGs in vivo, which correlates with stress resistance, thus showing the conserved role of S6K as a

Table 12.3 List of signal transduction proteins that translocate into SGs in yeast

Protein name	Description	References
Schizosaccharomyces pombe		
Hog1/Sty1	MAP kinase Sty1	Shiraishi et al. (2018)
Orb6	Serine/threonine-protein kinase Orb6	Magliozzi and Moseley (2021)
Pck2	Protein kinase C (PKC)-like Pck2	Kanda et al. (2021)
Pka1	cAMP-dependent protein kinase catalytic subunit Pka1	Nilsson and Sunnerhagen (2011)
Ppb1	Serine/threonine-protein phosphatase PP2B (calcineurin A) catalytic subunit Ppb1	Higa et al. (2015)
Rnc1	KH domain RNA-binding protein Rnc1	Satoh et al. (2018)
Shk1	PAK-related kinase Shk1	Magliozzi and Moseley (2021)
Saccharomyces cerevisiae		
ALD6	Magnesium-activated aldehyde dehydrogenase	Cherkasov et al. (2015)
BEM2	GTPase-activating protein BEM2/IPL2	Wallace et al. (2015)
CBK1	Serine/threonine-protein kinase CBK1	Shah et al. (2014)
DBP2	ATP-dependent RNA helicase DBP2	Cherkasov et al. (2015)
ESS1	Peptidyl-prolyl *cis-trans* isomerase ESS1	Zhu et al. (2020a)
FUS3	Mitogen-activated protein kinase FUS3	Shah et al. (2014)
HMT1	Protein arginine *N*-methyltransferase 1	Wallace et al. (2015)
HOG1	Mitogen-activated protein kinase involved in osmoregulation	Shiraishi et al. (2018)
KOG1	Target of rapamycin complex 1 subunit KOG1	Takahara and Maeda (2012)
KSS1	Mitogen-activated protein kinase (MAPK)	Shively et al. (2015)
MYO2	Myosin-2	Wallace et al. (2015)
NAB3	Nuclear polyadenylated RNA-binding protein 3	Wallace et al. (2015)
PBS2	MAP kinase kinase PBS2	Shah et al. (2014)
PKC1	Protein serine/threonine kinase	Amen and Kaganovich (2020)
PPH21	Serine/threonine-protein phosphatase PP2A-1 catalytic subunit	Shah et al. (2014)
PPT1	Serine/threonine-protein phosphatase T	Wallace et al. (2015)
RPB2	DNA-directed RNA polymerase II subunit RPB2	Jain et al. (2016)
RPC10	DNA-directed RNA polymerases I, II, and III subunit RPABC4	Cherkasov et al. (2015)
RPO21	DNA-directed RNA polymerase II subunit RPB1	Jain et al. (2016)
RPO26	DNA-directed RNA polymerases I, II, and III subunit RPABC2	Cherkasov et al. (2015)
RSP5	E3 ubiquitin-protein ligase RSP5	Wallace et al. (2015)
SCH9	Serine/threonine-protein kinase SCH9	Zhu et al. (2020a)
SST2	Protein SST2	Zhu et al. (2020a)
TIF4631	Eukaryotic initiation factor 4F subunit p150	Buchan et al. (2011)

(continued)

Table 12.3 (continued)

Protein name	Description	References
TIF4632	Eukaryotic initiation factor 4F subunit p130	Buchan et al. (2011)
TOR1	Serine/threonine-protein kinase TOR1	Takahara and Maeda (2012)
TPK1	cAMP-dependent protein kinase catalytic subunit	Barraza et al. (2017), Tudisca et al. (2012)
TPK2	cAMP-dependent protein kinase catalytic subunit	Barraza et al. (2017), Tudisca et al. (2012)
TPK3	cAMP-dependent protein kinase catalytic subunit	Barraza et al. (2017), Tudisca et al. (2012)
VMA2	V-type proton ATPase subunit B	Jain et al. (2016)
YDL124W	NADPH-dependent alpha-keto amide reductase	Cherkasov et al. (2015)
YPK1	Serine/threonine-protein kinase YPK1	Wallace et al. (2015)

downstream effector of mTORC1 signaling in SG assembly to promote survival in response to stressful conditions.

12.6.1.2 RAS Signaling Upregulates SGs

RAS proteins belong to the family of GTPases and are considered regulators of proliferation, migration, apoptosis, and survival (Santarpia et al. 2012). RAS is a key molecule that controls several tumorigenesis-related processes and genetic analyses found that RAS is one of the most deregulated oncogenes in human cancers (Sanchez-Vega et al. 2018). Mutant RAS proteins stimulate downstream signals, including the ERK and AKT signaling pathways, and tumor cells harboring mutant RAS exhibit more aggressive phenotypes (Sunaga et al. 2013). Almost one-third of all human cancers have RAS (KRAS-HRAS-NRAS) mutations (Cerami et al. 2012). KRAS is common in pancreatic ductal adenocarcinomas and colorectal cancer, and NRAS in melanoma, thyroid cancer, and leukemia (Gao et al. 2014). Cancer cells are exposed to various stresses, and the mutant RAS proteins help to adapt to tumor-associated stresses (Tao et al. 2014; Yang et al. 2018). Remarkably, the presence of SGs was observed in mutant KRAS pancreatic cancer cells as opposed to normal cells.

The pioneering work by Garabocka and Bar-Sagi allowed the discovery of the mutant KRAS-dependent capability of cancer cells to self-protect against stress-inducing stimuli and chemotherapeutic agents through the upregulation of SGs (Grabocka and Bar-Sagi 2016). Pancreatic ductal adenocarcinoma (PDAC), one of the most lethal solid tumors, has a 5-year survival rate that only surpassed 10% in the past year (Noda et al. 2022). Remarkably, SGs are markedly elevated in mutant KRAS pancreatic cancer cells and tumor samples from pancreatic adenocarcinoma patients, but not in surrounding normal tissue. The authors also demonstrated that mutant KRAS upregulates SGs by stimulating the production of the lipid signaling molecule 15-deoxy-delta 12,14 prostaglandin J2 (15-d-PGJ2). 15-d-PGJ2 was known to target cysteine 264 in eIF4A, destroying its interaction with eIF4G, the interaction required for the translation process (Kim et al. 2007). Mutant KRAS cancer cells were shown to have elevated levels of 15-d-PGJ2 as a result of enhanced prostaglandin (PG) synthesis and decreased PG degradation, due to upregulation of cyclooxygenase 2 (COX-2) expression and downregulation of hydroxyprostaglandin dehydrogenase 15-(NAD) (HPGD) expression, respectively. COX-2 inhibition suppressed SGs and sensitized cells to oxaliplatin. Importantly, mutant KRAS cells secreted 15-d-PGJ2, which enables paracrine upregulation of SGs in not only KRAS mutant cancer cells but also neighboring cells with normal KRAS, which rendered them more resistant to the cytotoxic effects of oxaliplatin. Importantly from clinical aspects, the authors further investigated the relationship between the levels of SG regulators and patient survival. As a result, positive regulators of SGs in patients of PDAC are linked to poor survival, thus proposing that SG upregulation may represent an important mechanism by which mutant KRAS confers reduced survival in PDAC.

12.6.2 SGs as a Therapeutic Target of Cancer Treatment

The studies presented above in the second section have provided the clinical potential of SGs as a target as well as a prognosis marker of cancer therapeutics. Notably, several components of SGs as represented by G3BP1 participate in cancer development through tumor-associated signaling pathways. Moreover, some conventional anticancer chemotherapy drugs as well as molecular-targeted drugs have been reported to induce SGs. In this section, we will introduce the relationship between SGs and cancer as well as the potential of targeting SGs as a therapeutic strategy and a prognosis marker of cancer therapy.

12.6.2.1 Targeting Components of SGs and Cancer

SGs are observed in many human cancers, and upregulation of several SG components has been reported in different kinds of tumors, such as pancreatic cancer, gastric cancer, thyroid cancer, breast cancer, sarcoma, colorectal cancer, head and neck cancer, and prostate cancer (Table 12.4). SG formation can be triggered not only by different stressful conditions but also by the overexpression of specific RNA-binding proteins (RBPs), including G3BP and TIA1, which play a role as a nucleating factor for SGs (Tourriere et al. 2003; Gilks et al. 2004; Hua and Zhou 2004b; Kedersha et al. 2005). Thus, these RBPs may also play a role to promote cancer progression, and targeting these key RBPs may be effective in cancer therapy.

Table 12.4 Cancer types with upregulated SG components

Cancer type	References
Breast cancer	French et al. (2002)
Esophageal squamous cell carcinoma	Zheng et al. (2022)
Gastric cancer	Lin et al. (2019), Zhao et al. (2021)
Head and neck cancers	Guitard et al. (2001), Wang et al. (2023)
Hepatocellular carcinoma	Dou et al. (2016)
Lung cancer	Zheng et al. (2019)
Melanoma	Liu et al. (2021)
Pancreatic cancer	Grabocka and Bar-Sagi (2016), Sim et al. (2019), Taniuchi et al. (2011)
Prostate cancer	Mukhopadhyay et al. (2021), Takayama et al. (2018)
Renal cell carcinomas	Wang et al. (2018)
Sarcoma	Somasekharan et al. (2015)
Thyroid cancer	Guitard et al. (2001)

12.6.2.2 SG Nucleating Factors

These factors largely belong to the family of RBPs that are involved in the early steps of SG formation and their overexpression alone can induce SG assembly without any environmental stimuli. Conversely, the loss of function of these factors can lead to the impairment of SG assembly.

G3BP: G3BP (Ras-GTPase-activating protein SH3 domain-binding protein 1) is one of the best-characterized components of SGs that initiates the assembly of SGs. G3BP overexpression induces SGs in a dose-dependent manner (Guillén-Boixet et al. 2020; Yang et al. 2020). G3BP1 and G3BP2 contain RNA-binding domains and are core components of SGs. G3BP is essential for SG formation as evidenced by the fact that the knockdown of G3BP1 reduces the assembly of SGs (Ghisolfi et al. 2012). The G3BP family consists of three homologous proteins (G3BP1, G3BP2a, and G3BP2b), of which G3BP1 is markedly overexpressed in various tumor tissues, especially in head and neck cancer, lung cancer, prostate cancer, and breast cancers (Guitard et al. 2001; French et al. 2002; Zheng et al. 2019; Mukhopadhyay et al. 2021). Treatment of cells with arsenite induces the dephosphorylation of G3BP1 at Ser-149, and this dephosphorylation stimulates the multimerization of G3BP1 and its interaction with HDAC6 (Tourriere et al. 2003). G3BP1 promotes cancer cell proliferation and metastasis, while loss of G3BP1 suppresses proliferation, migration, and invasion of various tumor cell lines (Table 12.5). Moreover, G3BP1 expression correlates with poor prognosis for patients of some types of cancers, including hepatocellular carcinoma (Zhang et al. 2007; Min et al. 2015; Somasekharan et al. 2015; Dou et al. 2016; Xiong et al. 2019; Zheng et al. 2019; Li et al. 2020, 2022; Mukhopadhyay et al. 2021; Zhao et al. 2021; Wang et al. 2022). Thus, G3BP1 expression may be a potential target for cancer treatment as well as an important biomarker of poor prognosis in multiple cancers.

Table 12.5 Roles of G3BPs in cancer progression

Cancer type	Function	References
Breast cancer	Proliferation	Winslow et al. (2013)
Renal cell carcinomas	Proliferation, metastasis	Wang et al. (2018)
Colon/colorectal	Proliferation, Invasion	Li et al. (2020), Cui et al. (2020)
Melanoma	proliferation, angiogenesis	Liu et al. (2021)
Gastric cancer	Proliferation, invasion, metastasis	Xiong et al. (2019)
Esophageal squamous cell carcinoma	Proliferation, invasion, metastasis	Zhang et al. (2019)
Pancreatic cancer	Invasion, metastasis	Taniuchi et al. (2011)
Sarcoma	Invasion, metastasis	Somasekharan et al. (2015)

Thus, reducing the expression of G3BP, or compounds targeting G3BP, such as compound C108 can be a promising new strategy for cancer therapy (Gupta et al. 2017). Small molecules such as resveratrol or the peptide GAP161 have been shown to reduce the expression of G3BP and prevent SG formation in colorectal cancer (Zhang et al. 2012). Further studies are needed on whether these compounds can affect tumor growth via SG regulation.

TIA1: Similar to G3BP1, T-cell intracellular antigen-1 (TIA1) is a key regulator of SGs. The overexpression of full-length TIA-1 is sufficient to cause the formation of SGs in the absence of stress (Gilks et al. 2004; Kedersha and Anderson 2007). TIA1 is an RNA-binding protein involved in many regulatory aspects of mRNA metabolism. TIA-1 has been reported to prevent cell proliferation, tumor growth, and invasion (Izquierdo et al. 2011). Notably, however, TIA1 was shown to exert tumor-promoting activity, which seems to be associated with its regulation of a set of cancer-related transcripts, in esophageal squamous cell carcinoma (ESCC). Moreover, TIA1 silencing inhibited proliferation, and the exogenous introduction of TIA1 promoted ESCC cell proliferation, thus demonstrating the clinical and functional significance of the TIA1 protein in ESCC tumorigenesis (Hamada et al. 2016). Whether or not, the role of TIA1 in SG assembly is functionally associated with the poor prognosis in ESCC patients remains unknown. However, the authors' Kaplan–Meier survival estimates showed that positive cytoplasmic TIA1 immunoreactivity was significantly associated with worse overall survival in all ESCC patients examined ($p = 0.0003$), but nuclear TIA1 immunoreactivity was not (Hamada et al. 2016). Furthermore, the authors identified putative binding mRNAs for cytoplasmic TIA1 and conclude that overexpressed cytoplasmic TIA1 promotes ESCC tumorigenesis at least partly through its binding with cancer-associated mRNAs such as SKP2 and CCNA2 mRNAs and induction of SKP2 and CCNA2 protein overexpression. Thus, cytoplasmic localized TIA1 seems to be involved in cancer-promoting RNA metabolism and contributes to the progression of ESCC.

12.6.2.3 Targeting Post-translational Modifications Related to SG Assembly

In addition to the strategy targeting the components of SGs, especially key factors nucleating the SGs, regulating the post-translational modifications related to SG assembly processes may be a promising approach to target cancer. These include ubiquitination, sumoylation, methylation, acetylation, O-GlcNAClaytion, and phosphorylation (Kedersha et al. 2013; Mahboubi and Stochaj 2017; Snead and Gladfelter 2019; Cao et al. 2020). These modifications are mediated by cellular signaling pathways. Therefore, enzymes that are responsible for these modifications are possible targets for blocking SG assembly and thus a target for cancer therapy (we will not go into detail and readers are recommended for the excellent review: Snead and Gladfelter 2019).

An interesting example from fission yeast suggested the MAPK-dependent phospho-regulation of a nucleating factor for SG assembly. A molecular genetic approach by Sugiura's lab using fission yeast identified Nrd1, an RRM-type RNA-binding protein as a regulator of the cell cycle by stabilizing the mRNA of the key cell-cycle regulator Cdc4, encoding myosin (Satoh et al. 2009). Moreover, the author's group showed that Pmk1 MAPK directly phosphorylates Nrd1, thereby negatively regulating the binding activity of Nrd1 to Cdc4 mRNA, thus showing the MAPK-mediated phospho-dependent regulation of the RNA-binding protein Nrd1. Importantly, Nrd1 was shown to play a role in SG formation and the phosphorylation of Nrd1 by MAPK enhances its localization to SGs (Satoh et al. 2012). The overexpression of Nrd1 induces SG assembly without any environmental stimuli, and the phosphorylated version of Nrd1 displayed an enhanced ability to stimulate Pabp-positive SG assembly. This study illustrates the importance of MAPK-dependent phosphorylation on the function and intracellular localization of an RNA-binding protein Nrd1. Intriguingly, Nrd1 shares significant sequence similarity and a preferred RNA-binding sequence (UCUU) with TIA-1 (Satoh et al. 2009). Since both TIA1 and Nrd1 play a key role to promote SG assembly, phosphorylation-dependent regulation of RNA-binding protein that acts to nucleate SG assembly may serve as a conserved mechanism to target SGs.

12.6.3 SGs and Cancer Chemotherapy

As described earlier, the KRAS mutant PDAC developed resistance to chemotherapy via the upregulation of SGs. Intriguingly, apart from the pathophysiological and genomic context, SGs can be induced by chemotherapy drugs for cancer (Table 12.6), and then formed SGs contribute to chemotherapy resistance. For example, chemotherapeutic drugs that influence the translation process or compounds targeting the translation elements can induce SGs. Thus, most of the anticancer drugs that can increase SG assembly can induce eIF2α phosphorylation

Table 12.6 List of anticancer drugs that form SGs and concomitant drugs that decrease SGs formation

Anticancer drug	SG components	Concomitant medication that promotes efficacy via SG inhibition	References
5-Fluorouracil	eIF2α↑(via PKR), ATXN2L, ATXN2, DCP1a, DDX6, eIF4G1, G3BP1, Musashi-1, TIAR		García et al. (2011), Kaehler et al. (2014), Chiou et al. (2017)
Bortezomib	eIF2α↑(via HRI), eIF4E, FMRP, FXR-1, G3BP1	Psammaplysin F	Fournier et al. (2010), Christen et al. (2019)
Cisplatin	eIF2α↑, eIF3η, G3BP1, TIAR	Thapsigargin, tunicamycin	Wang et al. (2019b), Pietras et al. (2022), Martins et al. (2011)
Etoposide	eIF2α↑, FMR1, G3BP1		Vilas-Boas et al. (2016)
Lapatinib	eIF2α↑(via PERK signaling), DDX3, eIF4G1, FMRP, G3BP1		Adjibade et al. (2020)
Oxaliplatin	eIF2α↑, ATXN2L, G3BP1		Martins et al. (2011), Lin et al. (2019), Zhao et al. (2021)
Paclitaxel	eIF2α↑, eIF3B, eIF4G, G3BP1, G3BP2, PABPC1, TIA-1		Shi et al. (2021), Szaflarski et al. (2016), Gupta et al. (2017)
Sorafenib	eIF2α↑(via PERK), eIF4E, eIF4G1, FXR-1, G3BP1	Psammaplysin F	Jiang and Wek (2005), Adjibade et al. (2015), Christen et al. (2019)
Temozolomide	eIF2α↑, G3BP1		Dastghaib et al. (2020), Bittencourt et al. (2019)
Vinblastine	eIF2α↑, eIF3B, eIF4G, G3BP1		Szaflarski et al. (2016)
Vincristine	eIF2α↑, eIF3B, eIF4G, G3BP1		Szaflarski et al. (2016)
Vinorelbine	eIF2α↑, eIF3B, eIF4B, eIF4G, G3BP1, TIA-1		Schwed-Gross et al. (2022), Szaflarski et al. (2016)

(Table 12.6). For example, 5-fluorouracil (5-FU), a uracil and thymine analog that is widely used in the treatment of a variety of solid tumors, also affects SG assembly based on RNA incorporation. 5-FU inhibits thymidylate synthase and its metabolites incorporate into RNA and DNA, thereby showing its antineoplastic effects by influencing the processing and maturation of rRNA, tRNA, and snRNA. The incorporation of 5-FU metabolite into RNA is a factor to trigger the SG assembly. 5-FU can activate PRK, which leads to eIF2α phosphorylation, thereby stimulating

SG assembly. Consistently, blocking the eIF2α phosphorylation was shown to enhance chemotherapeutically induced cell death in cancer cells such as gliomas for various drugs, including etoposide, bortezomib, or cisplatin (Vilas-Boas et al. 2016). Moreover, since oxidative stress is one of the conditions that induce SG assembly, compounds and drugs that stimulate reactive oxygen species (ROS), such as platinum or platinum-containing drugs, including oxaliplatin are capable to induce SGs. Interestingly, the relevance of Ataxin-2-like (ATXN2L), one of the nucleating factors for SG in chemoresistance was also reported. ATXN2L is a regulator of SGs based on the findings that ATXN2L overexpression induces the formation of SGs, while the reduced ATXN2L impairs SG formation. Importantly, silencing the ATXN2L expression (fewer SGs as compared with the control) can sensitize cancer cell lines against oxaliplatin by increasing ROS production and apoptosis (Lin et al. 2019). Thus, oxidative stress and SG formation synergize to facilitate cancer cells to acquire chemoresistance, which is negative for cancer treatment.

Intriguingly, not only conventional chemotherapy drugs, as shown above, but newly developed molecular-targeted drugs affect SG assembly (Table 12.6). These include Bortezomib (26S proteasome inhibitor), Lapatinib (tyrosine kinase inhibitor), and Sorafenib (tyrosine kinase inhibitor). Although these compounds exert antineoplastic effects by targeting distinct molecules, they commonly induce SG assembly by stimulating eIF2α phosphorylation. Upon different stress conditions, eIF2 phosphorylation is regulated by different four upstream kinases. Namely, HR1, PKR, PERK, and GCN2 have been reported to be responsible for inducing eIF2α phosphorylation. Bortezomib can stimulate eIF2α phosphorylation via HR1, while Sorafenib and Lapatinib induce eIF2α phosphorylation via PERK (Table 12.6).

12.7 Concluding Remarks and Future Perspectives

LLPS and SGs have emerged as a rapidly growing field with numerous biomedical scientific applications, at the intersection of molecular biology and cancer biology (Treeck et al. 2018; Alberti et al. 2019; Peran and Mittag 2020; Hofmann et al. 2021; Wiedner and Giudice 2021; Tong et al. 2022). LLPS compellingly plays a key role as a new layer of spatiotemporal regulation of cell signaling. Especially, increasing examples discussed above demonstrated that SGs have attracted considerable attention in the field of cancer biology, with broad impacts ranging from cancer-associated signaling systems to cancer development as well as the prognosis of cancer patients.

Although we focused on describing the relationship between SGs and cancer in this review, the importance of LLPS and its role in tumorigenesis is further expanding. For example, aberrant biomolecular condensates can induce transcriptional dysregulation, which is a key feature of cancer. Recent studies have implicated transcriptional condensates in the nucleus in the regulation of oncogenic transcriptional programs, as represented by YAP/TAZ condensates. The transcriptional coactivator Yes-associated protein (YAP) and its paralogue transcriptional coactivator with PDZ-binding motif (TAZ) are key players linked to cancer cell transcription via super-enhancers (Zanconato et al. 2016). These condensates appear to recruit RNA polymerase II to trigger the transcription of proliferative genes (Cai et al. 2019). Importantly, the transcriptional activity of YAP/TAZ is associated with their ability to form condensates. Thus, YAP and TAZ could work as signaling hubs for tumor development. Examples regarding the link between biomolecular condensates and other cellular protein quality control systems, such as proteasome degradation and autophagy are also emerging filed to gain the whole picture of the role of LLPS in cancer biology (see excellent reviews: Lu et al. 2021; Igelmann et al. 2022; Ren et al. 2022; Tong et al. 2022; Xie et al. 2023).

Although remarkable progress has been made regarding the role of biomolecular condensates, especially SGs as organizers of the signaling machinery, we are still at the early stage of unraveling the direct functional role of LLPS in organizing the dynamics of cancer signaling and its pathophysiological consequences. Several limitations remain in studying LLPS and its relevance to cell signaling. Disrupting key nucleating factors for the assembly or molecular interactions, or modifying posttranscriptional regulation essential for condensation often leads to disrupting entire SGs/condensates. This situation hampers the dissection of the loss of each component or the loss of condensates. Furthermore, although there is a close association between the prognosis of various types of cancer and SG upregulation, it is still difficult to visualize SGs in patients and compelling functional evidence will be needed to unequivocally assume the role of SG in poor prognosis of cancer. To overcome these hurdles, new techniques, especially genetically encoded tools are being employed to directly monitor or manipulate molecular processes within intact condensates in living cells (Huang et al. 2020; Quiroz et al. 2020; Zhang et al. 2020b; Freibaum et al. 2021; Tenner et al. 2021). We have established a chemical genetic screen to identify compounds targeting SGs by utilizing the cellular toxicity induced by the overexpression of key nucleating factors, such as Nrd1 and Rnc1. Given the remarkable conservation regarding the role of SGs as cellular organizers and signaling hubs, these approaches using model organisms may help to unravel the complex and conserved mechanisms of biomolecular condensates on cancer-associated signaling and its functional consequences on cell fate.

References

Adjibade P, St-Sauveur VG, Quevillon Huberdeau M et al (2015) Sorafenib, a multikinase inhibitor, induces formation of stress granules in hepatocarcinoma cells. Oncotarget. 6(41): 43927–43943

Adjibade P, Simoneau B, Ledoux N et al (2020) Treatment of cancer cells with Lapatinib negatively regulates general translation and induces stress granules formation. PLoS One. 15:e0231894. https://doi.org/10.1371/journal.pone.0231894

Alberti S, Gladfelter A, Mittag T (2019) Considerations and challenges in studying liquid-liquid phase separation and biomolecular condensates. Cell 176:419–434. https://doi.org/10.1016/j.cell.2018.12.035

Amen T, Kaganovich D (2020) Stress granules sense metabolic stress at the plasma membrane and potentiate recovery by storing active Pkc1. Sci Signal 13(623):eaaz6339. https://doi.org/10.1126/scisignal.aaz6339

Anderson P, Kedersha N (2002) Visibly stressed: the role of eIF2, TIA-1, and stress granules in protein translation. Cell Stress Chaperones 7(2):213–221

Andersson MK, Stahlberg A, Arvidsson Y et al (2008) The multifunctional FUS, EWS and TAF15 proto-oncoproteins show cell type-specific expression patterns and involvement in cell spreading and stress response. BMC Cell Biol 9:37. https://doi.org/10.1186/1471-2121-9-37

Arellano M, Valdivieso MH, Calonge TM et al (1999) *Schizosaccharomyces pombe* protein kinase C homologues, pck1p and pck2p, are targets of rho1p and rho2p and differentially regulate cell integrity. J Cell Sci 112(20):3569–3578

Arimoto K, Fukuda H, Imajoh-Ohmi S et al (2008) Formation of stress granules inhibits apoptosis by suppressing stress-responsive MAPK pathways. Nat Cell Biol 10(11):1324–1332

Asadi MR, Rahmanpour D, Moslehian MS et al (2021a) Stress granules involved in formation, progression and metastasis of cancer: a scoping review. Front Cell Dev Biol. 9:753. https://doi.org/10.3389/fcell.2021.745394

Asadi MR, Sadat Moslehian M, Sabaie H et al (2021b) Stress granules and neurodegenerative disorders: a scoping review. Front Aging Neurosci 13:650740. https://doi.org/10.3389/fnagi.2021.650740

Ash PE, Vanderweyde TE, Youmans KL et al (2014) Pathological stress granules in Alzheimer's disease. Brain Res 1584:52–58

Baek JH, Jang JE, Kang CM et al (2000) Hypoxia-induced VEGF enhances tumor survivability via suppression of serum deprivation-induced apoptosis. Oncogene. 19(40):4621–4631

Baez MV, Boccaccio GL (2005) Mammalian Smaug is a translational repressor that forms cytoplasmic foci similar to stress granules. J Biol Chem 280(52):43131–43140

Baguet A, Degot S, Cougot N et al (2007) The exon junction- complex-component metastatic lymph node 51 functions in stress-granule assembly. J Cell Sci 120(16):2774–2784

Balzer E, Moss EG (2007) Localization of the developmental timing regulator Lin28 to mRNP complexes, P-bodies and stress granules. RNA Biol 4(1):16–25

Barraza CE, Solari CA, Marcovich I et al (2017) The role of PKA in the translational response to heat stress in Saccharomyces cerevisiae. PLoS One. 12:e0185416. https://doi.org/10.1371/journal.pone.0185416

Bittencourt LFF, Negreiros-Lima GL, Sousa LP et al (2019) G3BP1 knockdown sensitizes U87 glioblastoma cell line to Bortezomib by inhibiting stress granules assembly and potentializing apoptosis. J Neurooncol. 144(3):463–473

Bowen ME, Boyden ED, Holm IA et al (2011) Loss-of-function mutations in PTPN11 cause metachondromatosis, but not Ollier disease or Maffucci syndrome. PLoS Genet 7(4):e1002050. https://doi.org/10.1371/journal.pgen.1002050

Brewster JL, Gustin MC (2014) Hog1: 20 years of discovery and impact. Sci Signal 7(343):re7. https://doi.org/10.1126/scisignal.2005458

Brown EJ, Albers MW, Shin TB et al (1994) A mammalian protein targeted by G1-arresting rapamycin-receptor complex. Nature 369(6483):756–758

Buchan JR (2014) mRNP granules. Assembly, function, and connections with disease. RNA Biol 11(8):1019–1030

Buchan JR, Yoon JH, Parker R (2011) Stress-specific composition, assembly and kinetics of stress granules in Saccharomyces cerevisiae. J Cell Sci 124(2):228–239

Buchan JR, Kolaitis RM, Taylor JP et al (2013) Eukaryotic stress granules are cleared by autophagy and Cdc48/VCP function. Cell 153(7):1461–1474

Buechler S (2009) Low expression of a few genes indicates good prognosis in estrogen receptor positive breast cancer. BMC Cancer 9:243. https://doi.org/10.1186/1471-2407-9-243

Burnett PE, Barrow RK, Cohen NA (1998) RAFT1 phosphorylation of the translational regulators p70 S6 kinase and 4E-BP1. Proc Natl Acad Sci USA 95(4):1432–1437

Burry RW, Smith CL (2006) HuD distribution changes in response to heat shock but not neurotrophic stimulation. J Histochem Cytochem 54(10):1129–1138

Cai D, Feliciano D, Dong P et al (2019) Phase separation of YAP reorganizes genome topology for long-term YAP target gene expression. Nat Cell Biol 21:1578–1589. https://doi.org/10.1038/s41556-019-0433-z

Cao X, Jin X, Liu B (2020) The involvement of stress granules in aging and aging-associated diseases. Aging Cell 19(4):e13136. https://doi.org/10.1111/acel.13136

Cargnello M, Roux PP (2011) Activation and function of the MAPKs and their substrates, the MAPK-activated protein kinases. Microbiol Mol Biol Rev. 75(1):50–83

Cerami E, Gao J, Dogrusoz U et al (2012) The cBio cancer genomics portal: an open platform for exploring multidimensional cancer genomics data. Cancer Discov 2:401–404. https://doi.org/10.1158/2159-8290.cd-12-0095

Chalupnikova K, Lattmann S, Selak N et al (2008) Recruitment of the RNA helicase RHAU to stress granules via a unique RNA-binding domain. J Biol Chem 283(50):35186–35198

Cherkasov V, Grousl T, Theer P et al (2015) Systemic control of protein synthesis through sequestration of translation and ribosome biogenesis factors during severe heat stress. FEBS Lett. 589(23):3654–3664

Chiou GY, Yang TW, Huang CC et al (2017) Musashi-1 promotes a cancer stem cell lineage and chemoresistance in colorectal cancer cells. Sci Rep. 7:2172. https://doi.org/10.1038/s41598-017-02057-9

Chitiprolu M, Jagow C, Tremblay V et al (2018) A complex of C9ORF72 and p62 uses arginine methylation to eliminate stress granules by autophagy. Nat Commun. 9:2794. https://doi.org/10.1038/s41467-018-05273-7

Cho YY (2017) RSK2 and its binding partners in cell proliferation, transformation and cancer development. Arch Pharm Res. 40(3):291–303

Cho YY, Lee MH, Lee CJ et al (2012) RSK2 as a key regulator in human skin cancer. Carcinogenesis. 33(12):2529–2537

Christen KE, Davis RA, Kennedy D (2019) Psammaplysin F increases the efficacy of bortezomib and sorafenib through regulation of stress granule formation. Int J Biochem Cell Biol 112:24–38

Clark DE, Errington TM, Smith JA et al (2005) The serine/threonine protein kinase, p90 ribosomal S6 kinase, is an important regulator of prostate cancer cell proliferation. Cancer Res. 65(8): 3108–3116

Courchet J, Buchet-Poyau K, Potemski A et al (2008) Interaction with 14-3-3 adaptors regulates the sorting of hMex-3B RNA-binding protein to distinct classes of RNA granules. J Biol Chem. 283(46):32131–32142

Cui X, Liang H, Hao C, Jing X (2020) Homer1 is a potential biomarker for prognosis in human colorectal carcinoma, possibly in association with G3BP1 signaling. Cancer Manage Res 12: 2899–2909. https://doi.org/10.2147/cmar.s240942

Dastghaib S, Shojaei S, Mostafavi-Pour Z et al (2020) Simvastatin induces unfolded protein response and enhances temozolomide-induced cell death in glioblastoma cells. Cells. 9:2339. https://doi.org/10.3390/cells9112339

De Leeuw F, Zhang T, Wauquier C et al (2007) The cold-inducible RNA binding protein migrates from the nucleus to cytoplasmic stress granules by a methylation dependent mechanism and acts as a translational repressor. Exp Cell Res 313(20):4130–4144

Deigendesch N, Koch-Nolte F, Rothenburg S (2006) ZBP1 subcellular localization and association with stress granules is controlled by its Z-DNA binding domains. Nucleic Acids Res. 34(18): 5007–5020

Detzer A, Engel C, Wünsche W et al (2011) Cell stress is related to re-localization of Argonaute 2 and to decreased RNA interference in human cells. Nucleic Acids Res. 39(7):2727–2741

Didiot MC, Subramanian M, Flatter E et al (2009) Cells lacking the fragile X mental retardation protein (FMRP) have normal RISC activity but exhibit altered stress granule assembly. Mol Biol Cell 20(1):428–437
Digilio MC, Conti E, Sarkozy A et al (2002) Grouping of multiple-lentigines/LEOPARD and Noonan syndromes on the PTPN11 gene. Am J Hum Genet 71(2):389–394
Doi A, Kita A, Kanda Y et al (2015) Geranylgeranyltransferase Cwg2-Rho4/Rho5 module is implicated in the Pmk1 MAP kinase-mediated cell wall integrity pathway in fission yeast. Genes Cells 20(4):310–323
Dou N, Chen J, Yu S et al (2016) G3BP1 contributes to tumor metastasis via upregulation of Slug expression in hepatocellular carcinoma. Am J Cancer Res 6:2641–2650
Eisinger-Mathason TS, Andrade J, Groehler AL et al (2008) Codependent functions of RSK2 and the apoptosis-promoting factor TIA-1 in stress granule assembly and cell survival. Mol Cell 31(5):722–736
Figley MD, Bieri G, Kolaitis RM et al (2014) Profilin 1 associates with stress granules and ALS-linked mutations alter stress granule dynamics. J Neurosci. 34(24):8083–8097
Fonteneau G, Redding A, Hoag-Lee H et al (2022) Stress granules determine the development of obesity-associated pancreatic cancer. Cancer Discov 12(8):1984–2005
Fournier MJ, Gareau C, Mazroui R (2010) The chemotherapeutic agent bortezomib induces the formation of stress granules. Cancer Cell Int. 10:12. https://doi.org/10.1186/1475-2867-10-12
Freibaum BD, Messing J, Yang P et al (2021) High-fidelity reconstitution of stress granules and nucleoli in mammalian cellular lysate. J Cell Biol 220:e202009079. https://doi.org/10.1083/jcb.202009079
French J, Stirling R, Walsh M, Kennedy HD (2002) The expression of Ras–GTPase activating protein SH3 domain-binding proteins, G3BPs, in human breast cancers. Histochem J 34:223–231. https://doi.org/10.1023/a:1021737413055
Fruehauf JP, Meyskens FL Jr (2007) Reactive oxygen species: a breath of life or death? Clin Cancer Res 13(3):789–794
Fujimura K, Katahira J, Kano F et al (2009) Selective localization of PCBP2 to cytoplasmic processing bodies. Biochim Biophys Acta 1793(5):878–887
Fujimura K, Suzuki T, Yasuda Y et al (2010) Identification of importin alpha1 as a novel constituent of RNA stress granules. Biochim Biophys Acta. 1803(7):865–871
Fukuda T, Naiki T, Saito M et al (2009) hnRNP K interacts with RNA binding motif protein 42 and functions in the maintenance of cellular ATP level during stress conditions. Genes Cells 14(2):113–128
Gallois-Montbrun S, Kramer B, Swanson CM et al (2007) Antiviral protein APOBEC3G localizes to ribonucleoprotein complexes found in P bodies and stress granules. J Virol 81(5):2165–2178
Gallouzi IE, Brennan CM, Stenberg MG et al (2000) HuR binding to cytoplasmic mRNA is perturbed by heat shock. Proc Natl Acad Sci U S A 97(7):3073–3078
Gao X, Ge L, Shao J et al (2010) Tudor-SN interacts with and co-localizes with G3BP in stress granules under stress conditions. FEBS Lett 584(16):3525–3532
Gao A, Yang J, Yang G et al (2014) Differential gene expression profiling analysis in workers occupationally exposed to benzene. Sci Total Environ 472:872–879. https://doi.org/10.1016/j.scitotenv.2013.11.089
García MA, Carrasco E, Aguilera M et al (2011) The chemotherapeutic drug 5-fluorouracil promotes PKR-mediated apoptosis in a p53-independent manner in colon and breast cancer cells. PLoS One 6(8):e23887. https://doi.org/10.1371/journal.pone.0023887
Gerwins P, Blank JL, Johnson GL (1997) Cloning of a novel mitogen-activated protein kinase kinase kinase, MEKK4, that selectively regulates the c-Jun amino terminal kinase pathway. J Biol Chem. 272(13):8288–8295
Ghisolfi L, Dutt S, McConkey ME et al (2012) Stress granules contribute to α-globin homeostasis in differentiating erythroid cells. Biochem Biophys Res Commun 420(4):768–774. https://doi.org/10.1016/j.bbrc.2012.03.070

Gilks N, Kedersha N, Ayodele M et al (2004) Stress granule assembly is mediated by prion-like aggregation of TIA-1. Mol Biol Cell 15(12):5383–5398. https://doi.org/10.1091/mbc.e04-08-0715

Goodier JL, Zhang L, Vetter MR et al (2007) LINE-1 ORF1 protein localizes in stress granules with other RNA-binding proteins, including components of RNAi RISC. Mol Cell Biol 27(18): 6469–6483

Gould CM, Newton AC (2008) The life and death of protein kinase C. Curr Drug Targets 9(8): 614–625

Grabocka E, Bar-Sagi D (2016) Mutant KRAS enhances tumor cell fitness by upregulating stress granules. Cell 167:1803–1813.e12. https://doi.org/10.1016/j.cell.2016.11.035

Grallert B, Kearsey SE, Lenhard M et al (2000) A fission yeast general translation factor reveals links between protein synthesis and cell cycle controls. J Cell Sci. 113(8):1447–1458

Grousl T, Opekarová M, Stradalova V et al (2015) Evolutionarily conserved 5′-3′ exoribonuclease Xrn1 accumulates at plasma membrane-associated eisosomes in post-diauxic yeast. PLoS One 10(3):e0122770. https://doi.org/10.1371/journal.pone.0122770

Gruber J, Harborth J, Schnabel J et al (2002) The mitotic-spindle-associated protein astrin is essential for progression through mitosis. J Cell Sci 115(21):4053–4059

Guil S, Long JC, Caceres JF (2006) hnRNP A1 relocalization to the stress granules reflects a role in the stress response. Mol Cell Biol. 26(15):5744–5758

Guillén-Boixet J, Kopach A, Holehouse AS et al (2020) RNA-induced conformational switching and clustering of G3BP drive stress granule assembly by condensation. Cell 181(2):346–361. https://doi.org/10.1016/j.cell.2020.03.049

Guitard E, Parker F, Millon R et al (2001) G3BP is overexpressed in human tumors and promotes S phase entry. Cancer Lett 162:213–221. https://doi.org/10.1016/s0304-3835(00)00638-8

Gupta N, Badeaux M, Liu Y et al (2017) Stress granule-associated protein G3BP2 regulates breast tumor initiation. Proc Natl Acad Sci U S A. 114(5):1033–1038

Gutiérrez-Venegas G, Arreguín-Cano JA, Arroyo-Cruz R (2010) Activation of ERK1/2 by protein kinase C-alpha in response to hydrogen peroxide-induced cell death in human gingival fibroblasts. Toxicol In Vitro 24(1):319–326

Hamada J, Shoda K, Masuda K et al (2016) Tumor-promoting function and prognostic significance of the RNA-binding protein T-cell intracellular antigen-1 in esophageal squamous cell carcinoma. Oncotarget 7(13):17111–17128. https://doi.org/10.18632/oncotarget.7937

Hammouda MB, Ford AE, Liu Y et al (2020) The JNK signaling pathway in inflammatory skin disorders and cancer. Cells. 9:857. https://doi.org/10.3390/cells9040857

Hara K, Maruki Y, Long X et al (2002) Raptor, a binding partner of target of rapamycin (TOR), mediates TOR action. Cell 110(2):177–189

Harwood FC, Klein Geltink RI, O'Hara BP et al (2018) ETV7 is an essential component of a rapamycin-insensitive mTOR complex in cancer. Sci Adv 4(9):eaar3938. https://doi.org/10.1126/sciadv.aar3938

Heberle AM, Razquin Navas P, Langelaar-Makkinje M et al (2019) The PI3K and MAPK/p38 pathways control stress granule assembly in a hierarchical manner. Life Sci Alliance 2(2): 201800257. https://doi.org/10.26508/lsa.201800257

Henao-Mejia J, He JJ (2009) Sam68 relocalization into stress granules in response to oxidative stress through complexing with TIA-1. Exp Cell Res. 315(19):3381–3395

Higa M, Kita A, Hagihara K et al (2015) Spatial control of calcineurin in response to heat shock in fission yeast. Genes Cells. 20(2):95–107

Hilliker A, Gao Z, Jankowsky E et al (2011) The DEAD-box protein Ded1 modulates translation by the formation and resolution of an eIF4F-mRNA complex. Mol Cell 43(6):962–972

Hofmann S, Kedersha N, Anderson P, Ivanov P (2021) Molecular mechanisms of stress granule assembly and disassembly. Biochim Biophys Acta 1868:118876. https://doi.org/10.1016/j.bbamcr.2020.118876

Hohmann S (2002) Osmotic stress signaling and osmoadaptation in yeasts. Microbiol Mol Biol Rev 66(2):300–372

Hoyer KK, Herling M, Bagrintseva K (2005) T cell leukemia-1 modulates TCR signal strength and IFN-gamma levels through phosphatidylinositol 3-kinase and protein kinase C pathway activation. J Immunol. 175(2):864–873
Hua Y, Zhou J (2004a) Rpp 20 interacts with SMN and is re-distributed into SMN granules in response to stress. Biochem Biophys Res Commun 314(1):268–276
Hua Y, Zhou J (2004b) Survival motor neuron protein facilitates assembly of stress granules. FEBS Lett 572(1-3):69–74
Huang C, Chen Y, Dai H et al (2020) UBAP2L arginine methylation by PRMT1 modulates stress granule assembly. Cell Death Differ 27:227–241. https://doi.org/10.1038/s41418-019-0350-5
Huynh MM, Jayanthan A, Pambid MR et al (2020) RSK2: a promising therapeutic target for the treatment of triple-negative breast cancer. Expert Opin Ther Targets. 24(1):1–5
Iannilli F, Zalfa F, Gartner A et al (2013) Cytoplasmic TERT associates to RNA granules in fully mature neurons: role in the translational control of the cell cycle inhibitor p15INK4B. PLoS One. 8(6):e66602. https://doi.org/10.1371/journal.pone.0066602
Igelmann S, Lessard F, Ferbeyre G (2022) Liquid–liquid phase separation in cancer signaling, metabolism and anticancer therapy. Cancers 14:1830. https://doi.org/10.3390/cancers14071830
Isakov N (2018) Protein kinase C (PKC) isoforms in cancer, tumor promotion and tumor suppression. Semin Cancer Biol 48:36–52
Jacinto E, Lorberg A (2008) TOR regulation of AGC kinases in yeast and mammals. Biochem J. 410(1):19–37
Jain S, Wheeler JR, Walters RW et al (2016) ATPase-modulated stress granules contain a diverse proteome and substructure. Cell. 164(3):487–498
Jedrusik-Bode M, Studencka M, Smolka C et al (2013) The sirtuin SIRT6 regulates stress granule formation in C. elegans and mammals. J Cell Sci. 126(22):5166–5177
Jiang HY, Wek RC (2005) Phosphorylation of the alpha-subunit of the eukaryotic initiation factor-2 (eIF2alpha) reduces protein synthesis and enhances apoptosis in response to proteasome inhibition. J Biol Chem. 280(14):14189–14202
Johnson DA, Akamine P, Radzio-Andzelm E et al (2001) Dynamics of cAMP-dependent protein kinase. Chem Rev 101(8):2243–2270
Johnson ME, Grassetti AV, Taroni JN et al (2016) Stress granules and RNA processing bodies are novel autoantibody targets in systemic sclerosis. Arthritis Res Ther 18:27. https://doi.org/10.1186/s13075-016-0914-4
Jung HY, Fattet L, Tsai JH et al (2019) Apical-basal polarity inhibits epithelial-mesenchymal transition and tumour metastasis by PAR-complex-mediated SNAI1 degradation. Nat Cell Biol 21(3):359–371
Kaehler C, Isensee J, Hucho T et al (2014) 5-Fluorouracil affects assembly of stress granules based on RNA incorporation. Nucleic Acids Res. 42(10):6436–6447
Kaleli HN, Ozer E, Kaya VO et al (2020) Protein kinase C isozymes and autophagy during neurodegenerative disease progression. Cells 9(3):553. https://doi.org/10.3390/cells9030553
Kanda Y, Satoh R, Matsumoto S et al (2016) Skb5, an SH3 adaptor protein, regulates Pmk1 MAPK signaling by controlling the intracellular localization of the MAPKKK Mkh1. J Cell Sci 129(16):3189–3202
Kanda Y, Satoh R, Takasaki T et al (2021) Sequestration of the PKC ortholog Pck2 in stress granules as a feedback mechanism of MAPK signaling in fission yeast. J Cell Sci 134(2): jcs250191. https://doi.org/10.1242/jcs.250191
Kato M, Han TW, Xie S et al (2012) Cell-free formation of RNA granules: low complexity sequence domains form dynamic fibers within hydrogels. Cell 149(4):753–767. https://doi.org/10.1016/j.cell.2012.04.017
Kawahara H, Imai T, Imataka H (2008) Neural RNA-binding protein Musashi1 inhibits translation initiation by competing with eIF4G for PABP. J Cell Biol 181(4):639–653
Kedersha N, Anderson P (2002) Stress granules: sites of mRNA triage that regulate mRNA stability and translatability. Biochem Soc Trans. 30(6):963–969

Kedersha N, Anderson P (2007) Mammalian stress granules and processing bodies. Methods Enzymol 431:61–81. https://doi.org/10.1016/S0076-6879(07)31005-7
Kedersha N, Anderson P (2009) Regulation of translation by stress granules and processing bodies. Prog Mol Biol Transl Sci 90:155–185
Kedersha NL, Gupta M, Li W et al (1999) RNA-binding proteins TIA-1 and TIAR link the phosphorylation of eIF-2 alpha to the assembly of mammalian stress granules. J Cell Biol. 147(7):1431–1442
Kedersha N, Chen S, Gilks N et al (2002) Evidence that ternary complex (eIF2-GTP-tRNA(i)(Met))-deficient preinitiation complexes are core constituents of mammalian stress granules. Mol Biol Cell. 13(1):195–210
Kedersha N, Stoecklin G, Ayodele M et al (2005) Stress granules and processing bodies are dynamically linked sites of mRNP remodeling. J Cell Biol 169(6):871–884
Kedersha N, Ivanov P, Anderson P (2013) Stress granules and cell signaling: more than just a passing phase? Trends Biochem Sci 38(10):494–506. https://doi.org/10.1016/j.tibs.2013.07.004
Kim EK, Choi EJ (2010) Pathological roles of MAPK signaling pathways in human diseases. Biochim Biophys Acta 1802(4):396–405
Kim DH, Sarbassov DD, Ali SM et al (2002) mTOR interacts with raptor to form a nutrient-sensitive complex that signals to the cell growth machinery. Cell 110(2):163–175
Kim DH, Sarbassov DD, Ali SM et al (2003) GbetaL, a positive regulator of the rapamycin-sensitive pathway required for the nutrient-sensitive interaction between raptor and mTOR. Mol Cell 11(4):895–904
Kim WJ, Back SH, Kim V et al (2005) Sequestration of TRAF2 into stress granules interrupts tumor necrosis factor signaling under stress conditions. Mol Cell Biol. 25(6):2450–2462
Kim SH, Dong WK, Weiler IJ et al (2006) Fragile X mental retardation protein shifts between polyribosomes and stress granules after neuronal injury by arsenite stress or in vivo hippocampal electrode insertion. J Neurosci 26(9):2413–2418
Kim WJ, Kim JH, Jang SK (2007) Anti-inflammatory lipid mediator 15d-PGJ2 inhibits translation through inactivation of eIF4A. EMBO J 26:5020–5032. https://doi.org/10.1038/sj.emboj.7601920
Kim JA, Jayabalan AK, Kothandan VK et al (2016) Identification of Neuregulin-2 as a novel stress granule component. BMB Rep. 49(8):449–454
Kobayashi T, Winslow S, Sunesson L et al (2012) PKCα binds G3BP2 and regulates stress granule formation following cellular stress. PLoS One 7(4):e35820. https://doi.org/10.1371/journal.pone.0035820
Kohno M, Pouyssegur J (2006) Targeting the ERK signaling pathway in cancer therapy. Ann Med. 38(3):200–211
Krisenko MO, Higgins RL, Ghosh S et al (2015) Syk is recruited to stress granules and promotes their clearance through autophagy. J Biol Chem. 290(46):27803–27815
Kwon S, Zhang Y, Matthias P (2007) The deacetylase HDAC6 is a novel critical component of stress granules involved in the stress response. Genes Dev. 21(24):3381–3394
Lai MC, Lee YH, Tarn WY (2008) The DEAD-box RNA helicase DDX3 associates with export messenger ribonucleoproteins as well as tip-associated protein and participates in translational control. Mol Biol Cell 19(9):3847–3858
Laplante M, Sabatini DM (2012) mTOR signaling in growth control and disease. Cell 149(2):274–293
Lee CH, Inoki K, Karbowniczek M (2007) Constitutive mTOR activation in TSC mutants sensitizes cells to energy starvation and genomic damage via p53. EMBO J 26(23):4812–4823
Leung AK, Calabrese JM, Sharp PA (2006) Quantitative analysis of Argonaute protein reveals microRNA-dependent localization to stress granules. Proc Natl Acad Sci USA 103(48):18125–18130
Leung AK, Vyas S, Rood JE et al (2011) Poly(ADP-ribose) regulates stress responses and microRNA activity in the cytoplasm. Mol Cell. 42(4):489–499

Li Z, Wang N, Fang J et al (2012) Role of PKC-ERK signaling in tamoxifen-induced apoptosis and tamoxifen resistance in human breast cancer cells. Oncol Rep 27(6):1879–1886

Li Y, Wang J, Zhong S et al (2020) Overexpression of G3BP1 facilitates the progression of colon cancer by activating β-catenin signaling. Mol Med Rep 22:4403–4411. https://doi.org/10.3892/mmr.2020.11527

Li M, Tang Y, Zuo X et al (2022) Loss of Ras GTPase-activating protein SH3 domain-binding protein 1 (G3BP1) inhibits the progression of ovarian cancer in coordination with ubiquitin-specific protease 10 (USP10). Bioengineered 13:721–734. https://doi.org/10.1080/21655979.2021.2012624

Lin Y, Protter DS, Rosen MK et al (2015) Formation and maturation of phase-separated liquid droplets by RNA-binding proteins. Mol Cell 60(2):208–219. https://doi.org/10.1016/j.molcel.2015.08.018

Lin L, Li X, Pan C et al (2019) ATXN2L upregulated by epidermal growth factor promotes gastric cancer cell invasiveness and oxaliplatin resistance. Cell Death Dis 10:173. https://doi.org/10.1038/s41419-019-1362-2

Liu GY, Sabatini DM (2020) mTOR at the nexus of nutrition, growth, ageing and disease. Nat Rev Mol Cell Biol 21(4):183–203

Liu Z, Yang Y, Gu A et al (2020) Par complex cluster formation mediated by phase separation. Nat Commun. 11(1):2266. https://doi.org/10.1038/s41467-020-16135-6

Liu S, Chen L, Chen H et al (2021) Circ_0119872 promotes uveal melanoma development by regulating the miR-622/G3BP1 axis and downstream signalling pathways. J Exp Clin Canc Res 40:66. https://doi.org/10.1186/s13046-021-01833-w

Loschi M, Leishman CC, Berardone N et al (2009) Dynein and kinesin regulate stress-granule and P-body dynamics. J Cell Sci. 122(21):3973–3982

Lotan R, Bar-On VG, Harel-Sharvit L et al (2005) The RNA polymerase II subunit Rpb4p mediates decay of a specific class of mRNAs. Genes Dev 19(24):3004–3016

Low WK, Dang Y, Schneider-Poetsch T et al (2005) Inhibition of eukaryotic translation initiation by the marine natural product pateamine A. Mol Cell 20(5):709–722

Lu J, Qian J, Xu Z et al (2021) Emerging roles of liquid–liquid phase separation in cancer: from protein aggregation to immune-associated signaling. Front Cell Dev Biol 9:631486. https://doi.org/10.3389/fcell.2021.631486

Luo G, Costanzo M, Boone C et al (2011) Nutrients and the Pkh1/2 and Pkc1 protein kinases control mRNA decay and P-body assembly in yeast. J Biol Chem 286(11):8759–8770

Luo Y, Na Z, Slavoff SA (2018) P-bodies: composition, properties, and functions. Biochemistry 57(17):2424–2431. https://doi.org/10.1021/acs.biochem.7b01162

Izquierdo JM, Alcalde J, Carrascoso I et al (2011) Knockdown of T-cell intracellular antigens triggers cell proliferation, invasion and tumour growth. Biochem J 435(2):337–344. https://doi.org/10.1042/BJ20101030

Madrid M, Vázquez-Marín B, Soto T (2017) Differential functional regulation of protein kinase C (PKC) orthologs in fission yeast. J Biol Chem 292(27):11374–11387

Magliozzi JO, Moseley JB (2021) Pak1 kinase controls cell shape through ribonucleoprotein granules. Elife 10:e67648. https://doi.org/10.7554/eLife.67648

Mahboubi H, Stochaj U (2014) Nucleoli and stress granules: connecting distant relatives. Traffic 15(10):1179–1193. https://doi.org/10.1111/tra.12191

Mahboubi H, Stochaj U (2017) Cytoplasmic stress granules: dynamic modulators of cell signaling and disease. Biochim Biophys Acta Mol Basis Dis 1863(4):884–895

Mahboubi H, Barise R, Stochaj U (2015) 5′-AMP-activated protein kinase alpha regulates stress granule biogenesis. Biochim Biophys Acta. 1853(7):1725–1737

Marin TM, Keith K, Davies B et al (2011) Rapamycin reverses hypertrophic cardiomyopathy in a mouse model of LEOPARD syndrome-associated PTPN11 mutation. J Clin Invest 121(3):1026–1043

Markmiller S, Soltanieh S, Server KL et al (2018) Context-dependent and disease-specific diversity in protein interactions within stress granules. Cell. 172(3):590–604

Marmor-Kollet H, Siany A, Kedersha N et al (2020) Spatiotemporal proteomic analysis of stress granule disassembly using APEX reveals regulation by SUMOylation and links to ALS pathogenesis. Mol Cell. 80(5):876–891
Marrocco V, Bogomolovas J, Ehler E et al (2019) PKC and PKN in heart disease. J Mol Cell Cardiol 128:212–226
Martins I, Kepp O, Schlemmer F et al (2011) Restoration of the immunogenicity of cisplatin-induced cancer cell death by endoplasmic reticulum stress. Oncogene. 30(10):1147–1158
Mazroui R, Huot ME, Tremblay S (2002) Trapping of messenger RNA by Fragile X Mental Retardation protein into cytoplasmic granules induces translation repression. Hum Mol Genet 11(24):3007–3017
Mediani L, Antoniani F, Galli V et al (2021) Hsp90-mediated regulation of DYRK3 couples stress granule disassembly and growth via mTORC1 signaling. EMBO Rep 22(5):e51740. https://doi.org/10.15252/embr.202051740
Meyerowitz J, Parker SJ, Vella LJ et al (2011) C-Jun N-terminal kinase controls TDP-43 accumulation in stress granules induced by oxidative stress. Mol Neurodegener. 6:57. https://doi.org/10.1186/1750-1326-6-57
Min L, Ruan Y, Shen Z et al (2015) Overexpression of Ras-GTPase-activating protein SH3 domain-binding protein 1 correlates with poor prognosis in gastric cancer patients. Histopathology 67: 677–688. https://doi.org/10.1111/his.12695
Molliex A, Temirov J, Lee J et al (2015) Phase separation by low complexity domains promotes stress granule assembly and drives pathological fibrillization. Cell 163(1):123–133. https://doi.org/10.1016/j.cell.2015.09.015
Moujalled D, James JL, Yang S et al (2015) Phosphorylation of hnRNP K by cyclin-dependent kinase 2 controls cytosolic accumulation of TDP-43. Hum Mol Genet. 24(6):1655–1669
Mukhopadhyay C, Yang C, Xu L et al (2021) G3BP1 inhibits Cul3SPOP to amplify AR signaling and promote prostate cancer. Nat Commun 12:6662. https://doi.org/10.1038/s41467-021-27024-x
Narayanan N, Wang Z, Li L et al (2017) Arginine methylation of USP9X promotes its interaction with TDRD3 and its anti-apoptotic activities in breast cancer cells. Cell Discov. 3:16048. https://doi.org/10.1038/celldisc.2016.48
Newton AC, Johnson JE (1998) Protein kinase C: a paradigm for regulation of protein function by two membrane-targeting modules. Biochim Biophys Acta 1376(2):155–172
Nilsson D, Sunnerhagen P (2011) Cellular stress induces cytoplasmic RNA granules in fission yeast. RNA. 17(1):120–133
Noda Y, Tomita H, Ishihara T et al (2022) Prediction of overall survival in patients with pancreatic ductal adenocarcinoma: histogram analysis of ADC value and correlation with pathological intratumoral necrosis. BMC Med Imaging 22:23. https://doi.org/10.1186/s12880-022-00751-3
Nonhoff U, Ralser M, Welzel F et al (2007) Ataxin-2 interacts with the DEAD/H-box RNA helicase DDX6 and interferes with P-bodies and stress granules. Mol Biol Cell 18(4):1385–1396
Nousch M, Reed V, Bryson-Richardson RJ et al (2007) The eIF4G-homolog p97 can activate translation independent of caspase cleavage. RNA 13(3):374–384
Omer A, Patel D, Lian XJ et al (2018) Stress granules counteract senescence by sequestration of PAI-1. EMBO Rep. 19:e44722. https://doi.org/10.15252/embr.201744722
Onishi H, Kino Y, Morita T et al (2008) MBNL1 associates with YB-1 in cytoplasmic stress granules. J Neurosci Res 86(9):1994–2002
Papadopoli D, Boulay K, Kazak L (2019) mTOR as a central regulator of lifespan and aging. F1000Res 8:F1000 Faculty Rev-998. https://doi.org/10.12688/f1000research.17196.1
Pare JM, Tahbaz N, Lopez-Orozco J et al (2009) Hsp90 regulates the function of argonaute 2 and its recruitment to stress granules and P-bodies. Mol Biol Cell. 20(14):3273–3284
Park C, Choi S, Kim YE et al (2017) Stress granules contain Rbfox2 with cell cycle-related mRNAs. Sci Rep. 7:11211. https://doi.org/10.1038/s41598-017-11651-w
Peran I, Mittag T (2020) Molecular structure in biomolecular condensates. Curr Opin Struct Biol 60:17–26. https://doi.org/10.1016/j.sbi.2019.09.007

Pietras P, Aulas A, Fay MM et al (2022) Translation inhibition and suppression of stress granules formation by cisplatin. Biomed Pharmacother 145:112382. https://doi.org/10.1016/j.biopha.2021.112382
Pintus G, Tadolini B, Posadino AM et al (2003) PKC/Raf/MEK/ERK signaling pathway modulates native-LDL-induced E2F-1 gene expression and endothelial cell proliferation. Cardiovasc Res 59(4):934–944
Quaresma AJ, Bressan GC, Gava LM (2009) Human hnRNP Q re-localizes to cytoplasmic granules upon PMA, thapsigargin, arsenite and heat-shock treatments. Exp Cell Res 315(6):968–980
Quiroz FG, Fiore VF, Levorse J et al (2020) Liquid-liquid phase separation drives skin barrier formation. Science 367:eaax9554. https://doi.org/10.1126/science.aax9554
Reineke LC, Tsai WC, Jain A et al (2017) Casein kinase 2 is linked to stress granule dynamics through phosphorylation of the stress granule nucleating protein G3BP1. Mol Cell Biol. 37: e00596. https://doi.org/10.1128/MCB.00596-16
Ren J, Zhang Z, Zong Z et al (2022) Emerging implications of phase separation in cancer. Adv Sci 9:2202855. https://doi.org/10.1002/advs.202202855
Rothe F, Gueydan C, Bellefroid E et al (2006) Identification of FUSE-binding proteins as interacting partners of TIA proteins. Biochem Biophys Res Commun 343(1):57–68
Roux PP, Blenis J (2004) ERK and p38 MAPK-activated protein kinases: a family of protein kinases with diverse biological functions. Microbiol Mol Biol Rev. 68(2):320–344
Rubinfeld H, Seger R (2005) The ERK cascade: a prototype of MAPK signaling. Mol Biotechnol. 31(2):151–174
Sanchez-Vega F, Mina M, Armenia J (2018) Oncogenic signaling pathways in the cancer genome atlas. Cell 173(2):321–337
Santarpia L, Lippman SM, El-Naggar AK (2012) Targeting the MAPK-RAS-RAF signaling pathway in cancer therapy. Expert Opin Ther Targets 16(1):103–119
Sassone-Corsi P (2012) The cyclic AMP pathway. Cold Spring Harb Perspect Biol 4(12):a011148. https://doi.org/10.1101/cshperspect.a011148
Satoh R, Morita T, Takada H et al (2009) Role of the RNA-binding protein Nrd1 and Pmk1 mitogen-activated protein kinase in the regulation of myosin mRNA stability in fission yeast. Mol Biol Cell 20(9):2473–2485
Satoh R, Tanaka A, Kita A et al (2012) Role of the RNA-binding protein Nrd1 in stress granule formation and its implication in the stress response in fission yeast. PLoS One. 7(1):e29683. https://doi.org/10.1371/journal.pone.0029683
Satoh R, Hara N, Kawasaki A et al (2018) Distinct modes of stress granule assembly mediated by the KH-type RNA-binding protein Rnc1. Genes Cells 23(9):778–785
Saxton RA, Sabatini DM (2017) mTOR signaling in growth, metabolism, and disease. Cell 168(6): 960–976
Scadden AD (2007) Inosine-containing dsRNA binds a stress-granule-like complex and downregulates gene expression in trans. Mol Cell 28(3):491–500
Schwed-Gross A, Hamiel H, Faber GP et al (2022) Glucocorticoids enhance chemotherapy-driven stress granule assembly and impair granule dynamics, leading to cell death. J Cell Sci. 135: jcs259629. https://doi.org/10.1242/jcs.259629
Sfakianos AP, Mellor LE, Pang YF et al (2018) The mTOR-S6 kinase pathway promotes stress granule assembly. Cell Death Differ. 25(10):1766–1780
Shah OJ, Wang Z, Hunter T (2004) Inappropriate activation of the TSC/Rheb/mTOR/S6K cassette induces IRS1/2 depletion, insulin resistance, and cell survival deficiencies. Curr Biol 14(18): 1650–1656
Shah KH, Nostramo R, Zhang B et al (2014) Protein kinases are associated with multiple, distinct cytoplasmic granules in quiescent yeast cells. Genetics. 198(4):1495–1512
Shi X, Si X, Zhang E et al (2021) Paclitaxel-induced stress granules increase LINE-1 mRNA stability to promote drug resistance in breast cancer cells. J Biomed Res. 35(6):411–424

Shigunov P, Sotelo-Silveira J, Stimamiglio MA et al (2014) Ribonomic analysis of human DZIP1 reveals its involvement in ribonucleoprotein complexes and stress granules. BMC Mol Biol. 15: 12. https://doi.org/10.1186/1471-2199-15-12

Shiraishi K, Hioki T, Habata A et al (2018) Yeast Hog1 proteins are sequestered in stress granules during high-temperature stress. J Cell Sci 131(1):jcs209114. https://doi.org/10.1242/jcs.209114

Shively CA, Kweon HK, Norman KL et al (2015) Large-scale analysis of kinase signaling in yeast pseudohyphal development identifies regulation of ribonucleoprotein granules. PLoS Genet. 11: e1005564. https://doi.org/10.1371/journal.pgen.1005564

Siegmund D, Wagner J, Wajant H (2022) TNF receptor associated factor 2 (TRAF2) signaling in cancer. Cancers. 14:4055. https://doi.org/10.3390/cancers14164055

Sim E, Irollo E, Grabocka E (2019) Pancreatic cancer, methods and protocols. Methods Mol Biol 1882:183–195. https://doi.org/10.1007/978-1-4939-8879-2_17

Snead WT, Gladfelter AS (2019) The control centers of biomolecular phase separation: how membrane surfaces, PTMs, and active processes regulate condensation. Mol Cell 76(2): 295–305

Solomon S, Xu Y, Wang B et al (2007) Distinct structural features of caprin-1 mediate its interaction with G3BP-1 and its induction of phosphorylation of eukaryotic translation initiation factor 2alpha, entry to cytoplasmic stress granules, and selective interaction with a subset of mRNAs. Mol Cell Biol 27(6):2324–2342

Somasekharan SP, El-Naggar A, Leprivier G et al (2015) YB-1 regulates stress granule formation and tumor progression by translationally activating G3BP1. J Cell Biol 208:913–929. https://doi.org/10.1083/jcb.201411047

Song MS, Grabocka E (2023) Stress granules in cancer. Rev Physiol Biochem Pharmacol. 185:25–52

Sosa V, Moliné T, Somoza R (2013) Oxidative stress and cancer: an overview. Ageing Res Rev 12(1):376–390

Stöhr N, Lederer M, Reinke C et al (2006) ZBP1 regulates mRNA stability during cellular stress. J Cell Biol 175(4):527–534. https://doi.org/10.1083/jcb.200608071

Stoecklin G, Stubbs T, Kedersha N et al (2004) MK2-induced tristetraprolin: 14-3-3 complexes prevent stress granule association and ARE-mRNA decay. EMBO J 23(6):1313–1324

Stohr N, Lederer M, Reinke C et al (2006) ZBP1 regulates mRNA stability during cellular stress. J Cell Biol. 175(4):527–534

Stratford AL, Fry CJ, Desilets C et al (2008) Y-box binding protein-1 serine 102 is a downstream target of p90 ribosomal S6 kinase in basal-like breast cancer cells. Breast Cancer Res. 10:R99. https://doi.org/10.1186/bcr2202

Sugiura R, Toda T, Shuntoh H et al (1998) *pmp1*$^+$, a suppressor of calcineurin deficiency, encodes a novel MAP kinase phosphatase in fission yeast. EMBO J 17(1):140–148

Sugiura R, Kita A, Shimizu Y et al (2003) Feedback regulation of MAPK signalling by an RNA-binding protein. Nature 424(6951):961–965

Sugiura R, Satoh R, Takasaki T (2021) ERK: a double-edged sword in cancer. ERK-dependent apoptosis as a potential therapeutic strategy for cancer. Cells 10(10):2509. https://doi.org/10.3390/cells10102509

Sunaga N, Kaira K, Imai H (2013) Oncogenic KRAS-induced epiregulin overexpression contributes to aggressive phenotype and is a promising therapeutic target in non-small-cell lung cancer. Oncogene 32(34):4034–4042

Szaflarski W, Fay MM, Kedersha N et al (2016) Vinca alkaloid drugs promote stress-induced translational repression and stress granule formation. Oncotarget. 7(21):30307–30322

Takada H, Nishimura M, Asayama Y et al (2007) Atf1 is a target of the mitogen-activated protein kinase Pmk1 and regulates cell integrity in fission yeast. Mol Biol Cell 18(12):4794–4802

Takahara T, Maeda T (2012) Transient sequestration of TORC1 into stress granules during heat stress. Mol Cell 47(2):242–252

Takahashi A, Tsutsumi R, Kikuchi I et al (2011) SHP2 tyrosine phosphatase converts parafibromin/Cdc73 from a tumor suppressor to an oncogenic driver. Mol Cell 43(1):45–56

Takasaki T, Hagihara K, Satoh R et al (2018) More than just an immunosuppressant: the emerging role of FTY720 as a novel inducer of ROS and apoptosis. Oxid Med Cell Longev 2018: 4397159. https://doi.org/10.1155/2018/4397159

Takayama K, Suzuki T, Fujimura T et al (2018) Association of USP10 with G3BP2 inhibits p53 signaling and contributes to poor outcome in prostate cancer. Mol Cancer Res 16:846. https://doi.org/10.1158/1541-7786.mcr-17-0471

Takekawa M, Posas F, Saito H (1997) A human homolog of the yeast Ssk2/Ssk22 MAP kinase kinase kinases, MTK1, mediates stress-induced activation of the p38 and JNK pathways. EMBO J. 16(16):4973–4982

Taniuchi K, Nishimori I, Hollingsworth MA (2011) The N-terminal domain of G3BP enhances cell motility and invasion by posttranscriptional regulation of BART. Mol Cancer Res 9:856–866. https://doi.org/10.1158/1541-7786.mcr-10-0574

Tao S, Wang S, Moghaddam SJ et al (2014) Oncogenic KRAS confers chemoresistance by upregulating NRF2. Cancer Res 74:7430–7441. https://doi.org/10.1158/0008-5472.can-14-1439

Tartaglia M, Gelb BD (2005) Noonan syndrome and related disorders: genetics and pathogenesis. Annu Rev Genomics Hum Genet 6:45–68

Tartaglia M, Niemeyer CM, Fragale A et al (2003) Somatic mutations in PTPN11 in juvenile myelomonocytic leukemia, myelodysplastic syndromes and acute myeloid leukemia. Nat Genet 34(2):148–150

Taylor SS, Zhang P, Steichen JM et al (2013) PKA: lessons learned after twenty years. Biochim Biophys Acta 1834(7):1271–1278

Tenner B, Zhang JZ, Kwon Y et al (2021) FluoSTEPs: fluorescent biosensors for monitoring compartmentalized signaling within endogenous microdomains. Sci Adv 7:eabe4091. https://doi.org/10.1126/sciadv.abe4091

Thedieck K, Holzwarth B, Prentzell MT et al (2013) Inhibition of mTORC1 by astrin and stress granules prevents apoptosis in cancer cells. Cell 154(4):859–874

Thomas MG, Tosar LJ, Loschi M et al (2005) Staufen recruitment into stress granules does not affect early mRNA transport in oligodendrocytes. Mol Biol Cell 16(1):405–420

Tong X, Tang R, Xu J et al (2022) Liquid–liquid phase separation in tumor biology. Signal Transduct Target Ther 7:221. https://doi.org/10.1038/s41392-022-01076-x

Tourriere H, Chebli K, Zekri L et al (2003) The RasGAP associated endoribonuclease G3BP assembles stress granules. J Cell Biol 160(6):823–831

Treeck BV, Protter DSW, Matheny T et al (2018) RNA self-assembly contributes to stress granule formation and defining the stress granule transcriptome. Proc National Acad Sci 115:2734–2739. https://doi.org/10.1073/pnas.1800038115

Tsai NP, Wei LN (2010) RhoA/ROCK1 signaling regulates stress granule formation and apoptosis. Cell Signal. 22(4):668–675

Tsai NP, Ho PC, Wei LN (2008) Regulation of stress granule dynamics by Grb7 and FAK signalling pathway. EMBO J. 27(5):715–726

Tsai NP, Tsui YC, Wei LN (2009) Dynein motor contributes to stress granule dynamics in primary neurons. Neuroscience. 159(2):647–656

Tudisca V, Simpson C, Castelli L et al (2012) PKA isoforms coordinate mRNA fate during nutrient starvation. J Cell Sci. 125(21):5221–5232

Turakhiya A, Meyer SR, Marincola G et al (2018) ZFAND1 recruits p97 and the 26S proteasome to promote the clearance of arsenite-induced stress granules. Mol Cell. 70(5):906–919

Uversky VN (2017) Intrinsically disordered proteins in overcrowded milieu: membrane-less organelles, phase separation, and intrinsic disorder. Curr Opin Struct Biol 44:18–30. https://doi.org/10.1016/j.sbi.2016.10.015

Välk K, Vooder T, Kolde R et al (2010) Metspalu, gene expression profiles of non-small cell lung cancer: survival prediction and new biomarkers. Oncology 79(3–4):283–292

Vessey JP, Vaccani A, Xie Y et al (2006) Dendritic localization of the translational repressor Pumilio 2 and its contribution to dendritic stress granules. J Neurosci 26(24):6496–6508

Vilas-Boas FA, da Silva AM, de Sousa LP et al (2016) Impairment of stress granule assembly via inhibition of the eIF2alpha phosphorylation sensitizes glioma cells to chemotherapeutic agents. J Neurooncol. 127(2):253–260

Wallace EW, Kear-Scott JL, Pilipenko EV et al (2015) Reversible, specific, active aggregates of endogenous proteins assemble upon heat stress. Cell. 162(6):1286–1298

Wang Y, Fu D, Chen Y et al (2018) G3BP1 promotes tumor progression and metastasis through IL-6/G3BP1/STAT3 signaling axis in renal cell carcinomas. Cell Death Dis 9:501. https://doi.org/10.1038/s41419-018-0504-2

Wang B, Maxwell BA, Joo JH et al (2019a) ULK1 and ULK2 regulate stress granule disassembly through phosphorylation and activation of VCP/p97. Mol Cell. 74(4):742–757

Wang S, Kwon SH, Su Y, Dong Z (2019b) Stress granules are formed in renal proximal tubular cells during metabolic stress and ischemic injury for cell survival. Am J Physiol Renal Physiol. 317(1):F116–F123

Wang D, Ao J, Xiong Y et al (2022) Systematic analysis of stress granule regulators-associated molecular subtypes predicts drug response, immune response, and prognosis in non-small cell lung cancer. Front Cell Dev Biol 10:868918. https://doi.org/10.3389/fcell.2022.868918

Wang N, Li T, Liu W et al (2023) USP7- and PRMT5-dependent G3BP2 stabilization drives de novo lipogenesis and tumorigenesis of HNSC. Cell Death Dis 14:182. https://doi.org/10.1038/s41419-023-05706-2

Wasserman T, Katsenelson K, Daniliuc S et al (2010) A novel c-Jun N-terminal kinase (JNK)-binding protein WDR62 is recruited to stress granules and mediates a nonclassical JNK activation. Mol Biol Cell. 21(1):117–130

Wiedner HJ, Giudice J (2021) It's not just a phase: function and characteristics of RNA-binding proteins in phase separation. Nat Struct Mol Biol 28:465–473. https://doi.org/10.1038/s41594-021-00601-w

Wilczynska A, Aigueperse C, Kress M et al (2005) The translational regulator CPEB1 provides a link between dcp1 bodies and stress granules. J Cell Sci 118(5):981–992

Winslow S, Leandersson K, Larsson C (2013) Regulation of PMP22 mRNA by G3BP1 affects cell proliferation in breast cancer cells. Mol Cancer 12:156. https://doi.org/10.1186/1476-4598-12-156

Wippich F, Bodenmiller B, Trajkovska MG et al (2013) Dual specificity kinase DYRK3 couples stress granule condensation/dissolution to mTORC1 signaling. Cell 152(4):791–805

Wolfson RL, Sabatini DM (2017) The dawn of the age of amino acid sensors for the mTORC1 pathway. Cell Metab 26(2):301–309

Wolozin B, Ivanov P (2019) Stress granules and neurodegeneration. Nat Rev Neurosci 20(11): 649–666

Wu HZ, Li LY, Jiang SL et al (2022) RSK2 promotes melanoma cell proliferation and vemurafenib resistance via upregulating cyclin D1. Front Pharmacol. 13:950571. https://doi.org/10.3389/fphar.2022.950571

Xie X, Matsumoto S, Endo A et al (2018) Deubiquitylases USP5 and USP13 are recruited to and regulate heat-induced stress granules through their deubiquitylating activities. J Cell Sci. 131: jcs210856. https://doi.org/10.1242/jcs.210856

Xie Q, Cheng J, Mei W et al (2023) Phase separation in cancer at a glance. J Transl Med 21:237. https://doi.org/10.1186/s12967-023-04082-x

Xiong R, Gao J, Yin T (2019) G3BP1 activates the TGF-β/Smad signaling pathway to promote gastric cancer. Oncotargets Ther 12:7149–7156. https://doi.org/10.2147/ott.s213728

Yang WH, Bloch DB (2007) Probing the mRNA processing body using protein macroarrays and "autoantigenomics". RNA. 13(5):704–712

Yang S, Liu G (2017) Targeting the Ras/Raf/MEK/ERK pathway in hepatocellular carcinoma. Oncol Lett. 13(3):1041–1047

Yang F, Peng Y, Murray EL (2006) Polysome-bound endonuclease PMR1 is targeted to stress granules via stress-specific binding to TIA-1. Mol Cell Biol 26(23):8803–8813

Yang Q, Huo S, Sui Y et al (2018) Mutation status and immunohistochemical correlation of KRAS, NRAS, and BRAF in 260 Chinese colorectal and gastric cancers. Front Oncol 8:487. https://doi.org/10.3389/fonc.2018.00487

Yang P, Mathieu C, Kolaitis RM et al (2020) G3BP1 is a tunable switch that triggers phase separation to assemble stress granules. Cell. 181(2):325–345

Youn JY, Dunham WH, Hong SJ et al (2018) High-density proximity mapping reveals the subcellular organization of mRNA-associated granules and bodies. Mol Cell. 69(3):517–532

Yu C, York B, Wang S et al (2007) An essential function of the SRC-3 coactivator in suppression of cytokine mRNA translation and inflammatory response. Mol Cell. 25(5):765–778

Zanconato F, Cordenonsi M, Piccolo S (2016) YAP/TAZ at the roots of cancer. Cancer Cell 29: 783–803. https://doi.org/10.1016/j.ccell.2016.05.005

Zhang W, Liu HT (2002) MAPK signal pathways in the regulation of cell proliferation in mammalian cells. Cell Res. 12(1):9–18

Zhang HZ, Liu JG, Wei YP et al (2007) Expression of G3BP and RhoC in esophageal squamous carcinoma and their effect on prognosis. World J Gastroentero 13:4126. https://doi.org/10.3748/wjg.v13.i30.4126

Zhang K, Daigle JG, Cunningham KM et al (2018) Stress granule assembly disrupts nucleocytoplasmic transport. Cell. 173(4):958–971

Zhang H, Zhang S, He H et al (2012) GAP161 targets and downregulates G3BP to suppress cell growth and potentiate cisplaitin-mediated cytotoxicity to colon carcinoma HCT116 cells. Cancer Sci 103(10):1848–1856. https://doi.org/10.1111/j.1349-7006.2012.02361.x

Zhang L, Zhao L, Yan X, Huang Y (2019) Loss of G3BP1 suppresses proliferation, migration, and invasion of esophageal cancer cells via Wnt/β-catenin and PI3K/AKT signaling pathways. J Cell Physiol 234:20469–20484. https://doi.org/10.1002/jcp.28648

Zhang JZ, Lu TW, Stolerman LM et al (2020a) Phase separation of a PKA regulatory subunit controls cAMP compartmentation and oncogenic signaling. Cell 182(6):1531–1544

Zhang X, Wang F, Hu Y et al (2020b) In vivo stress granule misprocessing evidenced in a FUS knock-in ALS mouse model. Brain 143:1350–1367. https://doi.org/10.1093/brain/awaa076

Zhao J, Fu X, Chen H et al (2021) G3BP1 interacts with YWHAZ to regulate chemoresistance and predict adjuvant chemotherapy benefit in gastric cancer. Br J Cancer. 124(2):425–436

Zheng H, Zhan Y, Zhang Y et al (2019) Elevated expression of G3BP1 associates with YB1 and p-AKT and predicts poor prognosis in nonsmall cell lung cancer patients after surgical resection. Cancer Med 8:6894–6903. https://doi.org/10.1002/cam4.2579

Zheng Y, Wu J, Deng R et al (2022) G3BP2 regulated by the lncRNA LINC01554 facilitates esophageal squamous cell carcinoma metastasis through stabilizing HDGF transcript. Oncogene 41:515–526. https://doi.org/10.1038/s41388-021-02073-0

Zhu M, Kuechler ER, Zhang J et al (2020a) Proteomic analysis reveals the direct recruitment of intrinsically disordered regions to stress granules in S. cerevisiae. J Cell Sci. 133:jcs244657. https://doi.org/10.1242/jcs.244657

Zhu G, Xie J, Kong W et al (2020b) Phase separation of disease-associated SHP2 mutants underlies MAPK hyperactivation. Cell 183(2):490–502

Chapter 13
The Multifaceted Regulation of TDP-43 Condensates at the Intersection of Physiology and Pathology: Implications for Neurodegenerative Diseases

Akihiro Sugai, Takuma Yamagishi, Shingo Koide, and Osamu Onodera

Abstract The intricate relationship between physiological and pathological aggregation in neurodegenerative diseases, such as amyotrophic lateral sclerosis (ALS), poses a significant research question. Comprehending the role of liquid–liquid phase separation (LLPS) and the toxicity of TDP-43 aggregation is of critical importance. This necessitates a comprehensive understanding of the multifarious factors involved in LLPS, including the regulation of TDP-43 intrinsically disordered regions (IDRs), isoforms, oligomers, aggregations, and interactions with RNA and other ALS-related proteins. Additionally, investigating the association between nervous system specificity and condensate formation in neurodegenerative diseases is essential. To develop effective therapies for neurodegenerative diseases, a thorough understanding of intracellular condensates and their roles is imperative.

Keywords Neurodegenerative diseases · ALS · TDP-43 · Intrinsically disordered region (IDR) · Alternative splicing · Liquid–liquid phase separation (LLPS)

13.1 Introduction

Neurodegenerative diseases, including amyotrophic lateral sclerosis (ALS), pose a considerable challenge to global health, as they lead to progressive decline in neurological function and ultimately result in death. Despite this, the pathogenesis of these diseases remains elusive, and effective treatments have not yet been established. Consequently, it is crucial to enhance our understanding of the molecular and cellular processes implicated in the pathogenesis of neurodegenerative diseases, particularly ALS.

A. Sugai (✉) · T. Yamagishi · S. Koide · O. Onodera
Department of Neurology, Brain Research Institute, Niigata University, Niigata, Japan
e-mail: akihiro.sugai@bri.niigata-u.ac.jp; tyama@bri.niigata-u.ac.jp; onodera@bri.niigata-u.ac.jp

R. Kurokawa (ed.), *Phase Separation in Living Cells*,
https://doi.org/10.1007/978-981-99-4886-4_13

Recent advancements in cellular biology have emphasized liquid–liquid phase separation (LLPS) as a noteworthy process that generates membraneless organelles or biomolecular condensates within cells, playing a pivotal role in organizing cellular components and modulating various biological processes. This phenomenon has garnered significant interest due to its potential association with the pathogenesis of diverse neurodegenerative diseases (Ryan and Fawzi 2019).

In the context of ALS, multiple proteins linked to the disease, such as TDP-43 and FUS have been demonstrated to induce LLPS, implying a potential connection between biomolecular condensate formation and function, and disease initiation and progression (Han et al. 2012; Kato et al. 2012; Molliex et al. 2015). Moreover, the intracellular condensate network perturbation has been hypothesized to contribute to the impairment of cellular processes and the emergence of neurodegenerative diseases.

This review endeavors to synthesize the existing knowledge regarding LLPS and its involvement in cellular processes, with an emphasis on ALS. Particular attention will be given to TDP-43, a protein emblematic of ALS pathology. In light of the disease's complexity, we will address the multifaceted nature of LLPS formation and its potential ramifications in disease pathogenesis.

13.2 The Complexities of ALS and FTLD: The Interplay of Genetics, Molecular Processes

ALS is characterized by muscle weakness and atrophy due to motor neuron damage and, in some cases, cognitive and behavioral changes. Another disorder, frontotemporal lobar degeneration (FTLD), is characterized by degeneration of the frontal and/or temporal lobes and is typified by alterations in personality, behavior, executive function, and progressive language impairment. There is a substantial overlap between ALS and FTLD, both genetically and pathologically. Disease progression is variable, with some individuals experiencing rapid functional decline, while others undergo a slower deterioration over several years. Epidemiological data indicate that the annual incidence of ALS and FTLD is approximately 2–3 cases per 100,000 individuals, respectively, with age being the greatest risk factor (Brown and Al-Chalabi 2017; Van Mossevelde et al. 2018).

The precise etiology of ALS and FTLD remains elusive, with genetic and environmental factors postulated to contribute: approximately 5–10% of ALS cases present a familial background (Byrne et al. 2011; Brown and Al-Chalabi 2017), while up to 50% of FTLD cases exhibit familial predisposition, albeit with ethnic disparities (Seelaar et al. 2011). Genetic investigations of ALS and FTLD reveal overlapping associations, with several genes implicated in both disorders (Table 13.1); notable genetic variants such as *C9ORF72*, *TARDBP* encoding TDP-43 protein, and *FUS* have been identified as causative factors for both ALS and FTLD, suggesting that the two diseases may share some underlying pathological

Table 13.1 Associations between ALS/FTLD-related genes, TDP-43 pathology, and LLPS

Functional classification	Gene	Phenotype	TDP-43 pathology	IDP	LLPS	Reference
RNA processing	*TARDBP*	ALS/FTD	+	+	+	Neumann et al. (2006), Arai et al. (2006), Molliex et al. (2015)
	FUS	ALS/FTD	–	+	+	Vance et al. (2009), Kato et al. (2012)
	C9ORF72	ALS/FTD	+	–	+	DeJesus-Hernandez et al. (2011), Renton et al. (2011), Hsiung et al. (2012), Mackenzie et al. (2013), Boeynaems et al. (2017)
	HNRNPA1	ALS/FTD, MSP	+	+	+	Kim et al. (2013), Molliex et al. (2015)
Proteostasis	*TBK1*	ALS/FTD	+	–	–	Freischmidt et al. (2015), Pottier et al. (2015)
	UBQLN2	ALS/FTD	+	+	+	Deng et al. (2011), Williams et al. (2012), Dao et al. (2018)
	SQSTM1	ALS/FTD, MSP	+	+	+	Rubino et al. (2012), Teyssou et al. (2013), Faruk et al. (2021)
Intracellular trafficking	*DCTN1*	ALS/FTD, Parkinsonism	+	–	–	Puls et al. (2003), Wider et al. (2009)
	PFN1	ALS/FTD	+	–	–	Wu et al. (2012)
	TUBA4A	ALS/FTD	+	–	–	Smith et al. (2014)
	ANXA11	ALS/FTD	+	+	+	Smith et al. (2017), Liao et al. (2019), Sainouchi et al. (2021)

This table features a selection of representative genes pertinent to the current review's discussion
IDP intrinsically disordered protein, *LLPS* liquid–liquid phase separation, *ALS/FTD* amyotrophic lateral sclerosis/frontotemporal dementia, *DPR* dipeptide repeat, *MSP* multiple system proteinopathy

mechanisms. ALS and FTLD, despite their heterogeneity and diverse pathological processes, can be categorized based on causative gene function, but with overlapping mechanisms (Brown and Al-Chalabi 2017).

Mutations in genes related to RNA metabolism, protein homeostasis, and intracellular trafficking are implicated in the pathogenesis of ALS and FTLD, indicating that these processes are crucial for understanding disease mechanisms (Table 13.1). RNA metabolism-related genes such as *TATDB*P, *FUS*, and *C9ORF72* have been demonstrated to be involved in various aspects of processing, emphasizing the significance of aberrant RNA metabolism in the pathogenesis of these diseases (Neumann et al. 2006; Arai et al. 2006; Vance et al. 2009; DeJesus-Hernandez et al. 2011). Furthermore, protein homeostasis-related genes such as *TBK1*, *UBQLN2*, and *SQSTM1* participate in protein folding and degradation processes, suggesting that disruption of protein homeostasis may be involved in ALS (Deng et al. 2011; Fecto et al. 2011; Freischmidt et al. 2015). Additionally, intracellular transport-related genes like *DCTN1*, *PFN1*, and *TUBA4A* are involved in intracellular transport and quality control (Puls et al. 2003; Wu et al. 2012; Smith et al. 2014). These ALS-associated genes emphasize the crucial role of stringent quality control of RNA and proteins, as well as their intracellular trafficking, in preserving neuronal health—particularly in the case of elongated motor neurons.

Understanding the functional classification of causative genes may offer insights into the underlying pathomechanisms of ALS and FTLD. Of particular interest are RNA metabolism-related proteins such as TDP-43, FUS, and hnRNP A/B. These genes also participate in the regulation of intracellular phase separation and the formation of membraneless organelles (Molliex et al. 2015; Boeynaems et al. 2017). Moreover, the mechanisms by which protein homeostasis-related genes and intracellular transport-related genes contribute to disease may also be involved in the formation of biomolecular condensates (Dao et al. 2018; Liao et al. 2019; Faruk et al. 2021) (Table 13.1).

Presuming that biomolecular condensates have a significant role in the onset and progression of ALS, genes associated with ALS may be central to the intricate and interactive milieu within cellular condensates. Meanwhile, despite the fact that familial ALS arises from various genetic mutations, a considerable proportion of these cases ultimately display TDP-43-related pathological characteristics (Table 13.1). Moreover, aberrations in TDP-43 are observed as pathological hallmarks in the vast majority of sporadic ALS cases (Neumann et al. 2006; Arai et al. 2006). Given these observations, a more comprehensive exploration of the condensates implicated in the pathogenesis of ALS and their association with TDP-43 is crucial.

13.3 Structural Organization of TDP-43 and Their Influence on Neurodegenerative Diseases

TDP-43 is a highly conserved RNA-binding protein implicated in a wide array of RNA processing events, encompassing splicing, stability, and transport. The protein's structure features an N-terminal domain (NTD), two RNA recognition motifs (RRM1 and RRM2), and a glycine-rich C-terminal domain (CTD) (François-Moutal et al. 2019). Due to the presence of nuclear localization signals (NLS) and nuclear export signals (NES), TDP-43 shuttles between the nucleus and cytoplasm, predominantly localized within the nucleus.

The NTD of TDP-43 plays a pivotal role in mediating oligomer formation, a process critical for the protein's biological activity (Afroz et al. 2017; Wang et al. 2018). This reversible TDP-43 oligomerization process is indispensable for the protein's alternative splicing function. Moreover, it has been posited that NTD-mediated binding may diminish the proximity of the C-terminal IDR, thereby impeding pathological aggregation (Afroz et al. 2017). The linker connecting the NTD and the RRM encompasses the NLS, which is recognized by importin-α for the active translocation of TDP-43 to the nucleus. A biochemical hallmark of ALS and FTLD brain tissue is the accumulation of cytoplasmic aggregates composed of TDP-43 C-terminal fragments. This fragmentation disrupts the NLS, eliminates the NTD, and results in the localization of TDP-43 C-terminal fragments within the cytoplasm, where they exhibit pronounced aggregation.

The RNA recognition motifs (RRMs) of TDP-43, specifically RRM1 and RRM2, are vital for nucleic acid binding and substantially impact TDP-43's interactions with RNA (Buratti 2001). Both RRM1 and RRM2 are involved in binding extended stretches of nucleic acids, with RRM1 assuming a predominant role and RRM2 functioning in a supplementary capacity (Kuo et al. 2014). Although RRMs are generally well-folded, certain regions within RRM1 and RRM2 have been pinpointed as amyloidogenic cores, potentially contributing to the nucleation or propagation of TDP-43 aggregation in ALS (Shodai et al. 2013). Furthermore, these regions serve as targets for an array of pathological post-translational modifications (Cohen et al. 2015), emphasizing their significance in the disease process.

The C-terminal domain (CTD) of TDP-43, known as a prion-like domain, harbors causative mutations for ALS and is significantly associated with aggregation. Structurally, the CTD is organized into distinct subdomains, including two Gly-Ser-rich regions (GaroS1/2), a hydrophobic amyloidogenic core (HP), and a Q/N-rich region (Mompeán et al. 2016). Two canonical α-helices reside in the HP segment, which overlaps a highly conserved region (CR) flanked by two intrinsically disordered regions (IDR1 and IDR2). The CTD dynamics and the equilibrium between fibril formation and dissolution are influenced by mutations and post-translational modifications (François-Moutal et al. 2019). Specific genetic mutations, particularly those located in conserved regions of the C-terminal domain, can attenuate LLPS formation (Conicella et al. 2016; Hallegger et al. 2021).

13.4 The Multifaceted Nature of LLPS: Biomolecular Interactions and Condensate Formation

The LLPS-mediated condensate formation is influenced by biomolecular interactions, component concentrations, and environmental factors, such as temperature, pH, and ionic strength (Banani et al. 2017). Cellular stress and macromolecular crowding also affect condensate properties. Comprehending the multifactorial nature of condensate formation is vital for studying the pathogenesis of neurodegenerative diseases, such as ALS.

The oligomerization of TDP-43, observed in motor neurons, is postulated to be involved in chronic pathology and intricately associated with the etiology and progression of ALS. Under physiological conditions, the NTD of TDP-43 mediates oligomerization, spatially distancing IDRs and thereby mitigating pathological aggregation (Afroz et al. 2017). Disrupting the delicate equilibrium between the NTD and the C-terminal IDRs may trigger a transition from physiological LLPS and condensate formation to pathological aggregation. RNA-binding proteins, such as TDP-43, interact with RNA molecules in a sequence-specific manner and participate in numerous cellular functions. Long, flexible, multivalent macromolecules like RNA assume critical structural roles, facilitating the formation of phase-separated, membraneless organelles and regulating TDP-43 stability, functionality, LLPS, and subcellular localization (Grese et al. 2021; Duan et al. 2022). In this context, alterations in the subcellular localization of nuclear proteins, including TDP-43 and FUS, diminish their interaction with RNA, consequently promoting aggregation formation in the cytoplasm (Mann et al. 2019).

Protein–protein interactions are critical for condensate formation, especially within the IDR of TDP-43, where a multitude of RNA-binding proteins, including ALS-related proteins such as hnRNP, engage in interactions (Buratti et al. 2005). These associations span a diverse range of intracellular functions, including LLPS related to RNA splicing, a primary function of TDP-43, and the formation of stress granules in response to cellular stress. Within stress granules, RNA-binding proteins like G3BP assume a central role in network formation, with the involvement of ALS-associated proteins such as TDP-43, FUS, TIA1, and ATXN2 (Yang et al. 2020). It has been posited that the quantitative balance of these interconnected proteins influences granule formation and disturbs its equilibrium, which, in turn, is associated with the transition to pathological aggregates. Collectively, these observations underscore the physiological importance of the interactions between ALS-related proteins and IDRs, and their potentially precarious balance in disease pathogenesis (Fig. 13.1).

13.5 Complex Regulatory Mechanisms of IDRs in RNA-Binding Proteins

The RNA-binding protein's IDR is unequivocally implicated in LLPS formation and pathological aggregation, rendering the regulation of its expression crucial. Intriguingly, IDR expression appears to be modulated through alternative splicing (Gueroussov et al. 2017), which entails a sophisticated network of ALS-associated RNA-binding proteins, including TDP-43, FUS, and hnRNP A/B, that participate in the alternative splicing of the IDR. The intricacy of this network is further augmented by the observation that the alternative splicing of RNA-binding proteins is itself mediated by membraneless organelles and LLPS, underscoring the complexity of these regulatory processes.

TDP-43 exemplifies the intricate relationship between IDRs and alternative splicing (Fig. 13.2). TDP-43 binds to the 3′ UTR of its own pre-mRNA and

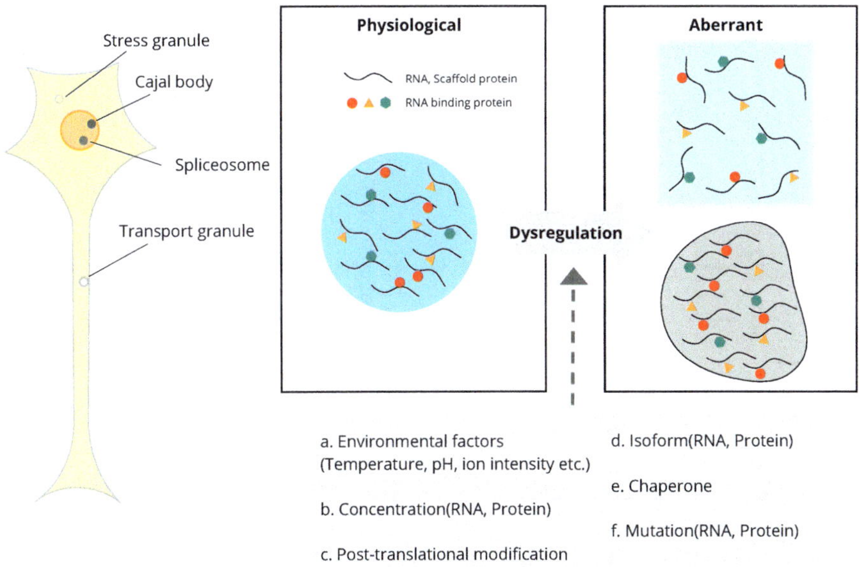

Fig. 13.1 Multifaceted nature of condensates in the nervous system. Condensates manifest in diverse aspects of cellular function and incorporate numerous factors. These perturbations may serve as precursors to not only disruptions in cellular processes but also to the formation of aggregates characteristic of neurodegenerative diseases

autoregulates its expression through alternative polyadenylation and alternative splicing (Ayala et al. 2011; Polymenidou et al. 2011), presumably via LLPS (Hallegger et al. 2021; Koehler et al. 2022). In this process, intron 6 and intron 7 are sequentially removed to generate NMD-sensitive isoforms (Koyama et al. 2016). The complexity of this autoregulation mechanism is further enhanced by the diversity of these splicing sites. As a result, isoforms in which only intron 6 is spliced may arise, but these isoforms are not, in principle, NMD-sensitive. Intron 6 is an intra-exon intron that completely encompasses the coding region of the TDP-43 IDR. Thus, this splicing isoform can produce a protein that lacks IDRs, suggesting that IDR expression is tightly auto-regulated.

HNRNPA1, an additional ALS-associated RNA-binding protein with an IDR, is pivotal in alternative splicing (Mayeda and Krainer 1992). TDP-43 interacts with HNRNPA1 through the IDR, modulating its splicing activity (Buratti et al. 2005), while the IDR-containing region of HNRNPA1 experiences alternative splicing, with TDP-43 involved in this process (Deshaies et al. 2018). The complex coordination of alternative splicing and its reciprocal effects on expression and function emphasizes the delicate equilibrium essential for cellular function.

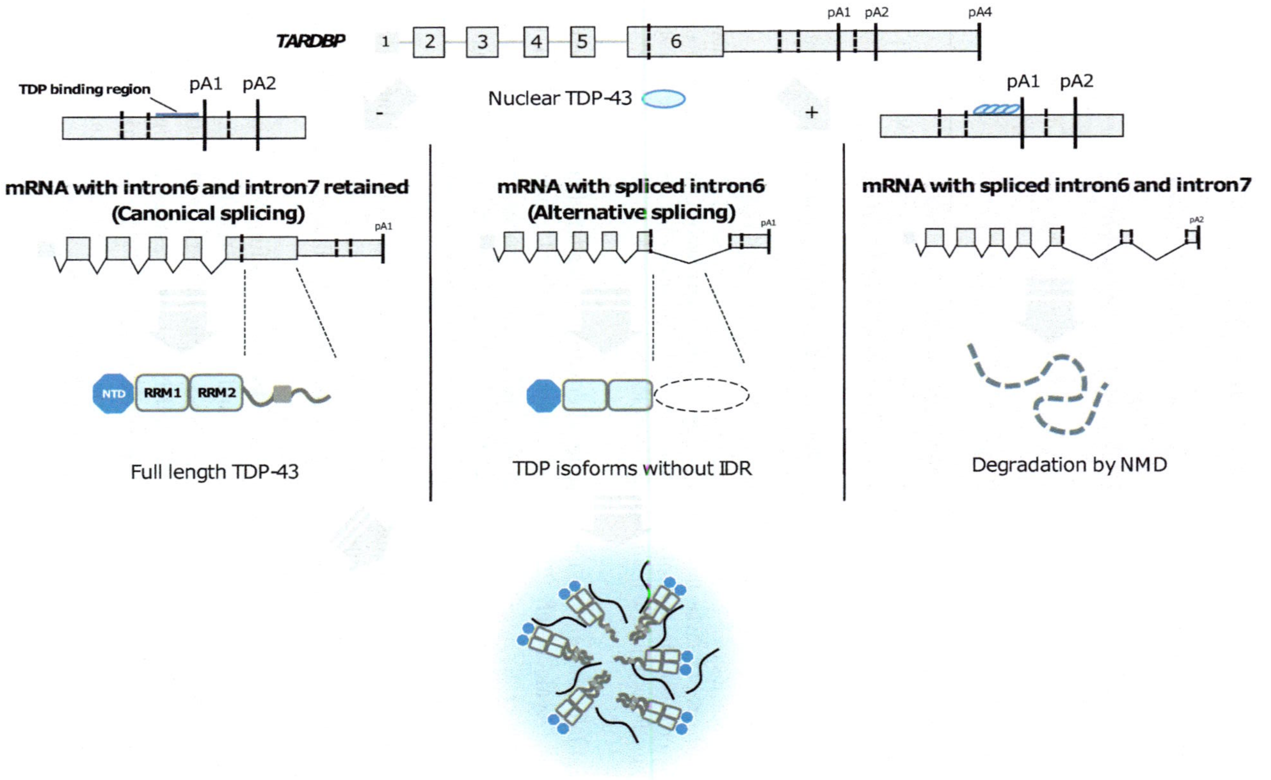

Fig. 13.2 TDP-43 autoregulatory mechanism and LLPS. TDP-43 interacts with the 3′ UTR of TARDBP pre-mRNA, putatively via liquid–liquid phase separation (LLPS). This association elicits alternative splicing events encompassing intron 6 and intron 7. When splicing exclusively involves intron 6, a short TDP isoform devoid of the IDR is generated. These specific isoforms may also participate in the formation of LLPS

13.6 Understanding TDP-43 Dysfunction and Its Impact on TDP-43 Isoforms on Condensate Formation

A final hallmark of ALS/FTLD pathogenesis is the loss of TDP-43 in the nucleus and its aggregation in the cytoplasm (Neumann et al. 2006; Arai et al. 2006). Since TDP-43 plays a key role in diverse mechanisms of RNA metabolism, the loss of its function is deeply implicated in pathogenesis. Indeed, the regulation of alternative splicing has been shown to be impaired in the brains of ALS and FTLD patients.

Of particular importance is its role in repressing the expression of cryptic exons within introns. The *STMN2* gene, which is involved in axon elongation in neurons, has a TDP-43-dependent cryptic exon, suggesting its involvement in neurodegeneration in motor neurons with reduced nuclear TDP-43 function (Klim et al. 2019). In addition, the *UNC13A* gene, a single nucleotide polymorphism (SNP) risk factor for ALS, also has a TDP-43-dependent cryptic exon. This SNP further exacerbates splicing abnormalities due to the loss of TDP-43 function (Brown et al. 2022; Ma et al. 2022). Importantly, many TDP-43-dependent genes with cryptic exons exhibit a neuro-specific expression pattern, implying deep involvement of TDP-43 and its loss of function in ALS pathogenesis.

Furthermore, the loss of nuclear TDP-43 function is associated with a decrease in GEM bodies, nuclear structures that serve as markers for survival motor neuron (SMN) proteins and are involved in splicing-related RNA maturation (Ishihara et al. 2013). GEM bodies are involved in mRNA splicing through the formation of spliceosomes. Decreased amounts of GEM bodies result in reduced expression of U small nuclear RNA (U snRNA) and disruption of normal mRNA splicing, further contributing to the complex pathogenesis of ALS and FTLD.

Nuclear TDP-43 dysfunction in ALS/FTLD also diminishes alternative splicing of TDP-43-encoding RNA (Koyama et al. 2016). Specifically, decreased splicing of intron 6 and intron 7 results in increased amounts of RNA encoding IDRs (Sugai et al. 2019) (Fig. 13.2). Increased IDR expression due to the loss of TDP-43 function could affect condensate formation dynamics and may exacerbate TDP-43 aggregation, a hallmark of these neurodegenerative diseases.

Intriguingly, the cytotoxicity of a short TDP-43 isoform lacking an IDR, generated by splicing intron 6, has been reported (Weskamp et al. 2019). This short TDP can form a heterodimer with full-length TDP-43 via the NTD, indicating that short TDP may affect condensate formation by an IDR-independent mechanism. Given this, the acquisition of toxicity of this isoform may be a debatable matter.

Recent findings have demonstrated a link between TDP-43 mutations causing ALS and TDP-43 autoregulation. In a mouse model of ALS with a single nucleotide mutation introduced into the mouse TDP-43 gene (TDP-43 Q331K), splicing resulting in the expression of short TDP is reduced. Consequently, TDP-43 expression is increased in these mice, leading to various RNA splicing alterations and abnormal RNA metabolism, even though nuclear TDP-43 function is preserved (White et al. 2018). In mice with knock-in for other ALS-causing mutations, RNA splicing targeted by TDP-43 exhibits changes in TDP-43 function that are enhanced, even if no clear increase in TDP-43 protein can be detected in the soluble fraction of Western blotting (Huang et al. 2020; Watanabe et al. 2020b).

The reported mechanism for decreased autoregulatory splicing due to TDP-43 mutations is related to the location of these gene mutations in a conserved region (CR) at the C-terminus of TDP-43. Mutations in the CR region have been associated with impaired LLPS formation (Conicella et al. 2016; Hallegger et al. 2021). Indeed, some studies suggest that LLPS may be involved in TDP-43 autoregulation (Hallegger et al. 2021; Koehler et al. 2022). However, it remains challenging to explain how RNA splicing of targets other than TDP-43 autoregulation could exhibit changes similar to TDP-43 excess. This complexity may be related to the sequence on the target RNA side of the LLPS, and the effects of TDP-43 mutations may have varying consequences (Hallegger et al. 2021). Alternatively, since ALS-causing mutations are not found only in the CR region, additional factors may be involved in the association between TDP-43 mutations and autoregulatory splicing.

Given all of this, under healthy conditions, appropriate expression of short TDP isoforms from alternative splicing may fine-tune TDP-43 condensate formation, contributing to cellular homeostasis (Fig. 13.2). This exemplifies the multifactorial nature of condensate formation and the complex network involving different isoforms. Comprehending TDP-43 isoform interactions and their impact on condensate formation is vital for elucidating ALS and FTLD pathogenesis (Fig. 13.1).

13.7 The Complex Relationship Between Physiological Condensates and Pathological Aggregates in Neurodegenerative Diseases

LLPS may serve as a precursor to pathological aggregations that typify neurodegenerative diseases, posing fundamental questions about the relationship between physiological and pathological aggregations. Physiological condensation and pathological aggregation differ in size, location, and composition. Some aggregations may form via LLPS, while others may not, and some might co-localize with microtubule-associated proteins in an aggresome formation-dependent pathway distinct from LLPS (Watanabe et al. 2020a). The roles of these various aggregation types in cytotoxicity and protection remain unclear.

The toxicity of TDP-43 aggregates remains controversial; studies using yeast models with random mutations in the C-terminus of TDP-43 have demonstrated that the resulting large TDP-43 aggregates are less toxic, while smaller aggregates are more toxic (Bolognesi et al. 2019). Consequently, the large aggregates may function as a protective mechanism to sequester potentially toxic oligomers. In transgenic mice with mutant TDP-43, pathologically, there is minimal TDP-43 aggregation as observed in human ALS autopsy brains (Todd and Petrucelli 2022). Nonetheless, motor neuron dropout occurs, suggesting that the presence of TDP-43 aggregates alone does not contribute to toxicity, but rather an excess of TDP-43 present as oligomers may be responsible for the toxicity (Sugai et al. 2019).

On the contrary, aggregation may impair TDP-43 function within the nucleus by sequestering newly formed functional TDP-43 in a prion-like manner, thus hindering its translocation to the nucleus. Furthermore, TDP-43 aggregates may impede nuclear-cytoplasmic trafficking (Woerner et al. 2015). The impaired nuclear function of TDP-43 leads to upregulation of TDP-43 expression, owing to its inherent capacity for autoregulation, which in turn exacerbates cytoplasmic aggregation and propagates the detrimental cycle of toxicity (Sugai et al. 2018). Overall, TDP-43 aggregation may reduce the excess of harmful oligomers locally, yet may simultaneously contribute to cytotoxicity through a secondary decline in functional TDP-43.

Further studies are needed to elucidate the mechanisms underlying the relationship between physiological condensates and pathological aggregation in neurodegenerative diseases. Additionally, research should focus on understanding the balance between toxic and protective effects mediated by both LLPS and non-LLPS aggregates.

13.8 Neuronal System Specificity in Neurodegenerative Diseases and the Multifaceted Impact of Condensates

Neurodegenerative diseases, despite the ubiquitous expression of their associated proteins, frequently display remarkable specificity within the affected nervous system. The fundamental mechanisms determining this nervous system specificity in neurodegenerative diseases remain elusive, in spite of extensive research efforts. Nevertheless, if condensates play a role in the initiation and progression of neurodegenerative diseases, a potential connection may exist between nervous system specificity and the multifaceted formation of condensates.

Neurons are particularly dependent on degradation systems, such as the ubiquitin-proteasome system and autophagy, for protein quality control, as aggregation cannot be eliminated by cell division (Schmidt et al. 2021). Moreover, exosome-mediated elimination contributes to the removal of pathogenic proteins associated with neurodegenerative diseases (Hung et al. 2023). Loss of these systems can lead to abnormal phase transitions and the formation of irreversible aggregates. ALS-causing genetic mutations have been found in genes involved in these degradation systems (Table 13.1). Aging, a risk factor for neurodegenerative diseases, may exacerbate the decline of homeostatic mechanisms in neurons, which are non-dividing cells.

Neurons are highly polarized, with axons and numerous branching dendrites. Consequently, the role of condensates in axonal transport and synapses is especially important. However, in neurodegenerative diseases, dysregulation of LLPS may result in the formation of abnormal condensates and impaired axonal transport

(Alami et al. 2014; Liu-Yesucevitz et al. 2014; Ryan and Fawzi 2019). Genes involved in the cytoskeleton and axonal transport have been identified as causative genes of ALS (Table 13.1). For example, mutations in ANXA11 disrupt RNA granule interaction with intracellular membranes in ALS, impairing axonal transport and leading to granule accumulation in the cell body and proximal axons, potentially contributing to motor neuron degeneration and disease progression (Liao et al. 2019).

Motor neurons, with their unique features and high energy demands, are particularly susceptible to energy deprivation, which may impair factors involved in condensate formation. In ALS, energy deficits can be attributed to mitochondrial dysfunction, which impairs ATP production (Stoica et al. 2014; Wang et al. 2019) and increases oxidative stress; defects in axonal transport hinder the delivery of essential cellular components and energy. Reduced intracellular ATP levels may compromise mechanisms controlling phase separation and RNP granule dynamics, highlighting the importance of understanding the interplay between energy metabolism, cellular homeostasis, and biomolecular condensate formation (Dang et al. 2021; Ren et al. 2022).

Neurons may undergo intricate post-transcriptional regulation, encompassing processes such as splicing, RNA editing, mRNA transport, and local translation in axons and dendrites. IDRs frequently overlap with exons subject to alternative splicing, potentially delineating tissue-specific and cell-specific phase separation and susceptibility to aggregation. Additionally, epigenetic factors, such as DNA methylation, influence tissue-dependent alternative splicing. For instance, in the motor cortex, alterations in DNA methylation states affecting TDP-43 alternative splicing occur with age, ultimately resulting in the accumulation of insoluble TDP-43 (Koike et al. 2021). This alternative splicing, generating shorter TDP forms, has been reported to occur more frequently in human spinal motor neurons, underscoring the significance of alternative splicing in neuronal specificity (Weskamp et al. 2019).

Human-specific motor neuron gene expression patterns, potentially linked to evolutionarily acquired functions like fine motor control of the hand, may contribute to motor neuron vulnerability in ALS (Henderson et al. 2019; Yadav et al. 2023; Pollen et al. 2023). The susceptibility of neurons responsible for advanced functions implies that human-specific gene expression and splicing patterns could play a vital role in ALS pathogenesis. These patterns might also affect the formation, regulation of biomolecular condensates in motor neurons, and the function of RNA-binding proteins such as TDP-43.

13.9 Challenges in Understanding Intracellular Condensates in Neurodegenerative Diseases

Notable advancements have been achieved in comprehending the formation and function of LLPS; however, deciphering the role of intracellular condensates in cellular processes and neurodegenerative diseases remains challenging. In vitro models serve as invaluable tools but might not adequately capture the intricacy of intracellular condensates. Enhanced experimental models could more accurately emulate the intracellular environment (Freibaum et al. 2021). Obtaining an integrated understanding of condensate involvement, with a focus on the nervous system's unique properties, is vital for devising targeted therapeutics. Therapeutic strategies that fail to consider the complex condensate network may yield unforeseen consequences.

Further investigations are necessary to address the complex, multifactorial pathogenesis of ALS. One crucial aspect to clarify is the etiology of TDP-43 pathology, which characterizes both sporadic ALS and many familial ALS cases. Based on the physiological functions of TDP-43, a comprehensive understanding of oligomer formation, splicing isoforms and their effects, interactions with other RNA-binding proteins, physiological condensation, pathological aggregation, and the impact of these interactions on cellular function and disease pathogenesis is required. Studies have extensively examined pathogenesis due to the loss or gain of TDP-43 function. Future research efforts will necessitate a holistic comprehension of the effects of ALS-causing gene mutations and the identification of upstream abnormalities that shape TDP-43 pathogenesis based on this understanding. LLPS, where physiological roles and pathology intersect, can be considered an upstream phenomenon in this context.

13.10 Conclusion

In conclusion, a holistic approach is imperative when investigating the role of physiological condensates in the specific nervous system, owing to the multifaceted nature of their participation in numerous intracellular processes. The potential involvement of physiological condensates in pathological aggregation formation signifies an evolutionary trade-off, the comprehension of which may elucidate the pathogenesis of neurodegenerative diseases, including ALS. A deeper understanding of the molecular mechanisms governing the delicate equilibrium between the advantageous physiological functions of condensates and their possible contribution to pathological aggregation is essential. This knowledge will foster the development of more effective therapeutic interventions targeting the pathogenesis of neurodegenerative diseases at the nexus of physiological and pathological phenomena.

References

Afroz T, Hock E-M, Ernst P et al (2017) Functional and dynamic polymerization of the ALS-linked protein TDP-43 antagonizes its pathologic aggregation. Nat Commun 8:45. https://doi.org/10.1038/s41467-017-00062-0

Alami NH, Smith RB, Carrasco MA et al (2014) Axonal transport of TDP-43 mRNA granules is impaired by ALS-causing mutations. Neuron 81:536–543. https://doi.org/10.1016/j.neuron.2013.12.018

Arai T, Hasegawa M, Akiyama H et al (2006) TDP-43 is a component of ubiquitin-positive tau-negative inclusions in frontotemporal lobar degeneration and amyotrophic lateral sclerosis. Biochem Biophys Res Commun 351:602–611. https://doi.org/10.1016/j.bbrc.2006.10.093

Ayala YM, De Conti L, Avendaño-Vázquez SE et al (2011) TDP-43 regulates its mRNA levels through a negative feedback loop. EMBO J 30:277–288. https://doi.org/10.1038/emboj.2010.310

Banani SF, Lee HO, Hyman AA, Rosen MK (2017) Biomolecular condensates: organizers of cellular biochemistry. Nat Rev Mol Cell Biol 18:285

Boeynaems S, Bogaert E, Kovacs D et al (2017) Phase separation of C9orf72 dipeptide repeats perturbs stress granule dynamics. Mol Cell 65:1044–1055.e5. https://doi.org/10.1016/j.molcel.2017.02.013

Bolognesi B, Faure AJ, Seuma M et al (2019) The mutational landscape of a prion-like domain. Nat Commun 10:4162. https://doi.org/10.1038/s41467-019-12101-z

Brown AL, Wilkins OG, Keuss MJ et al (2022) TDP-43 loss and ALS-risk SNPs drive mis-splicing and depletion of UNC13A. Nature 603:131–137. https://doi.org/10.1038/s41586-022-04436-3

Brown RH, Al-Chalabi A (2017) Amyotrophic lateral sclerosis. N Engl J Med 377:162–172. https://doi.org/10.1056/NEJMra1603471

Buratti E (2001) Nuclear factor TDP-43 and SR proteins promote in vitro and in vivo CFTR exon 9 skipping. EMBO J 20:1774–1784. https://doi.org/10.1093/emboj/20.7.1774

Buratti E, Brindisi A, Giombi M et al (2005) TDP-43 binds heterogeneous nuclear ribonucleoprotein A/B through its C-terminal tail. J Biol Chem 280:37572. https://doi.org/10.1074/jbc.m505557200

Byrne S, Walsh C, Lynch C et al (2011) Rate of familial amyotrophic lateral sclerosis: a systematic review and meta-analysis. J Neurol Neurosurg Psychiatry 82:623–627. https://doi.org/10.1136/JNNP.2010.224501

Cohen TJ, Hwang AW, Restrepo CR et al (2015) An acetylation switch controls TDP-43 function and aggregation propensity. Nat Commun 6:5845. https://doi.org/10.1038/ncomms6845

Conicella AE, Zerze GH, Mittal J, Fawzi NL (2016) ALS mutations disrupt phase separation mediated by α-helical structure in the TDP-43 low-complexity C-terminal domain. Structure 24:1537–1549. https://doi.org/10.1016/j.str.2016.07.007

Dang M, Lim L, Kang J, Song J (2021) ATP biphasically modulates LLPS of TDP-43 PLD by specifically binding arginine residues. Commun Biol 4:714. https://doi.org/10.1038/s42003-021-02247-2

Dao TP, Kolaitis R-M, Kim HJ et al (2018) Ubiquitin modulates liquid-liquid phase separation of UBQLN2 via disruption of multivalent interactions. Mol Cell 69:965–978.e6. https://doi.org/10.1016/j.molcel.2018.02.004

DeJesus-Hernandez M, Mackenzie IR, Boeve BF et al (2011) Expanded GGGGCC hexanucleotide repeat in noncoding region of C9ORF72 causes chromosome 9p-linked FTD and ALS. Neuron 72:245–256. https://doi.org/10.1016/j.neuron.2011.09.011

Deng HX, Chen W, Hong ST et al (2011) Mutations in UBQLN2 cause dominant X-linked juvenile and adult-onset ALS and ALS/dementia. Nature 477(7363):211–215. https://doi.org/10.1038/nature10353

Deshaies J-E, Shkreta L, Moszczynski AJ et al (2018) TDP-43 regulates the alternative splicing of hnRNP A1 to yield an aggregation-prone variant in amyotrophic lateral sclerosis. Brain 141:1320. https://doi.org/10.1093/brain/awy062

Duan L, Zaepfel BL, Aksenova V et al (2022) Nuclear RNA binding regulates TDP-43 nuclear localization and passive nuclear export. Cell Rep 40:111106. https://doi.org/10.1016/j.celrep.2022.111106

Faruk MO, Ichimura Y, Kageyama S et al (2021) Phase-separated protein droplets of amyotrophic lateral sclerosis-associated p62/SQSTM1 mutants show reduced inner fluidity. J Biol Chem 297:101405. https://doi.org/10.1016/j.jbc.2021.101405

Fecto F, Yan J, Vemula SP et al (2011) SQSTM1 mutations in familial and sporadic amyotrophic lateral sclerosis. Arch Neurol 68:1440–1446. https://doi.org/10.1001/ARCHNEUROL.2011.250

François-Moutal L, Perez-Miller S, Scott DD et al (2019) Structural insights into TDP-43 and effects of post-translational modifications. Front Mol Neurosci 12:1–22. https://doi.org/10.3389/fnmol.2019.00301

Freibaum BD, Messing J, Yang P et al (2021) High-fidelity reconstitution of stress granules and nucleoli in mammalian cellular lysate. J Cell Biol 220:e202009079. https://doi.org/10.1083/JCB.202009079

Freischmidt A, Wieland T, Richter B et al (2015) Haploinsufficiency of TBK1 causes familial ALS and fronto-temporal dementia. Nat neurosci 18(5):631–636. https://doi.org/10.1038/nn.4000

Grese ZR, Bastos AC, Mamede LD et al (2021) Specific RNA interactions promote TDP-43 multivalent phase separation and maintain liquid properties. EMBO Rep 22:e53632. https://doi.org/10.15252/embr.202153632

Gueroussov S, Weatheritt RJ, O'Hanlon D et al (2017) Regulatory expansion in mammals of multivalent hnRNP assemblies that globally control alternative splicing. Cell 170:324–339.e23. https://doi.org/10.1016/j.cell.2017.06.037

Hallegger M, Chakrabarti AM, Lee FCY et al (2021) TDP-43 condensation properties specify its RNA-binding and regulatory repertoire. Cell 184:4680–4696.e22. https://doi.org/10.1016/j.cell.2021.07.018

Han TW, Kato M, Xie S et al (2012) Cell-free formation of RNA granules: bound RNAs identify features and components of cellular assemblies. Cell 149:768. https://doi.org/10.1016/j.cell.2012.04.016

Henderson RD, Garton FC, Kiernan MC et al (2019) Human cerebral evolution and the clinical syndrome of amyotrophic lateral sclerosis. J Neurol Neurosurg Psychiatry 90:570–575. https://doi.org/10.1136/jnnp-2017-317245

Hsiung G-YR, DeJesus-Hernandez M, Feldman HH et al (2012) Clinical and pathological features of familial frontotemporal dementia caused by C9ORF72 mutation on chromosome 9p. Brain 135:709–722. https://doi.org/10.1093/brain/awr354

Huang S-L, Wu L-S, Lee M et al (2020) A robust TDP-43 knock-in mouse model of ALS. Acta Neuropathol Commun 8:3. https://doi.org/10.1186/s40478-020-0881-5

Hung S-T, Linares GR, Chang W-H et al (2023) PIKFYVE inhibition mitigates disease in models of diverse forms of ALS. Cell 186:786–802.e28. https://doi.org/10.1016/j.cell.2023.01.005

Ishihara T, Ariizumi Y, Shiga A et al (2013) Decreased number of Gemini of coiled bodies and U12 snRNA level in amyotrophic lateral sclerosis. Hum Mol Genet 22:4136–4147. https://doi.org/10.1093/hmg/ddt262

Kato M, Han TW, Xie S et al (2012) Cell-free formation of RNA granules: low complexity sequence domains form dynamic fibers within hydrogels. Cell 149:753. https://doi.org/10.1016/j.cell.2012.04.017

Kim HJ, Kim NC, Wang YD et al (2013) Mutations in prion-like domains in hnRNPA2B1 and hnRNPA1 cause multisystem proteinopathy and ALS. Nature 495:467–473. https://doi.org/10.1038/nature11922

Klim JR, Williams LA, Limone F et al (2019) ALS-implicated protein TDP-43 sustains levels of STMN2, a mediator of motor neuron growth and repair. Nat Neurosci 22:167–179. https://doi.org/10.1038/s41593-018-0300-4

Koehler LC, Grese ZR, Bastos ACS et al (2022) TDP-43 oligomerization and phase separation properties are necessary for autoregulation. Front Neurosci 16:503. https://doi.org/10.3389/FNINS.2022.818655/BIBTEX

Koike Y, Sugai A, Hara N et al (2021) Age-related demethylation of the TDP-43 autoregulatory region in the human motor cortex. Commun Biol 4(1):1107. https://doi.org/10.1038/s42003-021-02621-0

Koyama A, Sugai A, Kato T et al (2016) Increased cytoplasmic TARDBP mRNA in affected spinal motor neurons in ALS caused by abnormal autoregulation of TDP-43. Nucleic Acids Res 44: 5820–5836. https://doi.org/10.1093/nar/gkw499

Kuo P-H, Chiang C-H, Wang Y-T et al (2014) The crystal structure of TDP-43 RRM1-DNA complex reveals the specific recognition for UG- and TG-rich nucleic acids. Nucleic Acids Res 42:4712–4722. https://doi.org/10.1093/nar/gkt1407

Liao YC, Fernandopulle MS, Wang G et al (2019) RNA granules hitchhike on lysosomes for long-distance transport, using Annexin A11 as a molecular tether. Cell 179:147. https://doi.org/10.1016/j.cell.2019.08.050

Liu-Yesucevitz L, Lin AY, Ebata A et al (2014) ALS-linked mutations enlarge TDP-43-enriched neuronal RNA granules in the dendritic arbor. J Neurosci 34:4167–4174. https://doi.org/10.1523/JNEUROSCI.2350-13.2014

Ma XR, Prudencio M, Koike Y et al (2022) TDP-43 represses cryptic exon inclusion in the FTD–ALS gene UNC13A. Nature 603:124–130. https://doi.org/10.1038/s41586-022-04424-7

Mackenzie IR, Arzberger T, Kremmer E et al (2013) Dipeptide repeat protein pathology in C9ORF72 mutation cases: clinico-pathological correlations. Acta Neuropathol 126:859–879. https://doi.org/10.1007/s00401-013-1181-y

Mann JR, Gleixner AM, Mauna JC et al (2019) RNA binding antagonizes neurotoxic phase transitions of TDP-43. Neuron 0:1–18. https://doi.org/10.1016/J.NEURON.2019.01.048

Mayeda A, Krainer AR (1992) Regulation of alternative pre-mRNA splicing by hnRNP A1 and splicing factor SF2. Cell 68:365–375. https://doi.org/10.1016/0092-8674(92)90477-T

Molliex A, Temirov J, Lee J et al (2015) Phase separation by low complexity domains promotes stress granule assembly and drives pathological fibrillization. Cell 163:123–133. https://doi.org/10.1016/j.cell.2015.09.015

Mompeán M, Baralle M, Buratti E, Laurents DV (2016) An amyloid-like pathological conformation of TDP-43 is stabilized by hypercooperative hydrogen bonds. Front Mol Neurosci 9:125. https://doi.org/10.3389/fnmol.2016.00125

Neumann M, Sampathu DM, Kwong L et al (2006) Ubiquitinated TDP-43 in frontotemporal lobar degeneration and amyotrophic lateral sclerosis. Science 314:130–133. https://doi.org/10.1002/ana.21425.Phosphorylated

Pollen AA, Kilik U, Lowe CB, Camp JG (2023) Human-specific genetics: new tools to explore the molecular and cellular basis of human evolution. Nat Rev Genet. https://doi.org/10.1038/s41576-022-00568-4

Polymenidou M, Lagier-tourenne C, Hutt KR et al (2011) Long pre-mRNA depletion and RNA missplicing contribute to neuronal vulnerability from loss of TDP-43. Nat Neurosci 14:459–468. https://doi.org/10.1038/nn.2779

Pottier C, Bieniek KF, Finch N et al (2015) Whole-genome sequencing reveals important role for TBK1 and OPTN mutations in frontotemporal lobar degeneration without motor neuron disease. Acta Neuropathol 130:77–92. https://doi.org/10.1007/s00401-015-1436-x

Puls I, Jonnakuty C, LaMonte BH et al (2003) Mutant dynactin in motor neuron disease. Nat Genet 33(4):455–456. https://doi.org/10.1038/ng1123

Ren C-L, Shan Y, Zhang P et al (2022) Uncovering the molecular mechanism for dual effect of ATP on phase separation in FUS solution. Sci Adv 8:eabo7885. https://doi.org/10.1126/sciadv.abo7885

Renton AE, Majounie E, Waite A et al (2011) A hexanucleotide repeat expansion in C9ORF72 is the cause of chromosome 9p21-linked ALS-FTD. Neuron 72:257–268. https://doi.org/10.1016/j.neuron.2011.09.010

Rubino E, Rainero I, Chio A et al (2012) SQSTM1 mutations in frontotemporal lobar degeneration and amyotrophic lateral sclerosis. Neurology 79:1556–1562. https://doi.org/10.1212/WNL.0b013e31826e25df

Ryan VH, Fawzi NL (2019) Physiological, pathological, and targetable membraneless organelles in neurons. Trends Neurosci 42:693–708. https://doi.org/10.1016/j.tins.2019.08.005

Sainouchi M, Hatano Y, Tada M et al (2021) A novel splicing variant of ANXA11 in a patient with amyotrophic lateral sclerosis: histologic and biochemical features. Acta Neuropathol Commun 9:106. https://doi.org/10.1186/s40478-021-01202-w

Schmidt MF, Gan ZY, Komander D, Dewson G (2021) Ubiquitin signalling in neurodegeneration: mechanisms and therapeutic opportunities. Cell Death Differ 28:570–590. https://doi.org/10.1038/s41418-020-00706-7

Seelaar H, Rohrer JD, Pijnenburg YAL et al (2011) Clinical, genetic and pathological heterogeneity of frontotemporal dementia: a review. J Neurol Neurosurg Psychiatry 82:476–486. https://doi.org/10.1136/JNNP.2010.212225

Shodai A, Morimura T, Ido A et al (2013) Aberrant assembly of RNA recognition motif 1 links to pathogenic conversion of TAR DNA-binding protein of 43 kDa (TDP-43). J Biol Chem 288:14886–14905. https://doi.org/10.1074/jbc.M113.451849

Smith BN, Ticozzi N, Fallini C et al (2014) Exome-wide rare variant analysis identifies TUBA4A mutations associated with familial ALS. Neuron 84:324–331. https://doi.org/10.1016/j.neuron.2014.09.027

Smith BN, Topp SD, Fallini C et al (2017) Mutations in the vesicular trafficking protein annexin A11 are associated with amyotrophic lateral sclerosis. Sci Transl Med 9:eaad9157. https://doi.org/10.1126/scitranslmed.aad9157

Stoica R, De Vos KJ, Paillusson S et al (2014) ER-mitochondria associations are regulated by the VAPB-PTPIP51 interaction and are disrupted by ALS/FTD-associated TDP-43. Nat Commun 5:3996. https://doi.org/10.1038/ncomms4996

Sugai A, Kato T, Koyama A et al (2019) Non-genetically modified models exhibit TARDBP mRNA increase due to perturbed TDP-43 autoregulation. Neurobiol Dis 130:104534. https://doi.org/10.1016/J.NBD.2019.104534

Sugai A, Kato T, Koyama A et al (2018) Robustness and vulnerability of the autoregulatory system that maintains nuclear TDP-43 levels: a trade-off hypothesis for ALS pathology based on in silico data. Front Neurosci 12:28. https://doi.org/10.3389/FNINS.2018.00028

Teyssou E, Takeda T, Lebon V et al (2013) Mutations in SQSTM1 encoding p62 in amyotrophic lateral sclerosis: genetics and neuropathology. Acta Neuropathol 125:511–522. https://doi.org/10.1007/s00401-013-1090-0

Todd TW, Petrucelli L (2022) Modelling amyotrophic lateral sclerosis in rodents. Nat Rev Neurosci 23(4):231–251. https://doi.org/10.1038/s41583-022-00564-x

Van Mossevelde S, Engelborghs S, van der Zee J, Van Broeckhoven C (2018) Genotype–phenotype links in frontotemporal lobar degeneration. Nat Rev Neurol 14:363–378. https://doi.org/10.1038/s41582-018-0009-8

Vance C, Rogelj B, Hortobagyi T et al (2009) Mutations in FUS, an RNA processing protein, cause familial amyotrophic lateral sclerosis type 6. Science 323:1208–1211. https://doi.org/10.1126/science.1165942

Wang A, Conicella AE, Schmidt HB et al (2018) A single N-terminal phosphomimic disrupts TDP-43 polymerization, phase separation, and RNA splicing. EMBO J 37:e97452. https://doi.org/10.15252/embj.201797452

Wang P, Deng J, Dong J et al (2019) TDP-43 induces mitochondrial damage and activates the mitochondrial unfolded protein response. PLoS Genet 15:e1007947. https://doi.org/10.1371/journal.pgen.1007947

Watanabe S, Inami H, Oiwa K et al (2020a) Aggresome formation and liquid–liquid phase separation independently induce cytoplasmic aggregation of TAR DNA-binding protein 43. Cell Death Dis 11:909. https://doi.org/10.1038/s41419-020-03116-2

Watanabe S, Oiwa K, Murata Y et al (2020b) ALS-linked TDP-43M337V knock-in mice exhibit splicing deregulation without neurodegeneration. Mol Brain 13:8. https://doi.org/10.1186/s13041-020-0550-4

Weskamp K, Tank EM, Miguez R et al (2019) Shortened TDP43 isoforms upregulated by neuronal hyperactivity drive TDP43 pathology in ALS. J Clin Investig 130:1139. https://doi.org/10.1172/jci130988

White MA, Kim E, Duffy A et al (2018) TDP-43 gains function due to perturbed autoregulation in a Tardbp knock-in mouse model of ALS-FTD. Nat Neurosci 21:552–563. https://doi.org/10.1038/s41593-018-0113-5

Wider C, Dickson DW, Stoessl AJ et al (2009) Pallidonigral TDP-43 pathology in Perry syndrome. Parkinsonism Relat Disord 15:281–286. https://doi.org/10.1016/j.parkreldis.2008.07.005

Williams KL, Warraich ST, Yang S et al (2012) UBQLN2/ubiquilin 2 mutation and pathology in familial amyotrophic lateral sclerosis. Neurobiol Aging 33:2527.e3–2527.e10. https://doi.org/10.1016/j.neurobiolaging.2012.05.008

Woerner AC, Frottin F, Hornburg D, Feng LR (2015) Cytoplasmic protein aggregates interfere with nucleo-cytoplasmic transport of protein and RNA. Science 351:173–177. https://doi.org/10.1126/science.aad2033

Wu C, Fallini C, Ticozzi N et al (2012) Mutations in the profilin 1 gene cause familial amyotrophic lateral sclerosis. Nature 488:499–503. https://doi.org/10.1038/nature11280

Yadav A, Matson KJE, Li L et al (2023) A cellular taxonomy of the adult human spinal cord. Neuron 111:328–344.e7. https://doi.org/10.1016/j.neuron.2023.01.007

Yang P, Mathieu C, Kolaitis RM et al (2020) G3BP1 is a tunable switch that triggers phase separation to assemble stress granules. Cell 181:325–345.e28. https://doi.org/10.1016/j.cell.2020.03.046

Chapter 14
Functional Properties of Phase Separation and Intranuclear Complex of FUS in the Pathogenesis of ALS/FTLD

Shinsuke Ishigaki

Abstract FUS (fused in sarcoma), also known as TLS (translocated in liposarcoma), is an RNA-binding protein that mainly localizes to the nucleus and participates in various aspects of RNA metabolism such as transcription, alternative splicing, and RNA transport. It has been identified as a causative gene for amyotrophic lateral sclerosis (ALS) and a pathological molecule involved in ALS and frontotemporal lobar degeneration (FTLD). FUS undergoes liquid–liquid phase separation (LLPS), a phase transition phenomenon associated with membraneless organelle formation. Abnormal LLPS can result in irreversible aggregation of FUS. Restoring normal LLPS is a potential therapy for ALS and FTLD. 1,6-hexanediol can disrupt LLPS formation, so such small molecules could be therapeutic candidates. Another possibility is RNA molecules that bind to FUS or other related proteins, which modulate protein and/or protein complex formation essential for LLPS.

Keywords FUS · ALS · FTLD · LLPS · SFPQ

14.1 Introduction

Amyotrophic lateral sclerosis (ALS) has traditionally been viewed as a progressive neurodegenerative disorder primarily affecting the motor system (Hardiman et al. 2017). However, a subset of ALS patients may also exhibit clinical symptoms consistent with frontotemporal dementia (FTD) early on in their disease course (Tsujimoto et al. 2011; Masuda et al. 2016) Similarly, some patients with frontotemporal lobar degeneration (FTLD) may show both upper and lower motor symptoms (Riku et al. 2014). Notably, TAR DNA-binding protein of 43 kDa (TDP-43) and fused in sarcoma (FUS) are the major pathologic proteins implicated in both ALS and FTLD, and it is not uncommon for these two disorders to co-occur.

S. Ishigaki (✉)
Molecular Neuroscience Research Center, Shiga University of Medical Science, Otsu, Shiga, Japan
e-mail: shinsuke@belle.shiga-med.ac.jp

R. Kurokawa (ed.), *Phase Separation in Living Cells*,
https://doi.org/10.1007/978-981-99-4886-4_14

Indeed, recent studies have suggested that FTLD and ALS may represent different points on a continuous disease spectrum (Riku et al. 2014; Renton et al. 2014; Robberecht and Philips 2013).

Thus, ALS and FTLD are debilitating neurodegenerative disorders that can manifest either sporadically or through familial inheritance. These conditions exhibit distinct patterns of pathology, with ALS targeting the motor neuron systems and FTLD selectively affecting the frontotemporal cortices (Olney et al. 2017). Note that FTD is an umbrella term for several clinical syndromes of dementia that share similar symptoms and are characterized by atrophy of the frontal and anterior temporal lobes. In contrast, FTLD is a neuropathological term that refers to the underlying pathological features seen in the brains of individuals with FTD. In this article, the terminology will be consistent with FTLD.

FTLD is a group of clinically, pathologically, and genetically heterogeneous neurodegenerative disorders that involve the frontal and temporal lobes. Behavioral variant frontotemporal dementia, semantic dementia, and progressive non-fluent aphasia (PNFA) are three major clinical syndromes. TDP-43, FUS, and tau are three major pathogenetic proteins (Sobue et al. 2018). Progressive supranuclear palsy (PSP) and corticobasal degeneration (CBD) are known as 4R-tauopathies clinically characterized by Parkinsonism and various cognitive and motor symptoms. They are also recognized as related disorders with FTLD (Olney et al. 2017; Josephs et al. 2006).

PSP affects the globus pallidus, subthalamic nucleus, midbrain, pons, and cerebellar efferent system (dentate nucleus and upper cerebellar peduncle). It is characterized by neurofibrillary tangles and tufted astrocytes in neuronal cytoplasm, which can be detected through anti-4R-tau immunohistochemistry and silver staining. Tau aggregates in oligodendrocytes also form coiled bodies in white matter. CBD is characterized by asymmetrical para-sylvian atrophy and basal ganglia atrophy. The microscopic features include numerous immunopositive neuropil threads for hyperphosphorylated tau in the deep cortical layers and white matter, fuzzy pretangles in neuronal cytoplasm, and ballooned neurons. CBD is also identified by the presence of astrocytic plaques, which are tau aggregations in the distal portions of astrocytic processes, as opposed to PSP's tufted astrocytes that appear in the cytoplasm and proximal processes (Riku et al. 2022a).

Thus, FTLD itself encompasses a broad range of disease subtypes, ranging from FTLD-tau to FTLD-TDP, with the former involving tau pathology and the latter exhibiting the same disease entity as ALS (Fig. 14.1).

Previously, it was believed that FTLD-tau and FTLD-TDP had distinct pathogenic mechanisms due to their characteristic protein deposition in the pathology being different. However, our current study findings imply that similar sub-pathological molecular alterations may already occur in broadly defined FTLD cases, including those with PSP, CBD, and ALS (Ishigaki et al. 2020; Riku et al. 2022b).

The clinical and pathological features of sporadic cases of ALS and FTLD bear a striking resemblance to those of familial cases, which harbor mutations in various causative genes such as TDP-43, suggesting that those gene products play pivotal

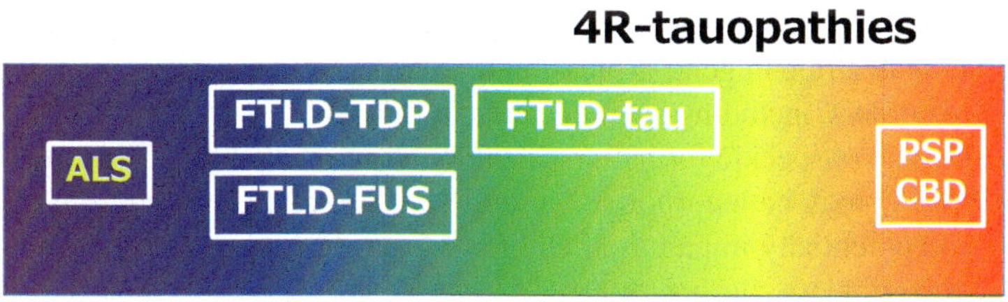

Fig. 14.1 Image of a wide-ranged disease spectrum of FTLD. ALS is one extreme of this spectrum, as shown by its clinical, pathological, and genetic features. The neuropathology can sort FTLD into three types: FTLD-TDP, FTLD-FUS, and FTLD-tau. Conversely, PSP and CBD are also sporadic forms of FTLD, since they exhibit the same 4R-tauopathy as FTLD-tau (Laforce Jr. 2013)

roles in the pathogenesis of these diseases (Lagier-Tourenne and Cleveland 2009). Similarly, FUS is implicated in familial ALS and FTLD, and in these disorders, both TDP-43 and FUS proteins undergo cytoplasmic redistribution and aggregation (Kwiatkowski et al. 2009; Vance et al. 2009; Ishigaki and Sobue 2018). Notably, TDP-43 and FUS share many common pathophysiological characteristics. Structurally, both proteins are heterogeneous ribonucleoproteins (hnRNPs) that harbor RNA recognition motifs (RRMs). Typically, these proteins are predominantly located within the nucleus and are involved in essential cellular processes such as transcription, alternative splicing, translation, and RNA transport. However, in pathological conditions, their altered forms tend to accumulate within the cytoplasm, leading to the formation of pathological aggregates (Lagier-Tourenne and Cleveland 2009; Svetoni et al. 2016).

Using a mouse model, we reported that FUS regulates alternative splicing of tau proteins in coordination with splicing factor, proline- and glutamine-rich (SFPQ). Under normal conditions, the two proteins form a high-molecular-weight complex in the nucleus. However, disease-associated mutations in FUS gene disrupt the formation of the complex, resulting in unregulated alternative splicing of tau, a disproportional increase in the 4R-tau/3R-tau ratio, and eventually neurodegeneration (Ishigaki et al. 2017). Moreover, the neuropathological study revealed spatial dissociation of SFPQ and FUS in the neuronal nuclei of ALS/FTLD-FUS, ALS/FTLD-TDP, PSP, and CBD. Furthermore, the ratio of 4R/3R-tau was elevated in cases with ALS/FTLD-TDP and PSP but was largely unaffected in cases with AD (Ishigaki et al. 2020; Riku et al. 2022b). Thus, impaired interactions between FUS and SFPQ and subsequent imbalanced tau isoform ratio constitute a common pathogenic mechanism across FTLD spectrum diseases.

Other reports also suggest that the loss of nuclear RNA-binding proteins FUS and SFPQ worsens the spatial separation of these proteins in FTLD spectrum disorders (Luisier et al. 2018; Tyzack et al. 2019). Those findings indicate that FUS and SFPQ may require interactions to maintain their stable nuclear localization and emphasize the need to examine the intranuclear spatial separation of functional complexes in the molecular pathogenesis of FTLD spectrum disorders.

Paraspeckles are non-membrane delimited entities formed on the long non-coding RNA *NEAT1_2* within the nucleus, consisting of TDP-43, FUS, and

SFPQ. While paraspeckles are typically absent in healthy motor neurons, they are believed to play a role in gene expression regulation when present. However, FTLD spectrum disorders, including sporadic and familial ALS cases, have been associated with increased paraspeckle formation (Shelkovnikova et al. 2018). This suggests that FUS and SFPQ may be incorporated into these ALS-related paraspeckle structures, which could potentially impair their ability to form functional complexes within the nucleus. Nevertheless, further research is needed to fully understand the potential significance of paraspeckle structures in FTLD spectrum disorders.

14.2 Functional Properties of FUS

FUS was first identified as a fusion protein (FUS-CHOP) that is associated with translocation in liposarcoma and other cancers (Crozat et al. 1993; Eneroth et al. 1990; Rabbitts et al. 1993). The fusion of the N-terminal FUS with various transcription factors such as the C-terminal CHOP gives it a very potent transcriptional activity and is thought to contribute to tumorigenesis (Bertolotti et al. 1999; Prasad et al. 1994). Besides this, FUS is a novel transcriptional regulator in cancer cell development (Kwiatkowski et al. 2009; Calvio et al. 1995). FUS was also identified as hnRNP P2, a subunit of the complex that is involved in pre-mRNA maturation (Calvio et al. 1995), and in 2009, two groups reported that it was the causative gene of familial ALS (Kwiatkowski et al. 2009; Vance et al. 2009), which attracted considerable attention in the field of neurological research. Since then, FUS, as well as many other RNA-binding proteins, is strongly related to the pathogenesis of ALS and its analog, frontotemporal lobar degeneration (FTLD).

In terms of protein structure, it has been reported that RRM binds to the stem-loop portion of RNA, and the zinc finger (ZnF) motif at the C-terminus recognizes the GGU sequence of RNA (Loughlin et al. 2019).

FUS belongs to the FET protein family, which also includes EWS and TAF15, all of which are causative genes of familial ALS (Svetoni et al. 2016; Bertolotti et al. 1999; Morohoshi et al. 1998). These three FET family proteins have an N-terminal QGSY-rich region, a highly conserved RRM, an RGG repeat with dimethylated arginine residues, and a C-terminal ZnF motif (Crozat et al. 1993; Prasad et al. 1994; Morohoshi et al. 1998; Iko et al. 2004). FUS is expressed in various tissues and mainly localized to the nucleus within cells. However, some FUS also localizes to the cytoplasm, such as the axons and dendrites of neurons. In vitro analyses show that FUS binds to RNA and single-stranded DNA, but rarely to double-stranded DNA (Crozat et al. 1993; Baechtold et al. 1999; Wang et al. 2008; Zinszner et al. 1997). The actual RNA-binding sites of FUS in vivo have been reported by several groups using CLIP-seq analyses of FUS. The FUS-recognizing RNA motifs are not clear, but they tend to be GU-rich sequences and have secondary structures (Coady and Manley 2015; Ishigaki et al. 2012; Lagier-Tourenne et al. 2012; Lerga et al. 2001; Masuda and Ohno 2016; Masuda et al. 2015; Rogelj et al. 2012; Schwartz et al. 2013; Sun et al. 2015; Udagawa et al. 2015; Yamazaki et al. 2012; Yokoi et al.

2017). FUS is involved in transcriptional regulation34 and reduces the transcription rate of RNA polymerase II by binding to RNA during synthesis. FUS also regulates polyadenylation through RNA binding and is deeply involved in RNA processing, mainly in mRNA length regulation. It is speculated that FUS regulates genes related to neuronal differentiation and synapse formation (Sun et al. 2015; Udagawa et al. 2015; Yamazaki et al. 2012; Yokoi et al. 2017; Kino et al. 2015).

Conversely, FUS is contained in the spliceosome and is known to bind many RNPs, including SFPQ, NONO, hnRNPA1, TDP-43, SMN, and other FET proteins, and is involved in selective splicing (Ishigaki et al. 2017; Coady and Manley 2015; Sun et al. 2015; An et al. 2019; Kahl et al. 2018; Tsuiji et al. 2013). Notably, FUS binds to microtubule associated protein tau (MAPT) and regulates its exon10 splicing. This affects the production of 3-repeat tau and 4-repeat tau, the main isoforms of tau protein. Several groups have reported that the loss of FUS function increases 4-repeat tau, the tau isoform (Ishigaki et al. 2012; Fujioka et al. 2013; Orozco and Edbauer 2013; Orozco et al. 2012). FUS has also been reported to be involved in some DNA repair functions. FUS is involved in DNA damage in neurons by binding to HDAC1. Its function in PARP-dependent DNA repair is impaired by ALS-related mutations (Naumann et al. 2018; Qiu et al. 2014; Singatulina et al. 2019; Wang et al. 2018).

FUS is present in the cytoplasm as well as the nucleus. It is found in dendrites and axons, where it regulates RNA transport, local translation, and axonal guidance (Errichelli et al. 2017; Fujii and Takumi 2005; Sephton et al. 2014).

Whole-body KO mice of FUS have different phenotypes depending on the background. C57B6 inbreds die soon after birth due to immune system abnormalities. Non-inbreds grow normally except for abnormal spermatogenesis (Hicks et al. 2000; Kuroda et al. 2000).

14.3 Relation to the Disease

FUS is a causative gene of familial ALS (ALS6) and is considered to be the second most frequent causative gene of familial ALS in Japan after SOD1 and is found mostly in the C-terminal region, mostly as a point mutation (Suzuki et al. 2023). In pathology, FUS is characterized by a change in localization from the nucleus to the cytoplasm and is found in neurons as a ubiquitin-positive cytoplasmic inclusion body (FUS). ALS is considered to be clinically, pathologically, and genetically on the same spectrum as FTLD, a form of dementia, and there is a group of patients with FTLD who also show FUS-positive pathology and are referred to as FTLD-FUS. FUS, TAF15, and EWSR belong to the FET family RNA binding proteins. Mutations in FET proteins, not only FUS but also TAF15 and EWSR have been causative for familial ALS (Svetoni et al. 2016).

Familial ALS caused by FUS mutations is associated with developmental disorders and psychiatric disorders, some cases present with nonspecific tauopathy due to FUS mutations, and FUS localization changes in the nucleus have been reported not only in ALS and FTLD but also in PSP and CBD, a subset of 4R tauopathies (Ishigaki et al. 2020; Baumer et al. 2010; Ferrer et al. 2015; Yamashita et al. 2012). These suggest that FUS may be involved in a broader range of neuropsychiatric mechanisms.

Similar to the TDP-43 model, FUS overexpression models have been shown to exhibit ALS-like phenotypes (Gao et al. 2017; Scekic-Zahirovic et al. 2016). Although simple overexpression of wild-type FUS may not produce the phenotype, FUS expression is self-regulated at the transcriptional and translational levels, and it is hypothesized that disruption of self-regulation leads to toxicity (Ling et al. 2019). Conversely, animal models of FUS loss in mice, zebrafish, and Drosophila have shown that loss of FUS in neurons of the cerebrum causes FTLD-like higher brain dysfunction (Ishigaki et al. 2017; Udagawa et al. 2015; Kino et al. 2015; Sasayama et al. 2012; Kabashi et al. 2011), although the involvement of loss of function in ALS is thought to be limited because motor neuron-specific KO of FUS does not result in ALS (Scekic-Zahirovic et al. 2016; Sharma et al. 2016). Considering the presence of reduced nuclear colocalization between FUS and SFPQ in the FTLD spectrum, including ALS/FTLD (Ishigaki et al. 2020; Riku et al. 2022b), it is likely that FUS function is also important for neurons in the cerebrum and may be involved in a wide range of neurological disease mechanisms.

14.4 LLPS

Of particular interest in the pathogenesis of FUS is its relationship to liquid–liquid phase separation (LLPS). FUS forms droplets as soon as it is purified and cleaved from the tag, and LLPS varies depending on the mutation, phosphorylation, RNA concentration, and other conditions. The relationship between FUS and stress granules, cytoplasmic inclusions, and FUS complexes in pathological conditions has also attracted attention, and it has been suggested that LLPS may also be involved in pathological conditions for many RNA-binding proteins other than FUS as well (Chong and Forman-Kay 2016; Monahan et al. 2017; Murakami et al. 2015; Murray et al. 2017; Patel et al. 2015; Shiina 2019; Yoshizawa et al. 2018); moreover, genes associated with ALS are more likely involved in LLPS suggesting that LLPS is one of the initial biochemical events to cause ALS pathophysiology (Pakravan et al. 2021; Orlando et al. 2019) (Fig. 14.2).

We have previously identified FUS-binding proteins in the nucleus of neurons by a global screening using immunoprecipitation followed by mass spectrometry (Ishigaki et al. 2017). FUS binds to many RNA-binding proteins which also contain ALS-associated molecules. The tendency to undergo LLPS of those proteins is quantified using the scores of PSPer (Orlando et al. 2019). We obtained many RNA-binding proteins such as several hnRNPs, SFPQ, NONO, PABP1, CPSF6, RU17, SNRPA, and SFRS5; all these LLPS scores are more than 0.3, indicating that FUS associates multiple RNA-binding proteins to make a large complex regulating RNA metabolism and the mechanism is strongly related to LLPS (Table 14.1).

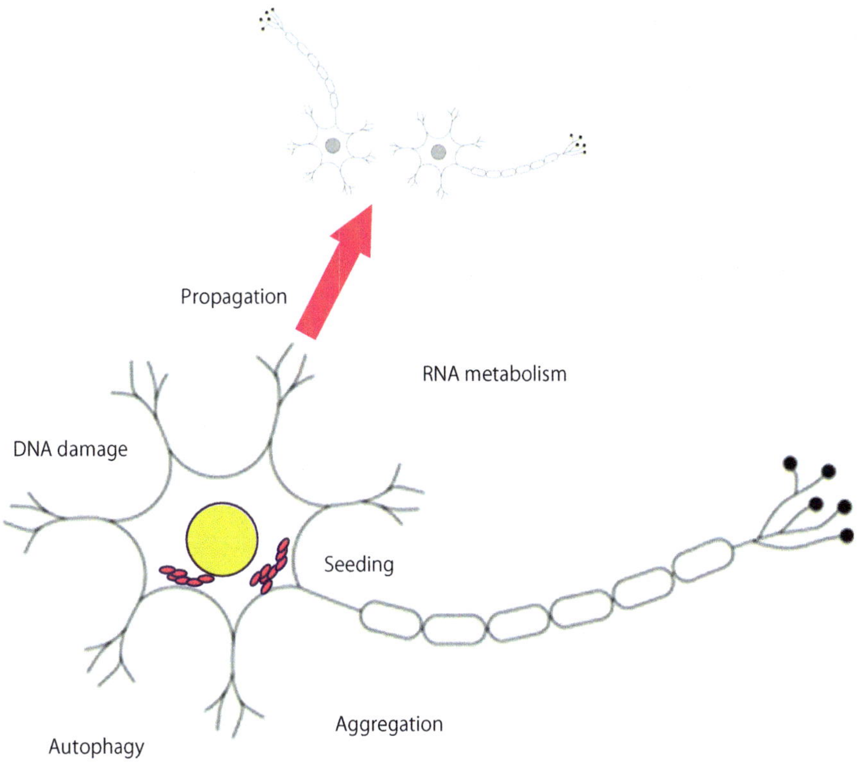

Fig. 14.2 Image of cellular mechanisms related to LLPS and neurodegeneration. LLPS involves many pathological processes, such as RNA metabolism, aggregation formation, autophagy, DNA damage, seeding, and propagation. (The image was adapted from a previous review article (Pakravan et al. 2021))

Moreover, several ALS-associated-RNA binding proteins exhibit high scores (more than 0.6), including FUS, EWS, TAF15, hnRNPA1, hnRNPA2B1, hnRNPA3, and ATXN2. It is noteworthy that the three FET family proteins, FUS, EWS, and TAF15 are in the top three on the list of LLPS. Other molecules with moderate scores (more than 0.3) also include RNA-binding proteins, TDP-43, Matrin3, TIA1, and hnRNPQ (Table 14.2) (Ishigaki et al. 2017; Pakravan et al. 2021).

14.5 Association with SFPQ

As mentioned above, we pathologically analyzed the roles of FUS using postmortem brain tissues. We found that FUS and SFPQ are colocalized in the nucleus of neurons under normal conditions. However, the affected brain lesions of FTLD spectrum disorders such as FTLD, ALS, PSP, and CBD showed impaired nuclear

Table 14.1 LLPS score of FUS-interacting proteins in the nucleus of neuronal cells

Accession #	MW [Da]	Description: Gene name [Gene ID]	FUS-binding score	LLPS score
Q8BX17	166,457	Gem-associated protein 5 [GEMI5]	241.61	0.148
Q921M3	135,465	Splicing factor 3B subunit 3 [SF3B3]	306.25	0.153
Q8VEK3	87,863	Heterogeneous nuclear ribonucleoprotein U [HNRPU]	163.88	0.388
Q9D0E1	77,597	Heterogeneous nuclear ribonucleoprotein M [HNRPM]	110.91	0.402
Q8VIJ6	75,394	Splicing factor, proline- and glutamine-rich [SFPQ]	754.66	0.484
P63017	70,827	Heat shock cognate 71 kDa protein [HSP7C]	84.27	0.151
P29341	70,598	Polyadenylate-binding protein 1 [PABP1]	211.83	0.468
Q7TMK9	69,590	Heterogeneous nuclear ribonucleoprotein Q [HNRPQ]	155.11	0.572
Q61545	68,376	RNA-binding protein EWS [EWS]	78.25	0.857
Q9WV25	60,211	Poly(U)-binding-splicing factor PUF60 [PUF60]	74.83	0.353
Q6NVF9	59,116	Cleavage and polyadenylation specificity factor subunit 6 [CPSF6]	248.24	0.669
Q99K48	54,506	Non-POU domain-containing octamer-binding protein [NONO]	920.84	0.46
P56959	52,642	RNA-binding protein FUS [FUS]	512.58	0.915
Q62376	51,961	U1 small nuclear ribonucleoprotein 70 kDa [RU17]	317.04	0.746
Q794E4	45,701	Heterogeneous nuclear ribonucleoprotein F [HNRPF]	52.91	0.415
Q921F2	44,519	TAR DNA-binding protein 43 [TADBP]	95.70	0.411
Q9Z204	34,364	Heterogeneous nuclear ribonucleoproteins C1/C2 [HNRPC]	186.73	0.419
Q62189	31,814	U1 small nuclear ribonucleoprotein A [SNRPA]	458.94	0.427
P97801	31,234	Survival motor neuron protein [SMN]	159.48	0.277
O35326	30,926	Splicing factor, arginine/serine-rich 5 [SFRS5]	66.95	0.624
P27048	23,640	Small nuclear ribonucleoprotein-associated protein B [RSMB]	78.91	0.358

FUS-binding score was modified from the previous report (Ishigaki et al. 2017). LLPS score was calculated using PSPer, an online tool (Orlando et al. 2019). LLPS scores indicated in red: likely to undergo LLPS, LLPS score > 0.3

colocalization of FUS and SFPQ (Ishigaki et al. 2020; Riku et al. 2022b). We reported that FUS regulates alternative splicing of tau proteins in coordination with SFPQ using a mouse model. Under normal conditions, the two proteins form a high-molecular-weight complex in the nucleus. However, disease-associated mutations in the FUS gene disrupt the complex formation. This leads to unregulated alternative splicing of tau, a disproportional increase in the 4R/3R-tau ratio, and neurodegeneration (Ishigaki et al. 2017). Loss of SFPQ causes neuronal loss in both the developmental and adult stages of mice (Takeuchi et al. 2018). SFPQ is an RBP that has multiple functions. It regulates mRNA processing, transcription, and DNA repair (Yarosh et al. 2015; Knott et al. 2016). SFPQ regulates the transcriptional

Table 14.2 LLPS score of ALS-associated proteins

Gene ID	LLPS score	Aggregation score
ADAR2	0.242	98.368
ANXA11	0.405	76.584
ATXN2	0.633	93.428
C9orf72	0.137	99.952
DNAJC7	0.289	62.24
EWS	0.857	4.521
FUS	0.915	4.473
FIG4	0.285	99.592
hnRNPA1	0.603	92.027
hnRNPA2B1	0.649	55.529
hnRNPA3	0.643	92.132
KIF5A	0.236	60.092
Matrin 3	0.422	7.724
NEK1	0.389	94.682
OPTN	0.197	79.727
SOD1	0.223	14.877
SGMR1	0.173	99.951
SYNCRIP (hnRNPQ)	0.571	74.295
SQSTM	0.321	6.455
TARDBP	0.376	97.627
TAF15	0.801	4.652
TIA1	0.422	47.614
TBK1	0.194	96.825
UBQLN2	0.241	98.684
VCP	0.288	61.728

The data was obtained from the previous report (Pakravan et al. 2021). LLPS scores indicated in red: likely to undergo LLPS, LLPS score > 0.3

elongation of very long genes (≥100 kb) in the brain and muscles (Takeuchi et al. 2018; Hosokawa et al. 2019).

Moreover, mutations in SFPQ occur in some familial ALS cases (Takeuchi et al. 2023; Thomas-Jinu et al. 2017), and loss of SFPQ from neuronal nuclei was observed in familial and sporadic ALS cases (Luisier et al. 2018). A report indicated that a proper LLPS state was necessary to recruit FUS and SFPQ together with other RNA-binding proteins in the complex during DNA damage response (Levone et al. 2021).

Consequently, dysfunction of FUS and SFPQ in neuronal nuclei may be a common pathomechanism across FTLD, ALS, and other FTLD spectrum diseases.

Thus, FUS and its associated RNA-binding proteins that undergo LLPS are involved in essential cellular processes such as alternative splicing, transcription, ncRNA metabolism, RNA transport, and DNA damage response. Deterioration of these mechanisms results in neuronal dysfunction and neurodegeneration.

14.6 ASO Therapeutics to Target the RBPs and the LLPS

There are ongoing research efforts to develop therapeutics that can regulate LLPS in the context of neurodegeneration, including ALS. One approach that has shown promise is targeting specific RNA-binding proteins (RBPs) that are known to drive LLPS in neurons. For example, in ALS, mutations in the RBP TDP-43 have been implicated in the formation of pathological protein aggregates and LLPS in neurons. Potential therapeutic approaches include targeting TDP-43 and other RBPs using small molecule inhibitors, antisense oligonucleotides (ASOs), and gene therapies (Pakravan et al. 2021). Chaperones such as heat shock proteins can be a candidate for LLPS-mediated therapeutics. Liu et al. found that Hsp27 inhibits LLPS of FUS by interacting as a chaperon when this protein is phosphorylated (Liu et al. 2020).

The therapeutic attempts using ASOs directly targeting gene expressions of FUS or TDP-43 showed rescue effects on the pathophysiology of ALS/FTLD model mice (Codron et al. 2022). FUS knock-in mouse lines that express the equivalent of ALS-associated mutant FUS-P525L and FUS-ΔEX14 protein show progressive, age-dependent motor neuron loss. The administration of an ASO targeting FUS gene expression, ION363 reduced postnatal levels of FUS protein in the brain and spinal cord, delaying motor neuron degeneration in mice (Korobeynikov et al. 2022). A Phase 3 safety and efficacy trial of ION363 (jacifusen) in familial ALS patients with mutations in the FUS gene is ongoing in the USA. Takeuchi et al. recently reported the use of gapmer-type ASOs against human TDP-43, modified with 2′-*O*,4′-C-ethylene nucleic acids (ENAs) that provide high stability. They tested the therapeutic potential of ENA-modified ASOs in reducing TDP-43 levels. The selected ENA-modified ASOs were administered intracerebroventricularly in a mouse model of ALS/FTLD expressing human TDP-43, resulting in an efficient reduction of TDP-43 levels in the brain and spinal cord. A single injection of ENA-modified ASOs in TDP-43 mice led to long-lasting improvement of behavioral abnormalities and suppression of cytoplasmic TDP-43 aggregation, even after TDP-43 levels returned to the initial levels (Takeuchi et al. 2023). ATXN2 is a gene encoding RNA-binding protein whose mutants with CAG repeats cause cerebrospinal ataxia. ATXN2 has been also known to be a risk factor for ALS. Moreover, the reduction of ATXN2 extended the lifespan and ALS phenotypes in TDP-43 mice (Becker et al. 2017). BIIB105 is an antisense oligonucleotide that functions by reducing the amount of ataxin-2 in cells. This is achieved by binding to the mRNA of ataxin-2, which is an intermediate molecule produced from the gene and required for protein production, causing its degradation and preventing the production of ataxin-2. Another target for ASO treatment in ALS is C9orf72, in which expansions of a G4C2 repeat yield polydipeptides cause aberrant LLPS status in the cell (Boeynaems et al. 2017; Boros et al. 2022; Tran et al. 2022). Several clinical trials including BIIB075, an ASO targeting the C9orf72 gene have been performed.

Those potential ASO therapeutics target root molecules of ALS and are expected to be effective for familial ALS/FTLD cases; however, there are concerns that the reduction of those gene expressions may cause side effects due to the loss of their innate functions. Although some reports have suggested that loss of FUS function in motor neurons may not contribute to motor neuron degeneration in ALS (Scekic-Zahirovic et al. 2016; Sharma et al. 2016), there is evidence to suggest that loss of FUS function in cerebral neurons can contribute to neuronal dysfunction and neurodegeneration in FTLD. FUS-deficient mice, generated either through silencing or FUS knockout, exhibit behavioral impairments (Udagawa et al. 2015; Kino et al. 2015). The behavioral phenotypes were rescued when wild-type FUS was restored in FUS-silenced mice, whereas a disease-associated mutant did not have the same effect (Ishigaki et al. 2017). Suppression of TDP-43 can cause a motor neuronal loss in mice and repress cryptic exons which may lead to neuronal dysfunction and degeneration (Iguchi et al. 2013; Ma et al. 2022; Liu et al. 2019; Highley et al. 2014; Ling et al. 2015). C9orf72 deficiency resulted in microglial activation and subsequent synaptic loss (Lall et al. 2021). Those findings alert the use of direct gene-silencing to root molecules that are thought to be toxic by a gain-of-toxicity manner.

Since FUS makes complexes with many RNA-binding proteins such as NONO, SFPQ, and hnRNPs, oligonucleotides targeting those RNA-binding proteins themselves are a possible candidate to modulate LLPS instead of targeting specific gene expressions. Unlike TDP-43, FUS does not exhibit definite sequence preferences for its RRM; however, the binding sequence of FUS is known to be a GU-rich sequence represented by GGUG and a sequence that has a stem-loop secondary structure (Fig. 14.3) (Ishigaki et al. 2012; Lerga et al. 2001; Masuda et al. 2015; Hoell et al. 2011).

Another approach is to develop small molecules that can modulate LLPS by altering the biophysical properties of protein condensates. 1,6-hexanediol is an agent that is known to resolve liquid-phase condensates in living cells (Sabari et al. 2018; Ulianov et al. 2021). The effects of 1,6-hexanediol are not protein specific; however, there are attempts to develop therapeutic small molecules which specifically target pathogenic proteins that undergo LLPS during the disease process. For example,

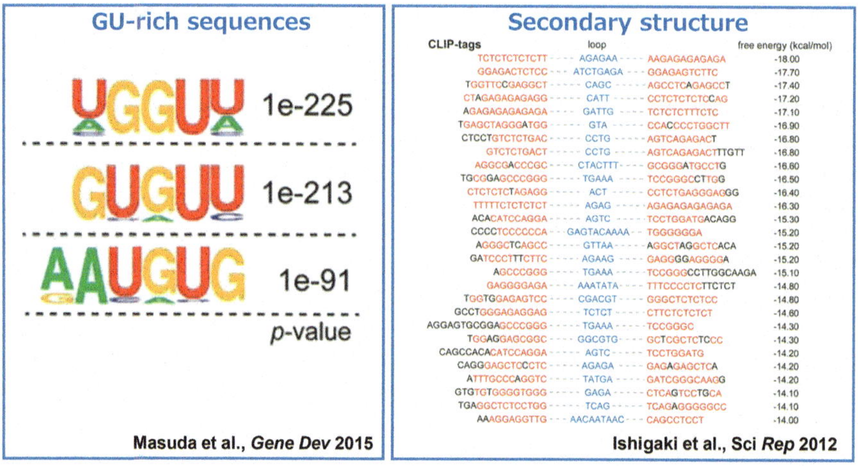

Fig. 14.3 The binding sequence of FUS is shown. GU-rich sequence represented by GGUG (Masuda et al. 2015) and a sequence that has a stem-loop secondary structure (Ishigaki et al. 2012)

researchers have identified several small molecules that can disrupt the LLPS of disease-associated proteins, such as TDP-43, in vitro (Babinchak et al. 2020). These molecules have the potential to be developed into drugs that can prevent or reverse the formation of pathological protein aggregates in neurodegenerative diseases. It has been known that RNA buffers the LLPS behavior of FUS (Maharana et al. 2018), indicating that certain nucleotides can be specific therapeutics to normalize aberrant forms of FUS or other disease-associated RNA-binding proteins. One of the possible approaches is to design candidate oligo sequences based on FUS preference sequences as shown in Fig. 14.3, with or without modification that enables long-lasting effects in the tissue such as LNA, MOE, or ENA.

14.7 Future Direction

Overall, while there is still much work to be done, targeting LLPS represents a promising avenue for the development of new therapeutics for neurodegenerative diseases, including ALS/FTLD in which many RNPs are involved. Recent reports have suggested that FUS is involved in a broader range of ALS/FTLD than the previously reported category of FUS-positive ALS/FTLD (Ishigaki et al. 2020; Tyzack et al. 2019; Fujimori et al. 2018; Ikenaka et al. 2020). The clarification of the direct relationship between the detailed liquid–liquid phase separation and disease pathogenesis is expected to lead to the discovery of new disease biomarkers and therapeutic targets focusing on the structure and functions of FUS and its associated RNPs.

References

An H et al (2019) ALS-linked FUS mutations confer loss and gain of function in the nucleus by promoting excessive formation of dysfunctional paraspeckles. Acta Neuropathol Commun 7:7. https://doi.org/10.1186/s40478-019-0658-x

Babinchak WM et al (2020) Small molecules as potent biphasic modulators of protein liquid-liquid phase separation. Nat Commun 11:5574. https://doi.org/10.1038/s41467-020-19211-z

Baechtold H et al (1999) Human 75-kDa DNA-pairing protein is identical to the pro-oncoprotein TLS/FUS and is able to promote D-loop formation. J Biol Chem 274:34337–34342. https://doi.org/10.1074/jbc.274.48.34337

Baumer D et al (2010) Juvenile ALS with basophilic inclusions is a FUS proteinopathy with FUS mutations. Neurology 75:611–618. https://doi.org/10.1212/WNL.0b013e3181ed9cde

Becker LA et al (2017) Therapeutic reduction of ataxin-2 extends lifespan and reduces pathology in TDP-43 mice. Nature 544:367–371. https://doi.org/10.1038/nature22038

Bertolotti A, Bell B, Tora L (1999) The N-terminal domain of human TAFII68 displays transactivation and oncogenic properties. Oncogene 18:8000–8010. https://doi.org/10.1038/sj.onc.1203207

Boeynaems S et al (2017) Phase separation of C9orf72 dipeptide repeats perturbs stress granule dynamics. Mol Cell 65:1044–1055.e5. https://doi.org/10.1016/j.molcel.2017.02.013

Boros BD, Schoch KM, Kreple CJ, Miller TM (2022) Antisense oligonucleotides for the study and treatment of ALS. Neurotherapeutics 19:1145–1158. https://doi.org/10.1007/s13311-022-01247-2

Calvio C, Neubauer G, Mann M, Lamond AI (1995) Identification of hnRNP P2 as TLS/FUS using electrospray mass spectrometry. RNA 1:724–733

Chong PA, Forman-Kay JD (2016) Liquid-liquid phase separation in cellular signaling systems. Curr Opin Struct Biol 41:180–186. https://doi.org/10.1016/j.sbi.2016.08.001

Coady TH, Manley JL (2015) ALS mutations in TLS/FUS disrupt target gene expression. Genes Dev 29:1696–1706. https://doi.org/10.1101/gad.267286.115

Codron P, Cassereau J, Vourc'h P (2022) InFUSing antisense oligonucleotides for treating ALS. Trends Mol Med 28:253–254. https://doi.org/10.1016/j.molmed.2022.02.006

Crozat A, Aman P, Mandahl N, Ron D (1993) Fusion of CHOP to a novel RNA-binding protein in human myxoid liposarcoma. Nature 363:640–644. https://doi.org/10.1038/363640a0

Eneroth M et al (1990) Localization of the chromosomal breakpoints of the t(12;16) in liposarcoma to subbands 12q13.3 and 16p11.2. Cancer Genet Cytogenet 48:101–107. https://doi.org/10.1016/0165-4608(90)90222-v

Errichelli L et al (2017) FUS affects circular RNA expression in murine embryonic stem cell-derived motor neurons. Nat Commun 8:14741. https://doi.org/10.1038/ncomms14741

Ferrer I et al (2015) Familial behavioral variant frontotemporal dementia associated with astrocyte-predominant tauopathy. J Neuropathol Exp Neurol 74:370–379. https://doi.org/10.1097/NEN.0000000000000180

Fujii R, Takumi T (2005) TLS facilitates transport of mRNA encoding an actin-stabilizing protein to dendritic spines. J Cell Sci 118:5755–5765. https://doi.org/10.1242/jcs.02692

Fujimori K et al (2018) Modeling sporadic ALS in iPSC-derived motor neurons identifies a potential therapeutic agent. Nat Med 24:1579–1589. https://doi.org/10.1038/s41591-018-0140-5

Fujioka Y et al (2013) FUS-regulated region- and cell-type-specific transcriptome is associated with cell selectivity in ALS/FTLD. Sci Rep 3:2388. https://doi.org/10.1038/srep02388

Gao FB, Almeida S, Lopez-Gonzalez R (2017) Dysregulated molecular pathways in amyotrophic lateral sclerosis-frontotemporal dementia spectrum disorder. EMBO J 36:2931–2950. https://doi.org/10.15252/embj.201797568

Hardiman O et al (2017) Amyotrophic lateral sclerosis. Nat Rev Dis Primers 3:17071. https://doi.org/10.1038/nrdp.2017.71

Hicks GG et al (2000) Fus deficiency in mice results in defective B-lymphocyte development and activation, high levels of chromosomal instability and perinatal death. Nat Genet 24:175–179. https://doi.org/10.1038/72842

Highley JR et al (2014) Loss of nuclear TDP-43 in amyotrophic lateral sclerosis (ALS) causes altered expression of splicing machinery and widespread dysregulation of RNA splicing in motor neurons. Neuropathol Appl Neurobiol 40:670–685. https://doi.org/10.1111/nan.12148

Hoell JI et al (2011) RNA targets of wild-type and mutant FET family proteins. Nat Struct Mol Biol 18:1428–1431. https://doi.org/10.1038/nsmb.2163

Hosokawa M et al (2019) Loss of RNA-binding protein Sfpq causes long-gene Transcriptopathy in skeletal muscle and severe muscle mass reduction with metabolic myopathy. iScience 13:229–242. https://doi.org/10.1016/j.isci.2019.02.023

Iguchi Y et al (2013) Loss of TDP-43 causes age-dependent progressive motor neuron degeneration. Brain 136:1371–1382. https://doi.org/10.1093/brain/awt029

Ikenaka K et al (2020) Characteristic features of FUS inclusions in spinal motor neurons of sporadic amyotrophic lateral sclerosis. J Neuropathol Exp Neurol 79:370–377. https://doi.org/10.1093/jnen/nlaa003

Iko Y et al (2004) Domain architectures and characterization of an RNA-binding protein, TLS. J Biol Chem 279:44834–44840. https://doi.org/10.1074/jbc.M408552200

Ishigaki S, Sobue G (2018) Importance of functional loss of FUS in FTLD/ALS. Front Mol Biosci 5:44. https://doi.org/10.3389/fmolb.2018.00044

Ishigaki S et al (2012) Position-dependent FUS-RNA interactions regulate alternative splicing events and transcriptions. Sci Rep 2:529. https://doi.org/10.1038/srep00529

Ishigaki S et al (2017) Altered tau isoform ratio caused by loss of FUS and SFPQ function leads to FTLD-like phenotypes. Cell Rep 18:1118–1131. https://doi.org/10.1016/j.celrep.2017.01.013

Ishigaki S et al (2020) Aberrant interaction between FUS and SFPQ in neurons in a wide range of FTLD spectrum diseases. Brain 143:2398–2405. https://doi.org/10.1093/brain/awaa196

Josephs KA et al (2006) Clinicopathologic analysis of frontotemporal and corticobasal degenerations and PSP. Neurology 66:41–48. https://doi.org/10.1212/01.wnl.0000191307.69661.c3

Kabashi E et al (2011) FUS and TARDBP but not SOD1 interact in genetic models of amyotrophic lateral sclerosis. PLoS Genet 7:e1002214. https://doi.org/10.1371/journal.pgen.1002214

Kahl A et al (2018) Cerebral ischemia induces the aggregation of proteins linked to neurodegenerative diseases. Sci Rep 8:2701. https://doi.org/10.1038/s41598-018-21063-z

Kino Y et al (2015) FUS/TLS deficiency causes behavioral and pathological abnormalities distinct from amyotrophic lateral sclerosis. Acta Neuropathol Commun 3:24. https://doi.org/10.1186/s40478-015-0202-6

Knott GJ, Bond CS, Fox AH (2016) The DBHS proteins SFPQ, NONO and PSPC1: a multipurpose molecular scaffold. Nucleic Acids Res 44:3989–4004. https://doi.org/10.1093/nar/gkw271

Korobeynikov VA, Lyashchenko AK, Blanco-Redondo B, Jafar-Nejad P, Shneider NA (2022) Antisense oligonucleotide silencing of FUS expression as a therapeutic approach in amyotrophic lateral sclerosis. Nat Med 28:104–116. https://doi.org/10.1038/s41591-021-01615-z

Kuroda M et al (2000) Male sterility and enhanced radiation sensitivity in TLS(-/-) mice. EMBO J 19:453–462. https://doi.org/10.1093/emboj/19.3.453

Kwiatkowski TJ Jr et al (2009) Mutations in the FUS/TLS gene on chromosome 16 cause familial amyotrophic lateral sclerosis. Science 323:1205–1208. https://doi.org/10.1126/science.1166066

Laforce R Jr (2013) Behavioral and language variants of frontotemporal dementia: a review of key symptoms. Clin Neurol Neurosurg 115:2405–2410. https://doi.org/10.1016/j.clineuro.2013.09.031

Lagier-Tourenne C, Cleveland DW (2009) Rethinking ALS: the FUS about TDP-43. Cell 136:1001–1004. https://doi.org/10.1016/j.cell.2009.03.006

Lagier-Tourenne C et al (2012) Divergent roles of ALS-linked proteins FUS/TLS and TDP-43 intersect in processing long pre-mRNAs. Nat Neurosci 15:1488–1497. https://doi.org/10.1038/nn.3230

Lall D et al (2021) C9orf72 deficiency promotes microglial-mediated synaptic loss in aging and amyloid accumulation. Neuron 109:2275–2291.e8. https://doi.org/10.1016/j.neuron.2021.05.020
Lerga A et al (2001) Identification of an RNA binding specificity for the potential splicing factor TLS. J Biol Chem 276:6807–6816. https://doi.org/10.1074/jbc.M008304200
Levone BR et al (2021) FUS-dependent liquid-liquid phase separation is important for DNA repair initiation. J Cell Biol 220:e202008030. https://doi.org/10.1083/jcb.202008030
Ling JP, Pletnikova O, Troncoso JC, Wong PC (2015) TDP-43 repression of nonconserved cryptic exons is compromised in ALS-FTD. Science 349:650–655. https://doi.org/10.1126/science.aab0983
Ling SC et al (2019) Overriding FUS autoregulation in mice triggers gain-of-toxic dysfunctions in RNA metabolism and autophagy-lysosome axis. Elife 8:e40811. https://doi.org/10.7554/eLife.40811
Liu EY et al (2019) Loss of nuclear TDP-43 is associated with decondensation of LINE retrotransposons. Cell Rep 27:1409–1421.e6. https://doi.org/10.1016/j.celrep.2019.04.003
Liu Z et al (2020) Hsp27 chaperones FUS phase separation under the modulation of stress-induced phosphorylation. Nat Struct Mol Biol 27:363–372. https://doi.org/10.1038/s41594-020-0399-3
Loughlin FE et al (2019) The solution structure of FUS bound to RNA reveals a bipartite mode of RNA recognition with both sequence and shape specificity. Mol Cell 73:490–504.e6. https://doi.org/10.1016/j.molcel.2018.11.012
Luisier R et al (2018) Intron retention and nuclear loss of SFPQ are molecular hallmarks of ALS. Nat Commun 9:2010. https://doi.org/10.1038/s41467-018-04373-8
Ma XR et al (2022) TDP-43 represses cryptic exon inclusion in the FTD-ALS gene UNC13A. Nature 603:124–130. https://doi.org/10.1038/s41586-022-04424-7
Maharana S et al (2018) RNA buffers the phase separation behavior of prion-like RNA binding proteins. Science 360:918–921. https://doi.org/10.1126/science.aar7366
Masuda A, Ohno K (2016) Regulation of mRNA length by FUS. Seikagaku 88:244–247
Masuda A et al (2015) Position-specific binding of FUS to nascent RNA regulates mRNA length. Genes Dev 29:1045–1057. https://doi.org/10.1101/gad.255737.114
Masuda M et al (2016) Involvement of the caudate nucleus head and its networks in sporadic amyotrophic lateral sclerosis-frontotemporal dementia continuum. Amyotroph Lateral Scler Frontotemporal Degener 17:571–579. https://doi.org/10.1080/21678421.2016.1211151
Monahan Z et al (2017) Phosphorylation of the FUS low-complexity domain disrupts phase separation, aggregation, and toxicity. EMBO J 36:2951–2967. https://doi.org/10.15252/embj.201696394
Morohoshi F et al (1998) Genomic structure of the human RBP56/hTAFII68 and FUS/TLS genes. Gene 221:191–198. https://doi.org/10.1016/s0378-1119(98)00463-6
Murakami T et al (2015) ALS/FTD mutation-induced phase transition of FUS liquid droplets and reversible hydrogels into irreversible hydrogels impairs RNP granule function. Neuron 88:678–690. https://doi.org/10.1016/j.neuron.2015.10.030
Murray DT et al (2017) Structure of FUS protein fibrils and its relevance to self-assembly and phase separation of low-complexity domains. Cell 171:615–627.e16. https://doi.org/10.1016/j.cell.2017.08.048
Naumann M et al (2018) Impaired DNA damage response signaling by FUS-NLS mutations leads to neurodegeneration and FUS aggregate formation. Nat Commun 9:335. https://doi.org/10.1038/s41467-017-02299-1
Olney NT, Spina S, Miller BL (2017) Frontotemporal dementia. Neurol Clin 35:339–374. https://doi.org/10.1016/j.ncl.2017.01.008
Orlando G et al (2019) Computational identification of prion-like RNA-binding proteins that form liquid phase-separated condensates. Bioinformatics 35:4617–4623. https://doi.org/10.1093/bioinformatics/btz274

Orozco D, Edbauer D (2013) FUS-mediated alternative splicing in the nervous system: consequences for ALS and FTLD. J Mol Med (Berl) 91:1343–1354. https://doi.org/10.1007/s00109-013-1077-2

Orozco D et al (2012) Loss of fused in sarcoma (FUS) promotes pathological tau splicing. EMBO Rep 13:759–764. https://doi.org/10.1038/embor.2012.90

Pakravan D, Orlando G, Bercier V, Van Den Bosch L (2021) Role and therapeutic potential of liquid-liquid phase separation in amyotrophic lateral sclerosis. J Mol Cell Biol 13:15–28. https://doi.org/10.1093/jmcb/mjaa049

Patel A et al (2015) A liquid-to-solid phase transition of the ALS protein FUS accelerated by disease mutation. Cell 162:1066–1077. https://doi.org/10.1016/j.cell.2015.07.047

Prasad DD, Ouchida M, Lee L, Rao VN, Reddy ES (1994) TLS/FUS fusion domain of TLS/FUS-erg chimeric protein resulting from the t(16;21) chromosomal translocation in human myeloid leukemia functions as a transcriptional activation domain. Oncogene 9:3717–3729

Qiu H et al (2014) ALS-associated mutation FUS-R521C causes DNA damage and RNA splicing defects. J Clin Invest 124:981–999. https://doi.org/10.1172/JCI72723

Rabbitts TH, Forster A, Larson R, Nathan P (1993) Fusion of the dominant negative transcription regulator CHOP with a novel gene FUS by translocation t(12;16) in malignant liposarcoma. Nat Genet 4:175–180. https://doi.org/10.1038/ng0693-175

Renton AE, Chio A, Traynor BJ (2014) State of play in amyotrophic lateral sclerosis genetics. Nat Neurosci 17:17–23. https://doi.org/10.1038/nn.3584

Riku Y et al (2014) Lower motor neuron involvement in TAR DNA-binding protein of 43 kDa-related frontotemporal lobar degeneration and amyotrophic lateral sclerosis. JAMA Neurol 71:172–179. https://doi.org/10.1001/jamaneurol.2013.5489

Riku Y et al (2022a) TDP-43 proteinopathy and tauopathy: do they have pathomechanistic links? Int J Mol Sci 23:15755. https://doi.org/10.3390/ijms232415755

Riku Y et al (2022b) Motor neuron TDP-43 proteinopathy in progressive supranuclear palsy and corticobasal degeneration. Brain 145:2769–2784. https://doi.org/10.1093/brain/awac091

Robberecht W, Philips T (2013) The changing scene of amyotrophic lateral sclerosis. Nat Rev Neurosci 14:248–264. https://doi.org/10.1038/nrn3430

Rogelj B et al (2012) Widespread binding of FUS along nascent RNA regulates alternative splicing in the brain. Sci Rep 2:603. https://doi.org/10.1038/srep00603

Sabari BR et al (2018) Coactivator condensation at super-enhancers links phase separation and gene control. Science 361:eaar3958. https://doi.org/10.1126/science.aar3958

Sasayama H et al (2012) Knockdown of the Drosophila fused in sarcoma (FUS) homologue causes deficient locomotive behavior and shortening of motoneuron terminal branches. PLoS One 7: e39483. https://doi.org/10.1371/journal.pone.0039483

Scekic-Zahirovic J et al (2016) Toxic gain of function from mutant FUS protein is crucial to trigger cell autonomous motor neuron loss. EMBO J 35:1077–1097. https://doi.org/10.15252/embj.201592559

Schwartz JC, Wang X, Podell ER, Cech TR (2013) RNA seeds higher-order assembly of FUS protein. Cell Rep 5:918–925. https://doi.org/10.1016/j.celrep.2013.11.017

Sephton CF et al (2014) Activity-dependent FUS dysregulation disrupts synaptic homeostasis. Proc Natl Acad Sci U S A 111:E4769–E4778. https://doi.org/10.1073/pnas.1406162111

Sharma A et al (2016) ALS-associated mutant FUS induces selective motor neuron degeneration through toxic gain of function. Nat Commun 7:10465. https://doi.org/10.1038/ncomms10465

Shelkovnikova TA et al (2018) Protective paraspeckle hyper-assembly downstream of TDP-43 loss of function in amyotrophic lateral sclerosis. Mol Neurodegener 13:30. https://doi.org/10.1186/s13024-018-0263-7

Shiina N (2019) Liquid- and solid-like RNA granules form through specific scaffold proteins and combine into biphasic granules. J Biol Chem 294:3532–3548. https://doi.org/10.1074/jbc.RA118.005423

Singatulina AS et al (2019) PARP-1 activation directs FUS to DNA damage sites to form PARG-reversible compartments enriched in damaged DNA. Cell Rep 27:1809–1821.e5. https://doi.org/10.1016/j.celrep.2019.04.031

Sobue G, Ishigaki S, Watanabe H (2018) Pathogenesis of frontotemporal lobar degeneration: insights from loss of function theory and early involvement of the caudate nucleus. Front Neurosci 12:473. https://doi.org/10.3389/fnins.2018.00473

Sun S et al (2015) ALS-causative mutations in FUS/TLS confer gain and loss of function by altered association with SMN and U1-snRNP. Nat Commun 6:6171. https://doi.org/10.1038/ncomms7171

Suzuki N, Nishiyama A, Warita H, Aoki M (2023) Genetics of amyotrophic lateral sclerosis: seeking therapeutic targets in the era of gene therapy. J Hum Genet 68:131–152. https://doi.org/10.1038/s10038-022-01055-8

Svetoni F, Frisone P, Paronetto MP (2016) Role of FET proteins in neurodegenerative disorders. RNA Biol 13:1089–1102. https://doi.org/10.1080/15476286.2016.1211225

Takeuchi A et al (2018) Loss of Sfpq causes long-gene transcriptopathy in the brain. Cell Rep 23:1326–1341. https://doi.org/10.1016/j.celrep.2018.03.141

Takeuchi T et al (2023) Sustained therapeutic benefits by transient reduction of TDP-43 using ENA-modified antisense oligonucleotides in ALS/FTD mice. Mol Ther Nucleic Acids 31:353–366. https://doi.org/10.1016/j.omtn.2023.01.006

Thomas-Jinu S et al (2017) Non-nuclear pool of splicing factor SFPQ regulates axonal transcripts required for normal motor development. Neuron 94:322–336.e5. https://doi.org/10.1016/j.neuron.2017.03.026

Tran H et al (2022) Suppression of mutant C9orf72 expression by a potent mixed backbone antisense oligonucleotide. Nat Med 28:117–124. https://doi.org/10.1038/s41591-021-01557-6

Tsuiji H et al (2013) Spliceosome integrity is defective in the motor neuron diseases ALS and SMA. EMBO Mol Med 5:221–234. https://doi.org/10.1002/emmm.201202303

Tsujimoto M et al (2011) Behavioral changes in early ALS correlate with voxel-based morphometry and diffusion tensor imaging. J Neurol Sci 307:34–40. https://doi.org/10.1016/j.jns.2011.05.025

Tyzack GE et al (2019) Widespread FUS mislocalization is a molecular hallmark of amyotrophic lateral sclerosis. Brain 142:2572–2580. https://doi.org/10.1093/brain/awz217

Udagawa T et al (2015) FUS regulates AMPA receptor function and FTLD/ALS-associated behaviour via GluA1 mRNA stabilization. Nat Commun 6:7098. https://doi.org/10.1038/ncomms8098

Ulianov SV et al (2021) Suppression of liquid-liquid phase separation by 1,6-hexanediol partially compromises the 3D genome organization in living cells. Nucleic Acids Res 49:10524–10541. https://doi.org/10.1093/nar/gkab249

Vance C et al (2009) Mutations in FUS, an RNA processing protein, cause familial amyotrophic lateral sclerosis type 6. Science 323:1208–1211. https://doi.org/10.1126/science.1165942

Wang X et al (2008) Induced ncRNAs allosterically modify RNA-binding proteins in cis to inhibit transcription. Nature 454:126–130. https://doi.org/10.1038/nature06992

Wang H et al (2018) Mutant FUS causes DNA ligation defects to inhibit oxidative damage repair in amyotrophic lateral sclerosis. Nat Commun 9:3683. https://doi.org/10.1038/s41467-018-06111-6

Yamashita S et al (2012) Sporadic juvenile amyotrophic lateral sclerosis caused by mutant FUS/TLS: possible association of mental retardation with this mutation. J Neurol 259:1039–1044. https://doi.org/10.1007/s00415-011-6292-6

Yamazaki T et al (2012) FUS-SMN protein interactions link the motor neuron diseases ALS and SMA. Cell Rep 2:799–806. https://doi.org/10.1016/j.celrep.2012.08.025

Yarosh CA, Iacona JR, Lutz CS, Lynch KW (2015) PSF: nuclear busy-body or nuclear facilitator? Wiley Interdiscip Rev RNA 6:351–367. https://doi.org/10.1002/wrna.1280

Yokoi S et al (2017) 3′UTR length-dependent control of SynGAP isoform alpha2 mRNA by FUS and ELAV-like proteins promotes dendritic spine maturation and cognitive function. Cell Rep 20:3071–3084. https://doi.org/10.1016/j.celrep.2017.08.100

Yoshizawa T et al (2018) Nuclear import receptor inhibits phase separation of FUS through binding to multiple sites. Cell 173:693–705.e22. https://doi.org/10.1016/j.cell.2018.03.003

Zinszner H, Sok J, Immanuel D, Yin Y, Ron D (1997) TLS (FUS) binds RNA in vivo and engages in nucleo-cytoplasmic shuttling. J Cell Sci 110(pt 15):1741–1750

Chapter 15
Microglia Lipid Droplets in Physiology and Neurodegeneration

Elizabeth West and Christopher Glass

Abstract Lipid droplets (LDs) are dynamic organelles found in organisms ranging from yeast to humans. Originally thought of as inert lipid-storing structures, a growing body of research suggests LDs have roles beyond supplying lipids for metabolism and membrane expansion. Lipid droplets are present under normal physiological conditions in certain cell types, such as hepatocytes and adipocytes. In these metabolically active cells, LDs are thought to primarily serve as lipid reservoirs (Walther and Farese, Annu Rev Biochem 81:687–714, 2012). In other cell types, such as macrophages and glia, lipid droplets primarily form under pathological conditions. For instance, macrophages form lipid laden foam cells during atherosclerosis, and microglia, which are brain resident macrophages, accumulate lipid droplets in many neurodegenerative diseases (Ralhan et al., J Cell Biol 220:e202102136, 2021; Gutierrez, Arq Bras Cardiol 119:542–543, 2022). Understanding the roles of lipid droplets across cell types in health and disease is an area of active research. The focus of this book chapter is to provide a brief overview of general lipid droplet biology followed by a discussion on our current knowledge of the formation and function of lipid droplets in microglia.

Keywords Lipid droplet · Microglia · Multiple sclerosis · Alzheimer's disease · Aging

E. West
Division of Biological Sciences, University of California, San Diego, La Jolla, CA, USA

Department of Cellular & Molecular Medicine, University of San Diego, La Jolla, CA, USA

C. Glass (✉)
Department of Cellular & Molecular Medicine, University of San Diego, La Jolla, CA, USA
e-mail: ckg@ucsd.edu

R. Kurokawa (ed.), *Phase Separation in Living Cells*,
https://doi.org/10.1007/978-981-99-4886-4_15

15.1 Introduction to Lipid Droplets

In the cell, neutral lipids are stored in phase separated lipid droplets, which consist of a phospholipid monolayer surrounding a neutral lipid core. Lipid droplet formation starts in the endoplasmic reticulum (ER) with the synthesis of neutral lipids. Triacylglycerides (TGs) and cholesterol esters (CEs) are the most abundant neutral lipids and are produced through the action of diacylglycerol acyltransferases (DGAT1 and DGAT2) and acyl-coenzyme A:cholesterol acyltransferase 1/sterol *O*-acyltransferase 1 (ACAT1/SOAT1), respectively (Buhman et al. 2001). Neutral lipids diffuse freely in the ER, but above a threshold concentration, spontaneous de-mixing occurs and the lipids condense into a lens. This process, called nucleation, is influenced by lipid composition. For example, increasing cholesterol or diacylglycerol concentrations favors nucleation (Thiam and Ikonen 2021). After nucleation, growth and budding occur. The budding process is influenced by phospholipid concentration and proteins, including the fat storage-inducing transmembrane proteins (FITs), perilipins, and seipin (Olzmann and Carvalho 2019). Exogenous lipids can also be incorporated into LD. For example, dietary and liver-derived lipids can be transferred via lipoproteins throughout the body for storage in adipocyte LDs. Additionally, lipid accumulation in macrophages can be driven by uptake of modified lipoproteins (Gonzales and Orlando 2007; Chistiakov et al. 2016).

Proteins associate with the lipid droplet surface and can localize to the LD surface from the ER and cytosol. Although LD proteomes differ between cell types, there is likely a core set of proteins involved in lipid metabolism, maintaining LD structure, and LD trafficking (Kory et al. 2016). Proteins that incorporate into lipid droplets are divided into two classes. Class I proteins localize to the ER and are incorporated into LD through an unknown mechanism. Class II proteins associate with the LD via the cytosol and are thought to associate with LDs primarily through amphipathic helices. How class II proteins preferentially associate with LDs versus other organelles is not fully understood. LD proteins mediate interactions with many different organelles. For example, lipid droplets can be degraded by lysosomes through chaperone mediated autophagy. They have also been shown to interact with the mitochondria, peroxisomes, and other lipid droplets. The proteins that mediate interactions with other organelles is an active area of research.

A growing number of functions have been ascribed to lipid droplets. In periods of cell growth or nutrient deprivation, lipids stored in LDs are released for use by lipolysis or lipophagy (Zechner et al. 2017). One of the proteins involved in lipolysis is hormone sensitive lipase (HSL) which can be activated by a number of signals, including insulin. Once activated, HSL associates with the lipid droplet surface where it can metabolize lipids for use as energy by peripheral tissues (Grabner et al. 2021). Beyond metabolic roles, lipid droplets may protect against pathogens and lipid toxicity by sequestering free fatty acids and peroxidated lipids, and evidence supports their roles in protecting against ER and mitochondrial stress. (Olzmann and Carvalho 2019; Bosch et al. 2020; Moulton et al. 2021). While lipid droplets mediate many beneficial processes, evidence suggests that in some cases, they may be detrimental. This is particularly the case for microglia in which lipid droplets form under numerous pathological conditions. For example, lipid droplet accumulating microglia have been described in aging and neurodegeneration where they appear to become proinflammatory and dysfunctional. Understanding the conditions and mechanisms of lipid droplet formation, in addition to their functional consequences is an active area of research. What follows is a discussion of our current knowledge of microglial lipid droplet biology.

15.2 Conditions in Which Lipid Droplet Accumulating Microglia Are Found

The brain is one of the most lipid rich organs in the body, owing in part to the cholesterol rich myelin produced by oligodendrocytes. As the brain's major phagocytes, microglia must be able to process engulfed lipids upon cell death and demyelinating injury. Under certain conditions, microglia accumulate lipids and form what are referred to as lipid droplet accumulating microglia or foamy microglia. The presence of lipid droplets in the brain was first described in 1907 by Alois Alzheimer who observed "glial cells showing adipose saccules" in the brains of deceased dementia patients (Alzheimer et al. 1995). However, it was only recently that these lipid laden cells became an active area of research. We now know microglia form lipid droplets under numerous conditions in vivo and in vitro, and we are beginning to understand their functions and contributions to diseases.

One of the most well characterized conditions in which lipid accumulating microglia occur is multiple sclerosis (MS). MS is an autoimmune disorder that leads to demyelination in the central nervous system. The disease is characterized by extensive immune cell infiltration, including T cells, B cells, and macrophages (Filippi et al. 2018). Postmortem human brain tissue from MS patients contains foamy macrophages and microglia in the lesion center and periphery. These lipid laden cells are enriched in free and esterified cholesterol (Boven et al. 2006; Bogie et al. 2020). They occur early in the demyelination process and are found near the periphery as the lesion becomes inactive (Kuhlmann et al. 2017). The extent of microglia and macrophage activation and lipid accumulation seems to depend on where the lesion occurs and whether the lesion is active or inactive (Filippi et al. 2018).

Mouse models of demyelination and MS have been crucial in expanding our understanding of lipid accumulating microglia. The two most common demyelinating models are cuprizone and lysolecithin which can be used to study the process of focal demyelination and remyelination. Lysolecithin injection activates phospholipase A_2 leading to demyelination at the injection site. Immune cells initially infiltrate the lesion, but inflammation resolves and remyelination occurs within 5–6 weeks. Cuprizone is a copper chelator delivered in the diet that causes demyelination in white matter areas of the brain. Upon cuprizone removal, remyelination occurs within 3–4 weeks. Prolonged cuprizone feeding can lead to oligodendrocyte depletion and chronic demyelination. To model the demyelination and neuroinflammatory aspects of MS, the experimental autoimmune encephalomyelitis (EAE) model is commonly used, where mice are injected with an adjuvant and peptides derived from myelin proteins. Myelin specific T cells infiltrate the brain and induce demyelination (Denic et al. 2011). Similar to what has been observed in tissue from human MS patients, foamy macrophages and microglia occur at lesion sites in mouse models of demyelination, MS, spinal cord injury, and traumatic brain injury (Wang et al. 2015; Cantuti-Castelvetri et al. 2018; Bogie et al. 2020; Haidar et al. 2022; Loix et al. 2022; Zambusi et al. 2022). Lipid accumulating microglia and macrophages can be induced and studied in vitro by treating cells with purified myelin (Bogie et al. 2012; Xu et al. 2021). These in vivo and in vitro models have helped identify receptors that mediate myelin uptake and lipid accumulation, including Fc, scavenger, complement, and other receptors that have been reviewed elsewhere (Grajchen et al. 2018).

Lipid accumulating microglia have also been described in the context of aging and aging related neurodegenerative diseases such as Alzheimer's and Parkinson's disease (Brekk et al. 2020; Claes et al. 2021; Smith et al. 2023). Several studies reported increased microglial LD accumulation in aged mice and humans compared to young individuals (Shimabukuro et al. 2016; Marschallinger et al. 2020; Arbaizar-Rovirosa et al. 2023). One study showed that plasma from aged mice induced lipid droplet formation in mouse microglia-like BV2 cells; however, the factors responsible for the increase were not identified (Marschallinger et al. 2020). Aged mice treated with stroke or demyelinating injury have enhanced lipid accumulation and worse disease outcomes compared to younger mice, raising the question of whether increased baseline lipid accumulation worsens disease outcomes and contributes to the onset of aging related degeneration, such as AD (Cantuti-Castelvetri et al. 2018; Arbaizar-Rovirosa et al. 2023). Although glial LD accumulation was documented over a century ago in AD, there are few reports about lipid accumulating microglia in mouse models of AD. One study that characterized their presence used a chimeric mouse model in which human microglia with an AD associated TREM2 mutation were implanted into a mouse model of AD (5xFAD). Whether more commonly used AD mouse models lack this pathology, or whether it has not been studied is unclear. Interestingly, studies using iPSC derived microglia generated from AD patients with the APOE4 risk alleles show increased lipid accumulation under baseline conditions. Evidence suggests that lipid accumulation in these cells could be due to defective lipid processing (Machlovi et al. 2022; Tcw et al. 2022; Victor et al. 2022). Another possible cause of increased lipid accumulation in AD could be due to increased myelin uptake from degenerating neurons, as observed in MS. However, an analysis of lipid droplet composition in aged mouse microglia found that cholesterol esters accounted for less than 1% of the total lipids. Triacylglycerides and diacylglycerides accounted for roughly a third of the lipids. Furthermore, the lipogenesis gene, *ACLY,* was slightly upregulated in LD high microglia from aged mice (Marschallinger et al. 2020). These results suggest that altered lipid metabolism, through either increased synthesis or reduced clearance, could contribute to age and AD related lipid accumulation.

Mitochondrial dysfunction and inflammation have also been linked to lipid droplet formation. Extensive work in *Drosophila* has shown that increasing neuronal reactive oxygen species (ROS) through mutations that disrupt neuronal mitochondrial function leads to glial LD formation (Liu et al. 2015, 2017; Muliyil et al. 2020; Moulton et al. 2021). The mechanism involves neuronal lipid synthesis and transfer to glial cells. Follow-up work in a mouse model that lacks the mitochondrial complex I subunit, Ndufs4, found that lipid droplets colocalize with microglia and astrocytes (Liu et al. 2015). In addition to ROS-induced LD formation, sterile inflammation, through LPS treatment induces LD formation in vitro and in vivo (Khatchadourian et al. 2012; Marschallinger et al. 2020). Microglia LDs induced by LPS treatment are enriched in phospholipids and glycerolipids. Treatment with the long-chain acyl-CoA synthase (ACSL) inhibitor, Triacsin C, prevents LPS induced lipid droplet production, suggesting de novo lipogenesis is responsible for lipid droplet accumulation. Additional studies are needed to understand the relative contributions of lipid uptake and de novo lipogenesis to LD formation in the many conditions in which they are found.

There is also evidence that lipid rich environments induce lipid droplet formation. BV2 cells form lipid droplets when treated with oleic acid (Khatchadourian et al. 2012, Marschallinger et al. 2020). This finding is corroborated by other studies that have used lipid treatment to induce lipid droplet formation in macrophages in vitro (van Dierendonck et al. 2020). There is also evidence for this in vivo. Mice fed a western type diet have a higher percentage of saturated lipids in the brain and an increased number of lipid droplets in microglia and astrocytes. Western diet fed mice treated with the demyelinating agent, lysolecithin, had higher numbers of myelin laden phagocytes than normal chow fed diet mice (Bosch-Queralt et al. 2021). Whether diet induces LD accumulation in human microglia has not been reported and warrants further investigation.

15.3 Transcriptional Regulation of Microglial Lipid Droplets

Several transcription factors have been implicated in the formation of lipid droplets. In the *Drosophila* model of ROS-induced lipid droplet formation, JNK and SREBP drive neuronal lipid synthesis in response to ROS. Synthesized lipids are transferred to glia where they form lipid droplets (Liu et al. 2015, 2017). Another study in *Drosophila* found that ROS can activate JNK in glia cells and lead to LD accumulation (Muliyil et al. 2020). In a mouse model of mitochondrial defect-induced ROS, pJNK levels are increased, suggesting a similar mechanism could account for the observed lipid accumulation in microglia and astrocytes (Liu et al. 2015). While the SREBP transcription factors are major regulators of cholesterol and fatty acid synthesis, there are few studies examining their role in lipid droplet accumulation in microglia. Other major regulators of lipid synthesis include members of the PPAR transcription factor family. Studies suggest PPARγ is involved in lipid droplet formation by regulating the expression of the lipid droplet associated protein, PLIN2 (Loving et al. 2021; Loix et al. 2022).

Besides lipid metabolism regulators, a study found that STAT1, a transcription factor that mediates responses to interferons, regulates lipid droplet formation in a mouse model of stroke. The authors found increased expression of interferon stimulated genes in postmortem brain tissue from stroke patients and in a mouse model of stroke (Arbaizar-Rovirosa et al. 2023). Interestingly, STAT1 knockout mice did not have increased lipid accumulation in microglia post-stroke. This finding is supported by in vitro studies in which macrophages treated with IFNγ develop lipid droplets (Rosas-Ballina et al. 2020). These results suggest that lipid droplet formation could be regulated by different transcriptional mechanisms depending on the environmental signal.

The LXR family of transcription factors plays important roles in regulating cholesterol and lipid efflux. A study found that myelin treated macrophages induce the LXR target genes *Abca1, Abcg1,* and *ApoE* in a manner largely dependent on LXRβ (Bogie et al. 2012). Reduced expression of LXR target genes has been implicated in impaired myelin clearance. In one study comparing recovery from lysolecithin injection in young versus old mice, the authors found increased myelin accumulation in aged mice. They attributed this to reduced expression of *Abca1, Abcg1,* and *ApoE,* which encode proteins needed for lipid efflux. Treatment with the LXR agonist, GW3965, restored *Abca1, Abcg1,* and *ApoE* expression, and reduced lipid and cholesterol accumulation (Cantuti-Castelvetri et al. 2018). Similarly, mice fed a western type diet were found to have increased myelin accumulation after lysolecithin injection compared to mice fed a normal chow diet. The authors measured decreased expression of *Abca1, Abcg1,* and *ApoE* and presented evidence that western type diet leads to reduce LXR target gene expression in a TBFβ dependent mechanism (Bosch-Queralt et al. 2021). Lipid accumulation due to reduced LXR target gene expression is also supported by prior work showing that microglia in *ApoE* knockout mice have extensive lipid accumulation (Mato et al. 1999; Nugent et al. 2020). These results highlight the importance of LXRs in promoting lipid efflux, but we still lack a mechanistic understanding how they influence lipid droplet dynamics.

15.4 Proteins Involved in Microglial Lipid Droplet Formation

While studies have begun to define lipid droplet proteomes, no such study has been conducted in microglia. Proteins in the perilipin (PLIN) family associate with lipid droplets and can be used to identify and quantify lipid droplet accumulation. In microglia, PLIN3 and PLIN2 have been used to detect LDs. In vivo myelin treatment induces *Plin2* gene and protein expression in human monocyte derived macrophages and mouse bone marrow derived macrophages (BMDMs). In mouse models of demyelination, macrophage and monocytes have increased PLIN2 expression, and human hippocampal brain sections from aged individuals have increased PLIN2 staining compared to young individuals.

One study found that PLIN2 knockout leads to decreased lipid droplets (Loix et al. 2022). *Plin2* $^{-/-}$ BMDMs treated with myelin accumulated LDs to a similar extent as WT BMDMs. However, when transferred to myelin-free media, *Plin2*$^{-/-}$ BMDMs cleared lipid droplets faster than WT BMDMs. They confirmed that microglia from *Plin2* knockout mice had reduced lipid droplet accumulation in response to lysolecithin and cuprizone treatment. These results suggest that PLIN2 is not needed for lipid droplet formation but prevents LD degradation. The authors present evidence in support of lipolysis as the mechanism driving lipid droplet degradation in *Plin2*$^{-/-}$ deficient microglia.

Proteins involved in lipid metabolism have also been shown to be important for lipid droplet formation in microglia. Stearyl-CoA desaturase-1 (SCD1) is the rate limiting enzyme in the production of mono-unsaturated fatty acids. Myelin treatment induces *Scd1* gene and protein expression in mouse BMDMs, resulting in increased levels of unsaturated fatty acids (Bogie et al. 2012, 2020). A study found that SCD1 inhibition in myelin treated BMDMs decreased lipid droplet accumulation. Through additional studies, the authors outlined a mechanism in which SCD1 activation leads to increased production of unsaturated fatty acid products that decrease surface levels of ABCA1, a lipid efflux transporter, in a PKCδ manner. Thus, inhibiting SCD1 retains ABCA1 surface levels and promotes lipid efflux (Bogie et al. 2020). In addition to SCD1, the lipid metabolism protein lipoprotein lipase (LPL) has been shown to regulate lipid droplet formation. LPL is a surface protein that breaks down triglycerides embedded in lipoproteins for cellular uptake. A study found that LPL knockdown in BV2 cells and LPL knockout in primary microglia resulted in increased lipid droplet accumulation. LPL knockdown BV2 cells had altered lipidomes, including increased levels of cholesterol esters. Treatment with a PPARγ agonist (rosiglitazone) or a pan PPAR agonist (bezafibrate) reduced LD accumulation in LPL KD cells. How LPL regulates lipid droplet formation and whether this finding is relevant in vivo remain unclear (Loving et al. 2021).

There is some evidence that TREM2, which is a known risk factor for Alzheimer's disease, could influence LD formation. In microglia, TREM2 plays an important role in environmental sensing by binding to a diverse repertoire of ligands, including lipids. TREM2 gene and protein expression is increased in primary microglia treated with myelin (Bosch-Queralt et al. 2021). In a study on mice treated with lysolecithin, researchers found that TREM2 KO mice lack foamy macrophages in the lesion site throughout the regeneration period (Gouna et al. 2021). In WT mice, the authors saw co-labeling of TREM2 and PLIN2 in the majority of microglial lipid droplets. In vitro, TREM2 KO primary microglia cultured with myelin did not form lipid droplets, whereas WT microglia accumulated lipid droplets upon myelin exposure. Lipidomics data confirmed that TREM2 KO microglia uptake myelin, but unlike in WT microglia, it remains free and unesterified. TREM2 KO microglia also had lower triglycerides and reduced expression of *soat1/Acat*, suggesting TREM2 deficiency impairs the storage of cholesterol and lipids in lipid droplets. Their results are at odds with a different study that treated TREM2 KO BMDMs and human iMG with myelin and saw neutral lipid accumulation, including several cholesterol ester species (Nugent et al. 2020). Another study transplanted iPSC derived microglia (iMG) into APP/PS1 mice and found that iMG carrying the hypofunctional TREM2-R47H mutation could form lipid droplets (Claes et al. 2021). Further studies are needed to clarify the involvement of TREM2 in lipid droplet formation.

Two studies have presented evidence for the involvement of granulin (GRN) in microglial lipid droplet formation. GRN was identified in a CRISPR screen to search for regulators of LPS induced lipid droplet formation (Marschallinger et al. 2020). BV2 cells with sgRNAs targeting *Grn* had increased lipid droplet accumulation compared to cells treated with control sgRNAs. *Grn*$^{-/-}$ *mice* had increased lipid droplet accumulating microglia in the hippocampus and thalamus compared to age matched WT microglia. This work is supported by another study examining recovery from traumatic brain injury (TBI) in zebrafish (Zambusi et al. 2022). In this model, lipid droplets accumulate in microglia at the injury site, but the lipid droplets resolve as regeneration progresses. When the authors knocked out granulins, the lipid droplet accumulating microglia persisted at the wound site and regeneration was attenuated. Through additional experiments, the authors present evidence that the TAR DNA binding protein, TDP-43, is responsible for the increased lipid droplet formation. Granulin knockout prevents TDP-43 clearance, leading to persistent lipid droplet accumulation. To support their finding, they presented data showing enrichment of PLIN2 and TDP-43 condensates in postmortem tissue from TBI patients. Proteins spanning a range of cellular functions are implicated in lipid droplet formation. Additional studies are needed to define the conditions and networks of interacting proteins that lead to lipid droplet formation.

15.5 Lipid Droplet Degradation in Microglia

In general, lipids are degraded through either lipolysis or lipophagy (Zechner et al. 2017). Both of these pathways have been shown to be involved in microglial lipid droplet degradation in various contexts. Studies in BV2 cells showed that disrupting autophagy through *Atg7* knockout increased lipid droplet formation (Xu et al. 2021). In support of this finding, a recent study showed that prolonged myelin exposure leads to defective autophagy and increased lipid droplet accumulation. Researchers treated BMDMs with myelin for either 24 or 72 h and found that extended myelin exposure led to defective autophagy that could be rescued with Trehalose (Haidar et al. 2022). Studies have also noted the close proximity of microglial lipid droplets to lysosomes, and lysosomal lipid accumulation in demyelination models, further supporting the role of lipophagy in lipid droplet degradation (Safaiyan et al. 2016;

Cantuti-Castelvetri et al. 2018; Marschallinger et al. 2020; Arbaizar-Rovirosa et al. 2023).

There is also evidence for lipolysis in lipid droplet degradation. While prolonged exposure to myelin impaired lipophagy in BMDMs, lipolysis increased. Researchers detected increased expression of the lipolytic genes, *Pnpla2* and *Mgll*, and increased levels of glycerol, a direct measure of lipolysis, upon prolonged myelin exposure. Despite increased lipolysis, there were still increased lipid droplets upon extended myelin exposure, suggesting lipophagy plays an important role in lipid droplet degradation in this context (Haidar et al. 2022). PLIN2 likely mediates lipid droplet lipolysis. In a study examining the effects of *Plin2* knockout on LDs, researchers found that the decreased LD formation observed in *Plin2*$^{-/-}$ macrophages was due to increased lipolysis. How PLIN2 regulates microglial LD dynamics, and the contributions of lipophagy and lipolysis to LD degradation warrant further investigation.

Proteins involved in lipid efflux have also been shown to regulate lipid droplet accumulation in microglia. As mentioned before, the LXR target genes, *Abca1, Abcg1,* and *ApoE* are important for lipid efflux. Decreased expression of these genes has been associated with increased lipid droplet accumulation (Cantuti-Castelvetri et al. 2018; Bogie et al. 2020; Bosch-Queralt et al. 2021). It is interesting to note that aged mice were reported to have decreased expression of these genes (Cantuti-Castelvetri et al. 2018). Whether decreased expression is also observed in humans and contributes to the age dependent increase in lipid droplets is unknown. It is also worth mentioning that mutations in *Apoe* and *Abca1* are associated with Alzheimer's disease. Mutations in *Apoe* are the most common cause of sporadic Alzheimer's disease, and individuals with two *Apoe4* alleles have a 12% increased risk of developing AD. iPSC derived astrocytes and microglia generated from *Apoe44* individuals have lipid defects, including increased lipid accumulation (Machlovi et al. 2022; Tcw et al. 2022; Victor et al. 2022). Additional studies are needed to understand the contribution of lipid accumulating microglia to aging and AD pathology. An effective approach to induce LXR target gene expression is through the use of the LXR agonist, GW3965. Multiple studies have shown that GW3965 treatment reduces microglial lipid droplet accumulation and may improve recovery from demyelinating injury (Cantuti-Castelvetri et al. 2018; Xu et al. 2021).

15.6 Functional Impacts of Lipid Droplet Accumulating Microglia

How lipid droplets influence microglia physiology and their relevance in different disease contexts is an active area of research. Lipid droplet accumulating microglia seem to emerge in conditions associated with inflammation. Some studies suggest they may be anti-inflammatory while other studies indicate they may promote inflammation. Lipid droplet accumulating microglia isolated from aged mouse hippocampi had increased ROS levels and increased production of proinflammatory cytokines at baseline, and after LPS treatment, compared to lipid droplet low microglia. Lipid droplet high microglia isolated from *Grn*$^{-/-}$ mice also had increased ROS and proinflammatory cytokine levels, compared to lipid droplet low microglia (Marschallinger et al. 2020).

In the context of remyelination, myelin laden microglia and macrophages in the early stages of recovery appear to be protective and necessary for remyelination (Boven et al. 2006; Bogie et al. 2013). However, lipid accumulating microglia that form in response to prolonged myelin exposure seem to be proinflammatory and detrimental for repair. As mentioned, one explanation for lipid accumulation upon prolonged myelin exposure is SCD1-mediated downregulation of ABCA1. Decreased surface ABCA1 impairs lipid efflux, leading to lipid accumulation. Extended treatment of BMDMs with myelin led to increased proinflammatory gene expression that was rescued by depleting cholesterol with methyl-beta-cyclodextrin, suggesting cholesterol accumulation can trigger proinflammatory gene expression. Mice with SCD-1 knocked out in macrophages and microglia had increased ABCA1 surface levels, reduced inflammation upon demyelination, and improved remyelination outcomes (Bogie et al. 2020). Plin2 knockout has also been used to reduce lipid droplet levels. Plin2 KO mice treated with the demyelinating agent, cuprizone, had reduced iNOS and proinflammatory gene expression levels, and remyelination was improved compared to WT mice (Loix et al. 2022). In support of this finding, another study showed that stimulating autophagy with Trehalose in cuprizone treated mice led to reduced lipid droplet formation and iNOS production and improved remyelination (Xu et al. 2021). Furthermore, targeting Granulin or TDP-43 in a zebrafish TBI model reduced lipid droplet accumulation, restored microglial homeostasis, and improved regeneration (Zambusi et al. 2022).

While there are numerous reports of reduced lipid droplet accumulation coinciding with improved remyelination, we lack a mechanistic understanding of whether and how lipid droplets drive a proinflammatory phenotype that limits repair. One possible explanation is increased inflammasome activation caused by cholesterol overload, as outlined in a study examining the cause of poor remyelination in aged mice (Cantuti-Castelvetri et al. 2018). Expression of lipid efflux genes is reduced in demyelinating lesions of aged mice, leading to cholesterol overload. The authors reported increased myelin overload and caspase-1 activation in *Apoe* KO macrophages and mice. *Nlrp3*$^{-/--}$ mice, which have impaired inflammasome activation, had improved remyelination, similar to mice treated with the LXR agonist, GW3965. These results demonstrate that decreasing cholesterol overload can limit inflammasome activation and improve remyelination.

Another possible way that lipid accumulating microglia limit repair is through reduced phagocytic function. An in vitro study found that myelin treated macrophages had reduced phagocytic uptake of polymorphonuclear neutrophils (PMNs), which infiltrate into spinal cord injuries. The authors speculate that reduced clearance of PMNs could promote inflammation and limit repair (Wang et al. 2015). Reduced phagocytic function has also been reported in lipid droplet accumulating

BV2 cells treated with LPS. Lipid droplet low BV2 cells phagocytosed zymosan particles, whereas phagocytosis was attenuated in lipid droplet high BV2 cells. Treatment with Triacsin C, which prevented LD formation by inhibiting the long-chain fatty acyl-CoA synthetase (ACSL), rescued phagocytosis (Marschallinger et al. 2020). Reduced phagocytosis in LD high microglia has also been observed in the context of aging (Arbaizar-Rovirosa et al. 2023). There are also reports of lipid accumulating microglia with enhanced phagocytic capacity. In a study on microglia isolated from APOE3 and APOE4 targeted replacement mice, the authors observed increased lipid accumulation and phagocytic uptake of multiple substrates in APOE4 microglia compared to APOE3 microglia. This demonstrates lipid accumulation may not always be associated with reduced phagocytosis (Machlovi et al. 2022).

Although many studies suggest lipid droplet accumulating microglia are detrimental, there is evidence supporting beneficial roles of lipid accumulation. One report found that myelin laden macrophages in lesion centers of postmortem MS brain tissue express anti-inflammatory markers (Boven et al. 2006). Another study examining the effects of TREM2 KO on lipid droplet formation in demyelinating injury found that TREM2 KO reduced lipid droplets. The reduced lipid droplet formation was associated with increased free cholesterol levels that led to ER stress. Reducing ER stress with 4-phenyl butyric acid increased lipid droplet formation and reduced $IBA1^{+}$ cells in the lesion center (Gouna et al. 2021). LDs are also beneficial for buffering against lipid toxicity. In *Drosophila*, neuronal ROS induce lipid synthesis and can lead to lipid peroxidation. Peroxidated lipids can be transferred and stored in glial lipid droplets that protect from neurodegeneration (Moulton et al. 2021). In mice, a study found that hyperactive neurons release fatty acids that can be transferred to astrocytes where they are stored in lipid droplets and subsequently metabolized (Ioannou et al. 2019). Additional studies are needed to understand whether microglia have roles in buffering toxic lipids, and in general, contexts in which microglial lipid droplets are beneficial.

15.7 Conclusion

Increasing evidence points to the roles of lipid droplets beyond inert storage organelles. There is growing interest in the mechanisms and functional impacts of lipid droplet accumulation in microglia. Lipid accumulation in microglia was originally described in the context of Alzheimer's disease and has since been described in many conditions, including multiple sclerosis, Parkinson's disease, stroke, traumatic brain injury, and aging. In neurodegenerative conditions, lipid droplets likely occur from myelin uptake and possibly through de novo lipogenesis. Reduced degradation through lipophagy and lipolysis could also contribute to LD accumulation. Proteins implicated in lipid droplet formation include PLIN2, LPL, ACAT/SOAT, TREM2, GRN, and TDP-43. The necessity of these proteins in the range of lipid droplet forming conditions warrants further investigation. Lipid droplet accumulating microglia are almost exclusively reported in the context of disease and inflammatory conditions. Many studies suggest they are dysfunctional and associated with worse disease outcomes. Studies have begun to address the mechanisms driving worsened outcomes, but we still lack mechanistic details. Moving forward, additional studies are needed to further define the transcriptional regulation and pathways involved in lipid droplet formation across the range of conditions in which they are found. A better understanding of how dynamic lipid droplets are and how they shape microglial phenotypes is also needed. Initial work to minimize microglial lipid accumulation has been beneficial in several disease contexts and could possibly pave the way for novel therapeutics.

References

Alzheimer A, Stelzmann RA, Schnitzlein HN, Murtagh FR (1995) An English translation of Alzheimer's 1907 paper, "Uber eine eigenartige Erkankung der Hirnrinde". Clin Anat 8(6): 429–431

Arbaizar-Rovirosa M, Pedragosa J, Lozano JJ, Casal C, Pol A, Gallizioli M, Planas AM (2023) Aged lipid-laden microglia display impaired responses to stroke. EMBO Mol Med 15(2): e17175

Bogie JF, Timmermans S, Huynh-Thu VA, Irrthum A, Smeets HJ, Gustafsson JA, Steffensen KR, Mulder M, Stinissen P, Hellings N, Hendriks JJ (2012) Myelin-derived lipids modulate macrophage activity by liver X receptor activation. PLoS One 7(9):e44998

Bogie JF, Jorissen W, Mailleux J, Nijland PG, Zelcer N, Vanmierlo T, Van Horssen J, Stinissen P, Hellings N, Hendriks JJ (2013) Myelin alters the inflammatory phenotype of macrophages by activating PPARs. Acta Neuropathol Commun 1:43

Bogie JFJ, Grajchen E, Wouters E, Corrales AG, Dierckx T, Vanherle S, Mailleux J, Gervois P, Wolfs E, Dehairs J, Van Broeckhoven J, Bowman AP, Lambrichts I, Gustafsson JA, Remaley AT, Mulder M, Swinnen JV, Haidar M, Ellis SR, Ntambi JM, Zelcer N, Hendriks JJA (2020) Stearoyl-CoA desaturase-1 impairs the reparative properties of macrophages and microglia in the brain. J Exp Med 217(5):e20191660

Bosch M, Sanchez-Alvarez M, Fajardo A, Kapetanovic R, Steiner B, Dutra F, Moreira L, Lopez JA, Campo R, Mari M, Morales-Paytuvi F, Tort O, Gubern A, Templin RM, Curson JEB, Martel N, Catala C, Lozano F, Tebar F, Enrich C, Vazquez J, Del Pozo MA, Sweet MJ, Bozza PT, Gross SP, Parton RG, Pol A (2020) Mammalian lipid droplets are innate immune hubs integrating cell metabolism and host defense. Science 370(6514):eaay8085

Bosch-Queralt M, Cantuti-Castelvetri L, Damkou A, Schifferer M, Schlepckow K, Alexopoulos I, Lutjohann D, Klose C, Vaculciakova L, Masuda T, Prinz M, Monroe KM, Di Paolo G, Lewcock JW, Haass C, Simons M (2021) Diet-dependent regulation of TGFbeta impairs reparative innate immune responses after demyelination. Nat Metab 3(2):211–227

Boven LA, Van Meurs M, Van Zwam M, Wierenga-Wolf A, Hintzen RQ, Boot RG, Aerts JM, Amor S, Nieuwenhuis EE, Laman JD (2006) Myelin-laden macrophages are anti-inflammatory, consistent with foam cells in multiple sclerosis. Brain 129(pt 2):517–526

Brekk OR, Honey JR, Lee S, Hallett PJ, Isacson O (2020) Cell type-specific lipid storage changes in Parkinson's disease patient brains are recapitulated by experimental glycolipid disturbance. Proc Natl Acad Sci U S A 117(44):27646–27654

Buhman KK, Chen HC, Farese RV Jr (2001) The enzymes of neutral lipid synthesis. J Biol Chem 276(44):40369–40372

Cantuti-Castelvetri L, Fitzner D, Bosch-Queralt M, Weil MT, Su M, Sen P, Ruhwedel T, Mitkovski M, Trendelenburg G, Lutjohann D, Mobius W, Simons M (2018) Defective

cholesterol clearance limits remyelination in the aged central nervous system. Science 359(6376):684–688
Chistiakov DA, Bobryshev YV, Orekhov AN (2016) Macrophage-mediated cholesterol handling in atherosclerosis. J Cell Mol Med 20(1):17–28
Claes C, Danhash EP, Hasselmann J, Chadarevian JP, Shabestari SK, England WE, Lim TE, Hidalgo JLS, Spitale RC, Davtyan H, Blurton-Jones M (2021) Plaque-associated human microglia accumulate lipid droplets in a chimeric model of Alzheimer's disease. Mol Neurodegener 16(1):50
Denic A, Johnson AJ, Bieber AJ, Warrington AE, Rodriguez M, Pirko I (2011) The relevance of animal models in multiple sclerosis research. Pathophysiology 18(1):21–29
van Dierendonck X, de la Rosa Rodriguez MA, Georgiadi A, Mattijssen F, Dijk W, van Weeghel M, Singh R, Borst JW, Stienstra R, Kersten S (2020) HILPDA uncouples lipid droplet accumulation in adipose tissue macrophages from inflammation and metabolic dysregulation. Cell Rep 30(6):1811–1822.e6
Filippi M, Bar-Or A, Piehl F, Preziosa P, Solari A, Vukusic S, Rocca MA (2018) Multiple sclerosis. Nat Rev Dis Primers 4(1):43
Gonzales AM, Orlando RA (2007) Role of adipocyte-derived lipoprotein lipase in adipocyte hypertrophy. Nutr Metab (Lond) 4:22
Gouna G, Klose C, Bosch-Queralt M, Liu L, Gokce O, Schifferer M, Cantuti-Castelvetri L, Simons M (2021) TREM2-dependent lipid droplet biogenesis in phagocytes is required for remyelination. J Exp Med 218(10):e20210227
Grabner GF, Xie H, Schweiger M, Zechner R (2021) Lipolysis: cellular mechanisms for lipid mobilization from fat stores. Nat Metab 3(11):1445–1465
Grajchen E, Hendriks JJA, Bogie JFJ (2018) The physiology of foamy phagocytes in multiple sclerosis. Acta Neuropathol Commun 6(1):124
Gutierrez PS (2022) Foam cells in atherosclerosis. Arq Bras Cardiol 119(4):542–543
Haidar M, Loix M, Vanherle S, Dierckx T, Vangansewinkel T, Gervois P, Wolfs E, Lambrichts I, Bogie JFJ, Hendriks JJA (2022) Targeting lipophagy in macrophages improves repair in multiple sclerosis. Autophagy 18(11):2697–2710
Ioannou MS, Jackson J, Sheu SH, Chang CL, Weigel AV, Liu H, Pasolli HA, Xu CS, Pang S, Matthies D, Hess HF, Lippincott-Schwartz J, Liu Z (2019) Neuron-astrocyte metabolic coupling protects against activity-induced fatty acid toxicity. Cell 177(6):1522–1535.e14
Khatchadourian A, Bourque SD, Richard VR, Titorenko VI, Maysinger D (2012) Dynamics and regulation of lipid droplet formation in lipopolysaccharide (LPS)-stimulated microglia. Biochim Biophys Acta 1821(4):607–617
Kory N, Farese RV Jr, Walther TC (2016) Targeting fat: mechanisms of protein localization to lipid droplets. Trends Cell Biol 26(7):535–546
Kuhlmann T, Ludwin S, Prat A, Antel J, Bruck W, Lassmann H (2017) An updated histological classification system for multiple sclerosis lesions. Acta Neuropathol 133(1):13–24
Liu L, Zhang K, Sandoval H, Yamamoto S, Jaiswal M, Sanz E, Li Z, Hui J, Graham BH, Quintana A, Bellen HJ (2015) Glial lipid droplets and ROS induced by mitochondrial defects promote neurodegeneration. Cell 160(1–2):177–190
Liu L, MacKenzie KR, Putluri N, Maletic-Savatic M, Bellen HJ (2017) The glia-neuron lactate shuttle and elevated ROS promote lipid synthesis in neurons and lipid droplet accumulation in glia via APOE/D. Cell Metab 26(5):719–737.e6
Loix M, Wouters E, Vanherle S, Dehairs J, McManaman JL, Kemps H, Swinnen JV, Haidar M, Bogie JFJ, Hendriks JJA (2022) Perilipin-2 limits remyelination by preventing lipid droplet degradation. Cell Mol Life Sci 79(10):515
Loving BA, Tang M, Neal MC, Gorkhali S, Murphy R, Eckel RH, Bruce KD (2021) Lipoprotein lipase regulates microglial lipid droplet accumulation. Cell 10(2):198
Machlovi SI, Neuner SM, Hemmer BM, Khan R, Liu Y, Huang M, Zhu JD, Castellano JM, Cai D, Marcora E, Goate AM (2022) APOE4 confers transcriptomic and functional alterations to primary mouse microglia. Neurobiol Dis 164:105615

Marschallinger J, Iram T, Zardeneta M, Lee SE, Lehallier B, Haney MS, Pluvinage JV, Mathur V, Hahn O, Morgens DW, Kim J, Tevini J, Felder TK, Wolinski H, Bertozzi CR, Bassik MC, Aigner L, Wyss-Coray T (2020) Lipid-droplet-accumulating microglia represent a dysfunctional and proinflammatory state in the aging brain. Nat Neurosci 23(2):194–208

Mato M, Ookawara S, Mashiko T, Sakamoto A, Mato TK, Maeda N, Kodama T (1999) Regional difference of lipid distribution in brain of apolipoprotein E deficient mice. Anat Rec 256(2): 165–176

Moulton MJ, Barish S, Ralhan I, Chang J, Goodman LD, Harland JG, Marcogliese PC, Johansson JO, Ioannou MS, Bellen HJ (2021) Neuronal ROS-induced glial lipid droplet formation is altered by loss of Alzheimer's disease-associated genes. Proc Natl Acad Sci U S A 118(52): e2112095118

Muliyil S, Levet C, Dusterhoft S, Dulloo I, Cowley SA, Freeman M (2020) ADAM17-triggered TNF signalling protects the ageing Drosophila retina from lipid droplet-mediated degeneration. EMBO J 39(17):e104415

Nugent AA, Lin K, van Lengerich B, Lianoglou S, Przybyla L, Davis SS, Llapashtica C, Wang J, Kim DJ, Xia D, Lucas A, Baskaran S, Haddick PCG, Lenser M, Earr TK, Shi J, Dugas JC, Andreone BJ, Logan T, Solanoy HO, Chen H, Srivastava A, Poda SB, Sanchez PE, Watts RJ, Sandmann T, Astarita G, Lewcock JW, Monroe KM, Di Paolo G (2020) TREM2 regulates microglial cholesterol metabolism upon chronic phagocytic challenge. Neuron 105(5): 837–854 e839

Olzmann JA, Carvalho P (2019) Dynamics and functions of lipid droplets. Nat Rev Mol Cell Biol 20(3):137–155

Ralhan I, Chang CL, Lippincott-Schwartz J, Ioannou MS (2021) Lipid droplets in the nervous system. J Cell Biol 220(7):e202102136

Rosas-Ballina M, Guan XL, Schmidt A, Bumann D (2020) Classical activation of macrophages leads to lipid droplet formation without de novo fatty acid synthesis. Front Immunol 11:131

Safaiyan S, Kannaiyan N, Snaidero N, Brioschi S, Biber K, Yona S, Edinger AL, Jung S, Rossner MJ, Simons M (2016) Age-related myelin degradation burdens the clearance function of microglia during aging. Nat Neurosci 19(8):995–998

Shimabukuro MK, Langhi LG, Cordeiro I, Brito JM, Batista CM, Mattson MP, Mello Coelho V (2016) Lipid-laden cells differentially distributed in the aging brain are functionally active and correspond to distinct phenotypes. Sci Rep 6:23795

Smith LJ, Bolsinger MM, Chau KY, Gegg ME, Schapira AHV (2023) The GBA variant E326K is associated with alpha-synuclein aggregation and lipid droplet accumulation in human cell lines. Hum Mol Genet 32(5):773–789

Tcw J, Qian L, Pipalia NH, Chao MJ, Liang SA, Shi Y, Jain BR, Bertelsen SE, Kapoor M, Marcora E, Sikora E, Andrews EJ, Martini AC, Karch CM, Head E, Holtzman DM, Zhang B, Wang M, Maxfield FR, Poon WW, Goate AM (2022) Cholesterol and matrisome pathways dysregulated in astrocytes and microglia. Cell 185(13):2213–2233 e2225

Thiam AR, Ikonen E (2021) Lipid droplet nucleation. Trends Cell Biol 31(2):108–118

Victor MB, Leary N, Luna X, Meharena HS, Scannail AN, Bozzelli PL, Samaan G, Murdock MH, von Maydell D, Effenberger AH, Cerit O, Wen HL, Liu L, Welch G, Bonner M, Tsai LH (2022) Lipid accumulation induced by APOE4 impairs microglial surveillance of neuronal-network activity. Cell Stem Cell 29(8):1197–1212.e8

Walther TC, Farese RV Jr (2012) Lipid droplets and cellular lipid metabolism. Annu Rev Biochem 81:687–714

Wang X, Cao K, Sun X, Chen Y, Duan Z, Sun L, Guo L, Bai P, Sun D, Fan J, He X, Young W, Ren Y (2015) Macrophages in spinal cord injury: phenotypic and functional change from exposure to myelin debris. Glia 63(4):635–651

Xu Y, Propson NE, Du S, Xiong W, Zheng H (2021) Autophagy deficiency modulates microglial lipid homeostasis and aggravates tau pathology and spreading. Proc Natl Acad Sci U S A 118(27):e2023418118

Zambusi A, Novoselc KT, Hutten S, Kalpazidou S, Koupourtidou C, Schieweck R, Aschenbroich S, Silva L, Yazgili AS, van Bebber F, Schmid B, Moller G, Tritscher C, Stigloher C, Delbridge C, Sirko S, Gunes ZI, Liebscher S, Schlegel J, Aliee H, Theis F, Meiners S, Kiebler M, Dormann D, Ninkovic J (2022) TDP-43 condensates and lipid droplets regulate the reactivity of microglia and regeneration after traumatic brain injury. Nat Neurosci 25(12):1608–1625

Zechner R, Madeo F, Kratky D (2017) Cytosolic lipolysis and lipophagy: two sides of the same coin. Nat Rev Mol Cell Biol 18(11):671–684

Chapter 16
Emerging Role of Phase Separation in COVID-19

Kenji Mizumura and Yasuhiro Gon

Abstract Viral infections have particularly been attracting attention after the spread of the novel coronavirus (SARS-CoV-2) in 2019. Viruses have a simple structure consisting of only a genome and structural proteins, and they cannot synthesize genomes or proteins, so they cannot propagate on their own. Therefore, they enter the cells of other organisms, gain control over the cell functions to synthesize their own genomes and structural proteins, form virus particles, and proliferate. Virus-infected cells contain membrane-less viral inclusion bodies; however, their functions remain unclear. It has recently been established that phase separation is involved in the formation of viral inclusion bodies and plays an important role in the utilization of intracellular mechanisms by viruses. Phase separation is also involved in the immune system; viruses reportedly change the function of phase separation according to the phase difference. In this review, we focus on SARS-CoV-2 and explain the role of phase separation in viral infection, proliferation, and immune evasion.

Keywords Phase separation · COVID-19 · SARS-CoV-2 · Viral infection

16.1 Introduction

On December 31, 2019, China reported an outbreak of a novel pneumonia in Wuhan. On January 24, 2020, Zhu et al. (2020) reported that a new coronavirus had been isolated in four patients. On the same day, Huang et al. (2020) published a

K. Mizumura (✉)
Division of Respiratory Medicine, Department of Internal Medicine, Nihon University School of Medicine, Itabashi-ku, Tokyo, Japan

Division of Biomedical Sciences, Research Center for Genomic Medicine, Saitama Medical University, Hidaka-shi, Saitama, Japan
e-mail: mizumura.kenji@nihon-u.ac.jp

Y. Gon
Division of Respiratory Medicine, Department of Internal Medicine, Nihon University School of Medicine, Itabashi-ku, Tokyo, Japan

R. Kurokawa (ed.), *Phase Separation in Living Cells*,
https://doi.org/10.1007/978-981-99-4886-4_16

clinical study on 41 cases in Wuhan, reporting that respiratory failure progressed rapidly after a prodromal period of approximately 1 week after symptom onset. On February 11, 2020, WHO announced that the name of the disease was coronavirus disease (COVID-19) (WHO 2020). The International Committee on Taxonomy of Viruses named the coronavirus severe acute respiratory syndrome coronavirus 2 (SARS-CoV-2) (Coronaviridae Study Group of the International Committee on Taxonomy of Viruses 2020). On March 11, 2020, WHO declared COVID-19 a pandemic. Unlike the severe acute respiratory syndrome (SARS) and Middle East respiratory syndrome (MERS), COVID-19 was asymptomatic or mild in many patients, but its spread was more extensive. As a result, the virus spread efficiently with the movement of people, rapidly increasing worldwide and threatening the stability of medical care due to large numbers of severely ill patients. Many cities went into emergency lockdowns as a means to counter the increase in the number of infections. However, this had a significant economic impact. According to Johns Hopkins University, as of May 8, 2022, the cumulative number of infections reported worldwide was 517.06 million, with 6.25 million reported deaths (Johns Hopkins University 2022).

This COVID-19 pandemic has led to vigorous research on coronaviruses in various fields, and phase separation has been identified as playing an important role in the proliferation of SARS-CoV-2 in infected cells and in the immune response of the host. This section describes the infection mechanism of SARS-CoV-2 and the role of phase separation in the pathology of COVID-19.

16.2 Structure of SARS-CoV-2

SARS-CoV-2 consists of four types of structural protein. These are envelope proteins (E) and membrane proteins (M) bound to the envelope, which is the membrane of the virus, as well as spike proteins (S) required for infection. These proteins are present in the membrane (Fig. 16.1). A single positive-strand RNA consisting of

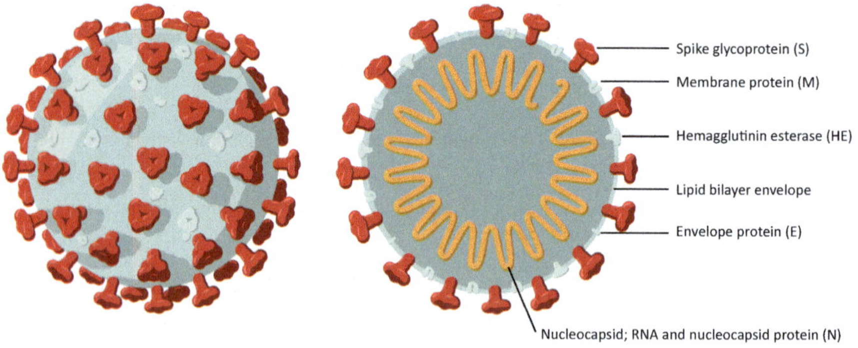

Fig. 16.1 Particle structure of SARS-CoV-2

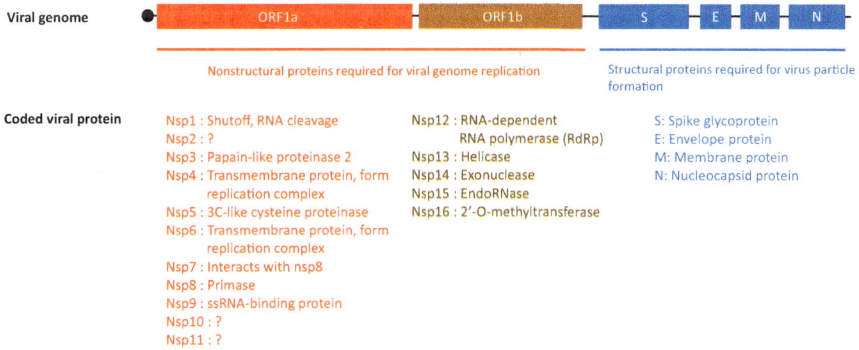

Fig. 16.2 Gene structure of SARS-CoV-2. *Nsp* nonstructural protein, *ORF* open reading frame

about 30,000 bases exists in the virus particle as a virus genome. The nucleocapsid protein (N) wraps around the viral genome and the nucleocapsid has a helical structure.

RNA viruses are divided into positive-strand RNA viruses and negative-strand RNA viruses. A positive-strand RNA virus has RNA with the same polarity as mRNA as its genome, and a negative-strand RNA virus has an RNA with the polarity of its complementary strand as its genome. SARS-CoV-2 is an enveloped positive-strand RNA virus. The viral genomic structure of SARS-CoV-2 has a long positive-strand RNA of approximately 30 kb as its genome. ORF1a and ORF1b, collectively referred to as ORF1ab, are encoded in the region approximately 2/3 from the 5′ end (Fig. 16.2). ORF1ab undergoes a frameshift to produce a total of 16 nonstructural proteins required for viral genome replication. Approximately one-third of ORF1ab encodes structural proteins required for S, M, E, and N viral particle formation.

16.3 Life Cycle of SARS-CoV-2

SARS-CoV-2 infection begins with the binding of SARS-CoV-2 spike protein to angiotensin-converting enzyme 2 (ACE2) on the host cell membrane (Fig. 16.3) (Zhou et al. 2020). Viral RNA invades the cytoplasm of the host cell by membrane fusion of SARS-CoV-2 with the host cell. In order for membrane fusion to occur, ACE2 must be cleaved. This cleavage is carried out by the transmembrane protease serine 2 (TMPRSS2) which is a host cell protease (Hoffmann et al. 2020). An alternative pathway has been proposed in which membrane fusion occurs by endocytosis after the virus has been taken up by the host cell.

In order for the virus to multiply, the viral genome and viral proteins need to be synthesized in the infected cells. When SARS-CoV-2 viral RNA invades cells, the nonstructural proteins encoded by ORF1ab are directly translated by human ribosomes, because the genome of SARS-CoV-2, a positive-strand RNA virus, has the

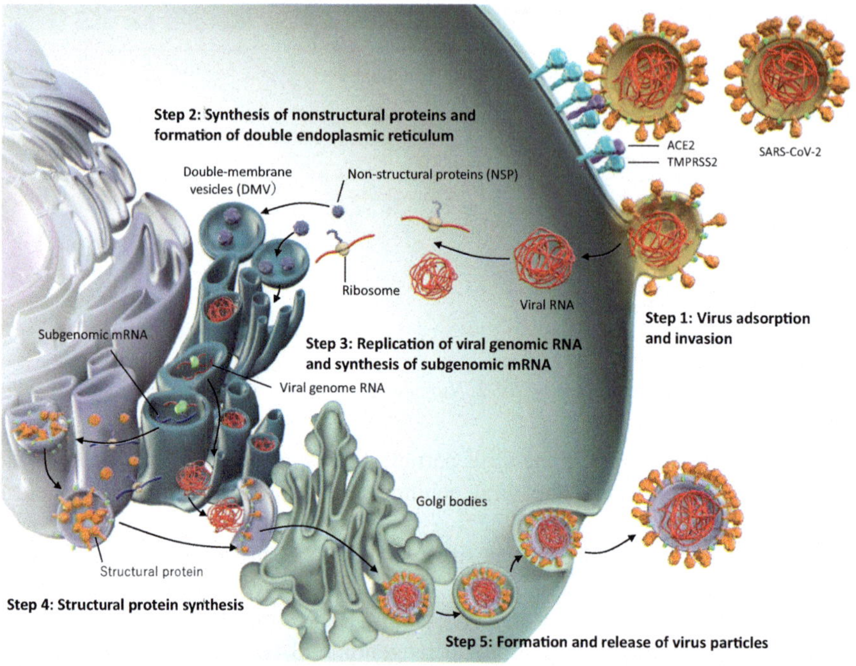

Fig. 16.3 Life cycle of SARS-CoV-2. Step 1: Virus adsorption and invasion. The SARS-CoV-2 spike protein receptor binding domain (RBD) binds to angiotensin-converting enzyme 2 (ACE2) receptors on the cell surface of the human throat and lungs. When the transmembrane protease serine 2 (TMPRSS2) on the cell surface cleaves a part of the spike protein, fusion of the virus envelope and the cell membrane begins, and the virus invades the cell. Step 2: Synthesis of nonstructural proteins and formation of double endoplasmic reticulum. Human ribosomes synthesize nonstructural proteins (NSPs), including RNA-dependent RNA polymerase (RdRp). NSP modifies the structure of the endoplasmic reticulum into double-membrane vesicles (DMVs) and also incorporates RdRp to establish replication-transcription complexes (RTCs) that can hide from the innate immune system and stably replicate and synthesize RNA. Step 3: Replication of viral genomic RNA and synthesis of subgenomic mRNA.Within the DMVs, RdRp replicates viral genomic RNA and synthesizes subgenomic mRNA that encodes structural proteins. Step 4: Structural protein synthesis. Viral genome and subgenomic mRNA synthesized in the DMVs move to the cytoplasm through the molecular tunnel composed of nonstructural protein 3 (Nsp3) on the DMV. Subgenomic mRNA that has migrated to the cytoplasm is translated into structural proteins by human ribosomes. Step 5: Formation and release of virus particles. The replicated viral genomic RNA and structural proteins aggregate to form particles that are released extracellularly via the Golgi complex or lysosomes

same polarity as mRNA (Figs. 16.2 and 16.3). There are 16 nonstructural proteins of SARS-CoV-2, including RNA-dependent RNA polymerase (RdRp) which is required to synthesize genomic RNA. The synthesized nonstructural protein begins to remodel the cell structure, inflating long, thin endoplasmic reticulum like a soap bubble, called double-membrane vesicles (DMVs). The DMVs use RdRp to establish replication-transcription complexes (Scudellari 2021; Alsaadi and Jones 2019).

DMVs may also serve as a refuge in which viral RNA escapes detection by the innate immune system inside the cell.

In the DMVs, RdRp synthesizes negative-strand anti-genomic RNA from positive-strand genomic RNA and continuously synthesizes positive-strand genomic RNA using de novo negative-strand anti-genomic RNA as a template to promote viral genome replication (Fig. 16.3). RdRp simultaneously synthesizes negative-strand anti-genomic RNA that encodes structural proteins of different lengths (called subgenomic mRNA). The synthesized viral genome is thought to move to the cytoplasm through a molecular tunnel composed of nonstructural protein 3 (Nsp3) on the DMV (Wolff et al. 2020). Subgenomic mRNA that has migrated to the cytoplasm is translated into structural proteins by human ribosomes. The replicated viral genomic RNA and structural proteins aggregate to form particles that are released extracellularly via the Golgi complex or lysosomes.

16.4 SARS-CoV-2 Proliferation and Phase Separation

It has been reported that SARS-CoV-2 nucleocapsid protein and viral RNA cause liquid–liquid phase separation in host cells (Iserman et al. 2020; Savastano et al. 2020). Liquid–liquid phase separation creates dense protein/viral RNA aggregates in the host cell. RdRp is incorporated into these aggregates, suggesting that liquid–liquid phase separation is involved in viral replication in SARS-CoV-2 infection (Savastano et al. 2020). In Sect. 16.3, it is described that RdRp replicates viral genomic RNA and synthesizes subgenomic mRNA in DMVs, but it has been reported that there is almost no protein in DMVs (Klein et al. 2020). It appears that RNA synthesis does not occur in the DMVs because they contain minimal protein; as such, it is assumed that DMVs do not contain RdRp, which is a protein that is required for RNA synthesis. DMVs function as containers that take in viral RNA from the cytoplasm and hide it from the immune system (Klein et al. 2020). These findings suggest that droplets by liquid–liquid phase separation rather than DMVs may be the actual site of genomic RNA synthesis. However, further studies are needed to determine the respective roles of DMVs and liquid–liquid phase separation in SARS-CoV-2 proliferation.

The SARS-CoV-2 nucleocapsid protein coordinates two major functions: transcription of viral genes and compression of the RNA genome into viral particles. Nucleocapsid proteins may regulate these effects by phase separation (Carlson et al. 2020). When the nucleocapsid protein is phosphorylated, it becomes a phase close to liquid, and transcription by RdRp proceeds without compression of the RNA genome (Fig. 16.4) (Carlson et al. 2020). Conversely, when the nucleocapsid protein is unmodified, the phase becomes closer to a gel, promoting the formation of viral RNA into nucleocapsids, and compression of the RNA genome occurs. Furthermore, the contents of aggregates may regulate switching between transcription of viral genes and compression of the genome. The SARS-CoV-2 membrane protein surrounds the outer edge of nucleocapsid and binds the viral particle membrane to

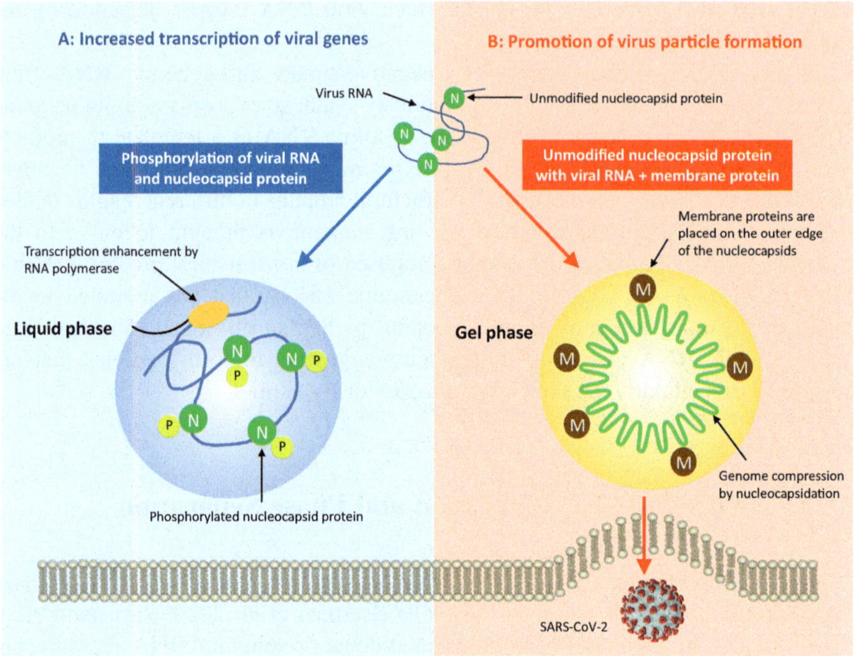

Fig. 16.4 SARS-CoV-2 proliferation and phase separation. (**a**) The nucleocapsid protein and viral RNA form a binary coexistence aggregate. When the nucleocapsid protein is phosphorylated, it becomes a phase close to liquid, and transcription by RNA-dependent RNA polymerase (RdRp) proceeds without compression of the RNA genome. (**b**) When the membrane protein, the nucleocapsid protein, and the viral RNA coexist by phase separation, the membrane protein is arranged on the outer edge so as to surround the outer edge of the nucleocapsid protein and the viral RNA. Furthermore, when the nucleocapsid protein is unmodified, the phase becomes closer to a gel, promoting the formation of viral RNA into nucleocapsids, and compression of the RNA genome occurs

nucleocapsid to construct a viral structure (Fig. 16.1). When the membrane protein, nucleocapsid protein, and viral RNA coexist by phase separation, the membrane protein is placed on the outer edge so as to surround the outer edge of the nucleocapsid protein and viral RNA (Fig. 16.4) (Lu et al. 2021). The ternary coexistence aggregate of membrane protein, nucleocapsid protein, and viral RNA is immiscible with the nucleocapsid protein and viral RNA binary coexistence aggregate. This suggests that the ternary coexistence aggregate of membrane protein, nucleocapsid protein, and viral RNA may be involved in the synthesis of viral particles, and that the binary coexistence aggregate of nucleocapsid protein and viral RNA may be involved in the transcription of viral genes (Fig.16.4).

16.5 Innate Immune Response to SARS-CoV-2 and Phase Separation

Mitochondrial antiviral-signaling protein (MAVS) on the surface of the mitochondria is an innate component of the immune response system that detects infecting RNA viruses. In SARS-CoV-2 infection, like other viral infections, this innate immunity responds first, but the infection sometimes progresses because the type I interferon response is inhibited. Determining the mechanism of innate immune suppression by SARS-CoV-2 infection and promoting the innate immune response are important in developing new treatments for COVID-19.

Stress granules, which are one of the stress adaptation mechanisms of cells, are phase-separated granules in which mRNA arrested during translation, 40S ribosome, and RNA-binding protein are aggregated. Recently, stress granules have been reported to function as a site of the immune response to viral infections (Fig. 16.5) (McCormick and Khaperskyy 2017). Stress granules containing the viral RNA sensors, retinoic acid-inducible gene I (RIG-I) and tripartite motif-containing protein 25 (TRIM25), are formed during viral infection (Sanchez-Aparicio et al. 2017). Following this, the stress granules in close proximity to mitochondria form a complex of RIG-1 and MAVS, which produces interferon to protect against viral infection.

Recently, it has been reported that the stress granule protein, Ras-GTPase-activating protein SH3 domain-binding protein 1 (G3BP1) acts as a switch that induces phase separation and causes stress granules to be assembled (Yang et al. 2020). SARS-CoV-2 nucleocapsid protein undergoes liquid–liquid phase separation with RNA to form droplets (Fig. 16.5). SARS-CoV-2 nucleocapsid protein recruits G3BP1 (Lu et al. 2021). The isolation of G3BP1 suppresses the antiviral effect of stress granules, thus serving as a proliferation strategy. Furthermore, droplets consisting of SARS-CoV-2 nucleocapsid protein and RNA suppress innate immunity to the virus by inhibiting MAVS polyubiquitination and aggregation (Fig. 16.5) (Wang et al. 2021). This process involves the dimer-forming region of the nucleocapsid protein. An experiment was conducted in which the interfering peptides targeting the dimer-forming region were synthesized and administered to SARS-CoV-2-infected mice in order to block the MAVS inhibitory effect (Wang et al. 2021). The interfering peptides significantly reduced spike protein and nucleocapsid protein in the lungs of the SARS-CoV-2-infected mice compared to the control mice, suggesting that they inhibited the growth of SARS-CoV-2 (Wang et al. 2021).

16.6 Conclusion

It is clear that phase separation is intricately involved not only in viral proliferation, but also in suppression of the immune response in COVID-19. Phase separation of nucleocapsid protein and viral RNA in SARS-CoV-2-infected cells increases at

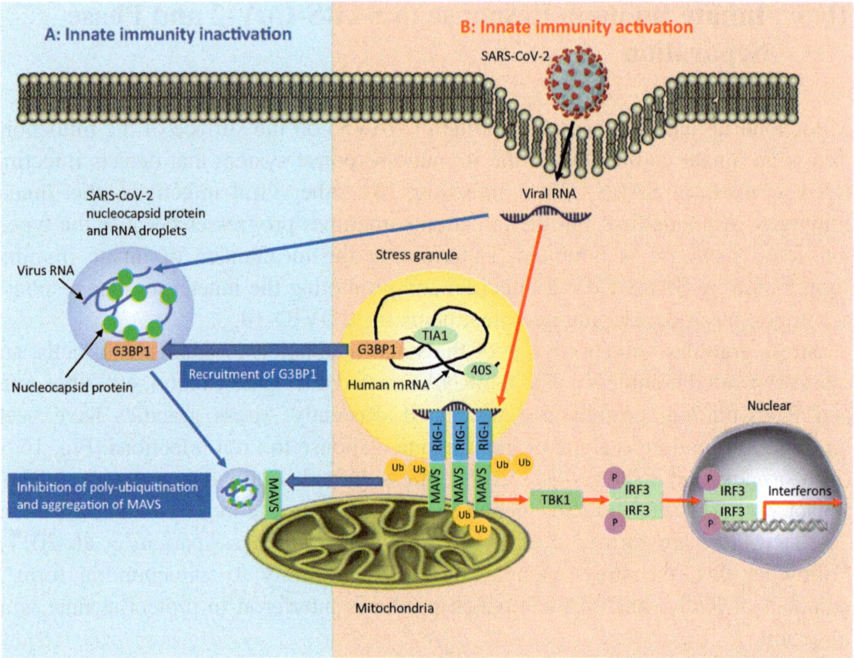

Fig. 16.5 Innate immune response to SARS-CoV-2 and phase separation. (**a**) *Innate immunity inactivation:* The SARS-CoV-2 nucleocapsid protein and RNA droplets suppress innate immunity to the virus by inhibiting mitochondrial antiviral-signaling protein (MAVS) polyubiquitination and aggregation. Furthermore, the nucleocapsid protein inhibits the formation of stress granules by binding to the stress granule protein, Ras-GTPase-activating protein SH3 domain-binding protein (G3BP1). *Ub* ubiquitin, *P* phosphorylation. (**b**) *Innate immunity activation:* When viral infection occurs, mRNA translation is arrested, and 40S ribosome and RNA-binding protein undergo liquid–liquid phase separation to form stress granules. The stress granules contain the RNA sensors, retinoic acid-inducible gene I (RIG-I) and tripartite motif-containing protein 25 (TRIM25). The stress granules in close proximity to mitochondria form a complex of RIG-1 and MAVS, leading to the production of interferon. *IRF3* interferon regulatory factor 3, *TBK1* TANK-binding kinase 1

human body temperature (33–37 °C), but is suppressed at higher temperatures (40 ° C) (Iserman et al. 2020). Fever during viral infection may be a defense mechanism targeting phase separation in living organisms. However, it is currently unclear whether SARS-COV-2 RNA synthesis takes place in the DMV or inside the droplets generated from liquid–liquid phase separation. Furthermore, the relationship between phase separation and the Golgi bodies and lysosomes in the process of virus particle formation and virus shedding is also unclear. Further investigations are needed to investigate the role of these structures in SARS-CoV-2 proliferation.

Currently, various therapeutic agents targeting the molecular mechanism of SARS-CoV-2 infection are being developed worldwide, and phase separation is considered to be a promising therapeutic target.

References

Alsaadi EAJ, Jones IM (2019) Membrane binding proteins of coronaviruses. Future Virol 14(4): 275–286

Carlson CR, Asfaha JB, Ghent CM et al (2020) Phosphoregulation of phase separation by the SARS-CoV-2 N protein suggests a biophysical basis for its dual functions. Mol Cell 80(6): 1092–1103.e4

Coronaviridae Study Group of the International Committee on Taxonomy of Viruses (2020) The species severe acute respiratory syndrome-related coronavirus: classifying 2019-nCoV and naming it SARS-CoV-2. Nat Microbiol 5(4):536–544

Hoffmann M, Kleine-Weber H, Schroeder S et al (2020) SARS-CoV-2 cell entry depends on ACE2 and TMPRSS2 and is blocked by a clinically proven protease inhibitor. Cell 181(2):271–280.e8

Huang C, Wang Y, Li X et al (2020) Clinical features of patients infected with 2019 novel coronavirus in Wuhan, China. *Lancet* 395(10223):497–506

Iserman C, Roden CA, Boerneke MA et al (2020) Genomic RNA elements drive phase separation of the SARS-CoV-2 nucleocapsid. Mol Cell 80(6):1078–1091.e6

Johns Hopkins University (2022) COVID-19 Map – Johns Hopkins Coronavirus Resource Center. https://coronavirus.jhu.edu/map.html

Klein S, Cortese M, Winter SL et al (2020) SARS-CoV-2 structure and replication characterized by in situ cryo-electron tomography. Nat Commun 11(1):5885

Lu S, Ye Q, Singh D et al (2021) The SARS-CoV-2 nucleocapsid phosphoprotein forms mutually exclusive condensates with RNA and the membrane-associated M protein. Nat Commun 12(1): 502

McCormick C, Khaperskyy DA (2017) Translation inhibition and stress granules in the antiviral immune response. Nat Rev Immunol 17(10):647–660

Sanchez-Aparicio MT, Ayllon J, Leo-Macias A, Wolff T, Garcia-Sastre A (2017) Subcellular localizations of RIG-I, TRIM25, and MAVS complexes. J Virol 91(2):e01155

Savastano A, Ibanez de Opakua A, Rankovic M, Zweckstetter M (2020) Nucleocapsid protein of SARS-CoV-2 phase separates into RNA-rich polymerase-containing condensates. Nat Commun 11(1):6041

Scudellari M (2021) How the coronavirus infects cells - and why Delta is so dangerous. Nature 595(7869):640–644

Wang S, Dai T, Qin Z et al (2021) Targeting liquid-liquid phase separation of SARS-CoV-2 nucleocapsid protein promotes innate antiviral immunity by elevating MAVS activity. Nat Cell Biol 23(7):718–732

WHO (2020) WHO Director-General's remarks at the media briefing on 2019-nCoV on 11 February 2020. https://www.who.int/director-general/speeches/detail/who-director-general-s-remarks-at-the-media-briefing-on-2019-ncov-on-11-february-2020

Wolff G, Limpens R, Zevenhoven-Dobbe JC et al (2020) A molecular pore spans the double membrane of the coronavirus replication organelle. Science 369(6509):1395–1398

Yang P, Mathieu C, Kolaitis RM et al (2020) G3BP1 is a tunable switch that triggers phase separation to assemble stress granules. Cell 181(2):325–345.e28

Zhou P, Yang XL, Wang XG et al (2020) A pneumonia outbreak associated with a new coronavirus of probable bat origin. Nature 579(7798):270–273

Zhu N, Zhang D, Wang W et al (2020) A novel coronavirus from patients with pneumonia in China, 2019. N Engl J Med 382(8):727–733

Index

R. Kurokawa (ed.), *Phase Separation in Living Cells*,
https://doi.org/10.1007/978-981-99-4886-4

The manufacturer's authorised representative in the EU is Springer Nature Customer Service Centre GmbH, Europaplatz 3, 69115 Heidelberg, Germany. If you have any concerns regarding our products, please contact ProductSafety@springernature.com

Printed and bound by CPI Group (UK) Ltd, Croydon, CR0 4YY
01/10/2026
02225866-0011